ALLE ZEIT WACH
1842

W. Salomons   U. Förstner (Eds.)

Environmental Management of Solid Waste

Dredged Material and Mine Tailings

With 118 Figures

Springer-Verlag
Berlin Heidelberg New York
London Paris Tokyo

Dr. WIM SALOMONS
Delft Hydraulics Laboratory
Institute for Soil Fertility
P.O. Box 30003
NL-9750 RA Haren (Gr)
The Netherlands

Professor Dr. ULRICH FÖRSTNER
Arbeitsbereich Umweltschutztechnik
Technische Universität Hamburg-Harburg
Harburger Schloßstraße 20
Postfach 90 14 03
D-2100 Hamburg 90
Fed. Rep. of Germany

ISBN 3-540-18232-2 Springer-Verlag Berlin Heidelberg New York
ISBN 0-387-18232-2 Springer-Verlag New York Berlin Heidelberg

Library of Congress Cataloging-in-Publication Data. Environmental management of solid waste–dredged material and mine tailings. Includes bibliographies and index. 1. Mineral industries–Waste disposal–Environmental aspects. 2. Tailings (Metallurgy)–Environmental aspects. 3. Dredging spoil–Environmental aspects. 4. Reclamation of land. I. Salomons, W. (Willem), 1945-. II. Förstner, Ulrich. III. Title: Dredged material and mine tailings. TD899.M47E58 1988 628.4'4 88-2016

Printed in the United States of America

Typesetting: Overseas Typographers, Inc., Makati, Philippines
2132/3130-543210

Preface

Mine tailings and dredged materials are solid wastes, which are produced at single sites in huge quantities. Costs are dramatically increasing both for the installation of new disposal sites and for the restoration of former deposits, which had been planned and operated in an inappropriate manner. The problems and questions these waste materials pose with regard to safe environmental deposition are similar: aquatic versus terrestrial disposal, revegetation, leaching of contaminants. Larger projects in the fields of both mine tailings reclamation and dredged material disposal are increasingly requiring a multidisciplinary team approach. Scientists with many different backgrounds – engineers, hydrologists, geologists analytical chemists, biologists, ecologists, environmental planners and economists – are searching for long-term solution, which involve minimum harm to nature, but can still be paid by society.

A typical project (and one of the motives for this book) was the planning of the "sludge island" off Rotterdam. After one decade of preparatory research, which was continuously interacted by public discussions, the Port of Rotterdam and the Netherlands Waterways Administration has now started to construct a containment for approximately 150 million m^3 of moderately to strongly polluted dredged sludge from the harbor area. The various "ecological" advantages will be discussed in several contributions to the present work. This solution also seems to be economically competitive with the old inland sites and even to the sea disposal of dredged materials (!). It is noteworthy, however, that the Municipality of Rotterdam has definitely stated that there is no intention of creating further large-scale sites after 2002. Measures have to be undertaken to improve the quality of the sediments, particularly by municipal and industrial dischargers in the Rhine River catchment area.

Mine tailings, compared to the more recent environmental implications with dredged materials, have been recognized as problematic materials for a long time. A report from 1868 of the River Pollution Commission in Britain described the situation in mid-Wales: "All these streams are turbid, whitened by the waste of the lead mines in their course; and flood waters bring down poisonous slime which, spreading over the adjoining flats, either befoul or destroy grass, and thus injure cattle and horses grazing on the dried herbage, or, by killing the plants whose roots have held the land together, render the shores more liable to abrasion and destruction on the next occasion of high water". Until now, problems arising from acid mine effluents are not satisfactorily solved in any part of the world. The major emphasis devoted to this aspect in the present book clearly demonstrates that the prediction and prevention of acid mine drainage are key elements of a strategy to control pollution from mining operations.

A major part of mineral reserves are in less-developed countries. A disproportionate fraction of resource development is expected to take place in the areas where environmental protection measures may be limited. Such experience will imply far-going demands from the host countries: (1) Reclamation should be carried out, as far as possible, during the life of the mine; (2) technology to ameliorate long-term effects should be as self-supporting as possible; (3) simple, reliable, low-energy techniques for minimizing deleterious effects of mining should be developed. The latter requirement can be summarized in the short expression "working with and not against nature". Initial efforts will be described in this book.

With 30 contributions, the two-volume book is mainly addressed to two major groups of potential users: to environmental chemists, biologists and geochemists working for mining companies, consultant agencies and universities, and to managers and planners in both industry and governmental agencies.

Chemistry and Biology of Dredged Materials and Mine Tailings, is introduced by review articles on three scientific disciplines, which seem to be particularly relevant for the present subject: solution/solid interactions of metal (A. Bourg), microbial processes (O.H. Tuovinen) and behavior of vegetation (W.H.O. Ernst). These reviews are followed by in-depth contributions on biological and chemical characteristics of the two types of solid wastes. They include case histories as well as laboratory (experimental) assessment of actual or potential environmental impact of both organic and inorganic priority pollutants.

Environmental Management of Dredged Materials and Mine Tailings will certainly fulfill expectations of a wide spectrum of practitioners, in that latest results are presented from management plans and decision-making processes in both fields. Examples are given from new mining operations in both developed and developing countries (I. Ritchie; R. Higgins), and the most advanced approaches to dredged material handling by the Municipality of Rotterdam and the U.S. Government (H. Nijssen; C.R. Lee and R.K. Peddicord). With respect to the future development in these areas, the reader should particularly refer to the articles on "ecological engineering" (K. Kalin and R.O. van Everdingen), "biological engineering" (D.V. Ellis and L. Taylor), use of "integrated biological systems" (H. von Michaelis). Several other contributions describing methods of "geochemical engineering" emphasize the increasing efforts of using natural resources available at the disposal site for reducing negative environmental effects of all types of solid waste materials.

Last but not least we would like to express our gratitude to all contributors for their enthusiastic and cooperative response to this project. We are very thankful to Dr. Engel, Springer-Verlag, for his constant encouragement, and for the much appreciated assistance of the publisher in the preparation of these volumes.

W. Salomons
U. Förstner

Contents

Part I Prediction of Effects and Treatment

The Predictive Assessment of the Migration of Leachate in the Subsoils Surrounding Mine Tailing and Dredging Spoil Sites

M. LOXHAM[1]

1 Introduction

Mine tailings and the very similar dredging spoil form just one example of a wide category of bulk inorganic wastes that can be disposed of by surface storage or landfilling. Other wastes include pulverised fuel ash, waste phosphate gypsum, cement fly ash, desulphurisation sludges and domestic waste incinerator ash. The disposal can be simply passive or have a geotechnical function such as land make-up or embankment fill.

This category of wastes differs from conventional chemical waste primarily in its volume and stability and in the fact that the largest proportion of the waste is in no way hazardous and differs only marginally, if at all, from naturally occurring minerals.

Unfortunately, they do contain trace levels of leachable toxics such as heavy metals, amphoterics, metalloids, fluorides and in the case of dredging spoil and much incinerator ash, persistent organic residues. In some cases radionuclides can be released such as radium, the actinides or the gas radon. The concentrations in the leachate are invariably low but none the less environmentally significant. Furthermore, the release can take place over very long time spans and potentially pollute large areas.

This long-term impact is to be distinguished from short-term impact by the displacement of the associated pore water or transport water. The pore water very often contains comparatively high levels of processing fluids such as acids, organics or bases as well as soluble non-toxic salts. In fact, many of these wastes are characterised by a short-term peak release from these fluids followed by a long-term steady state one, by ordinary leaching processes.

This release of toxic material can significantly degrade the ground- and surface waters in the surrounding environment. It is usually feasible and cost effective to control the releases to surface waters and some measures can be taken against the short-term release into the groundwater. However, the long-term threat to the groundwater is more difficult to address.

The most significant mechanism leading to the spread of the toxic components from the immediate vicinity of the site into the environment is that of advection with the groundwater. The quantitive assessment of this is the key to the overall safety analysis of many disposal sites and to the rational, cost-effective design of any necessary environmental protection countermeasures.

[1]Delft Soil Mechanics Laboratory, P.O. 69, 2600 AB Delft, The Netherlands

In recent years great strides have been made in predictive advection modelling techniques and this is fortunate as the impact time factors for both the dispersion of the toxics and any countermeasures taken against it, is so long that direct experimental simulation is out of the question. Typically, environmental engineers are required to assess the effects of leachate migration on the quality of the surrounding aquifers on time scales of up to thousands of years.

In the following, a description will be given of a practical methodology for the quantification of these advection effects without going into the mathematical physics behind them. For these, the reader is referred to standard texts on groundwater pollution (Bear 1972; Fried 1975; Kirkham and Powers 1972). The methodology has been found useful and adequate for a wide variety of problems and some examples will be given. Emphasis is placed on the optimalisation of investigatory effort to reach a given conclusion and in particular that of keeping to a minimum the highly expensive field and laboratory work.

2 Source-Path-Target Methodology

Before going into computational details it is worthwhile to consider the overall methodology of site assessment. In almost all cases it is possible to identify three elements in the safety assessment problem. These are:

1. A source;
2. A target;
3. A pathway connecting them.

This is illustrated in Fig 1.

In practice multiple targets and associated pathways can be identified and the source-path-target diagrams can become very complicated.

The source is characterised by its nature and its emission strength. This can be an advective leachate flux and concentration or if this flux is very small, a diffusive flux. In general, the emission will change with both time and spatially over the source site. The emission flux is often referred to as the "aquifer loading". This reflects the convention that the unsaturated zone under the site down to the groundwater table and any artificial barriers such as liners under the site are included as part of the source description rather than that of the path. The release scenarios considered are more often than not the result of risk analysis studies of the (composite) source.

The target is characterised by a maximum allowable impact value. Typical targets are streams, abstraction wells and other outflows. However, more diffuse targets can be imagined such as a crop-rooting zone, a nature reserve or even a whole aquifer. In practice the target is specified as a maximum allowable concentration value (MAC value) and the actual numerical value of these are the subject of much controversy. Drinking water standards are often specified and the regulatory authorities usually do not allow credit to be taken for in-target dilution effects as for example the mixing of an outflow into the bulk water flow of a river.

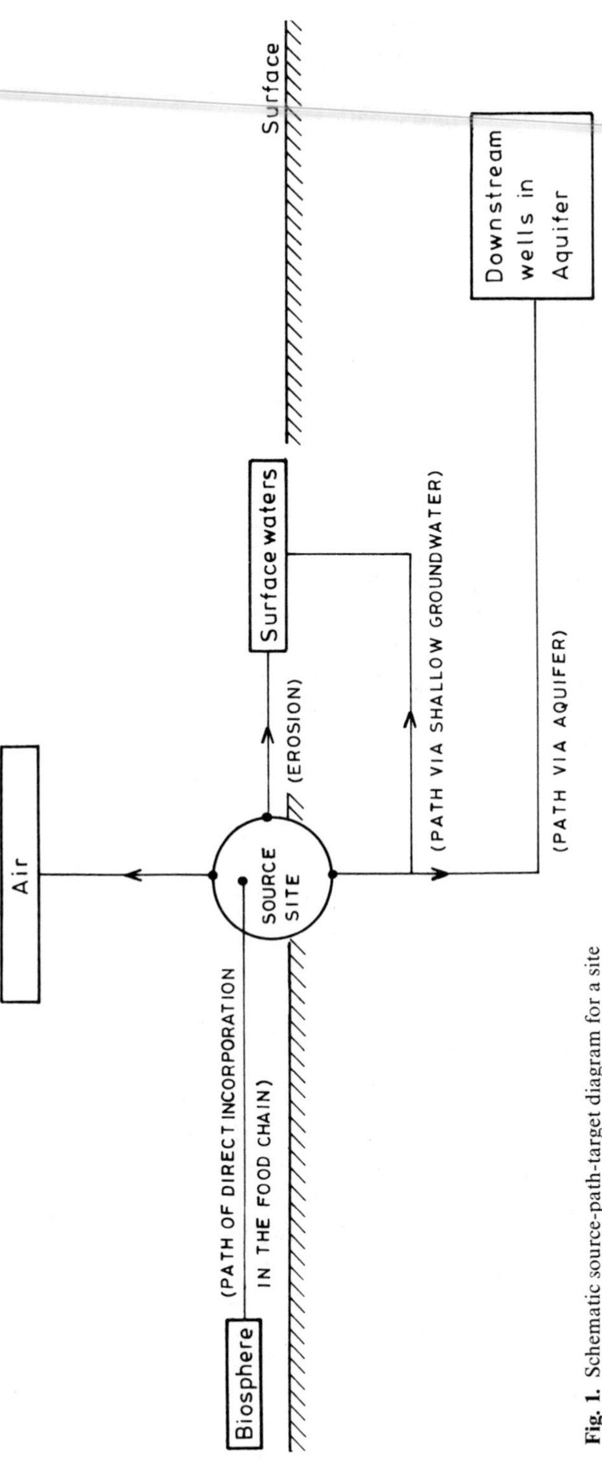

Fig. 1. Schematic source-path-target diagram for a site

The source and target are connected along a path. This path is characterised by its dynamics. Typically, the concentrations decrease along the path by dispersion and dilution effects, but also by radioactive or biological decay. The quantification of the phenomena along the path is the objective of most site-assessment studies.

If the expected impact of a particular site at the target is found to be unacceptable, then countermeasures have to be considered. These can be directed at any one of the three components of the S-P-T system:

1. The source can be reduced (or removed);
2. The target can be moved;
3. The path can be manipulated.

An example of this manipulation is illustrated in Fig. 2 for the case of a cut-off wall backed up by a guard well.

Figure 2 also illustrates two important features of the engineering approach to the problem. Firstly, the difference between the MAC value and the predicted concentration at the target can be taken as a measure of the impact safety factor being carried in the design. This factor is usually specified from both economic and other non-technical considerations. This value has a range of uncertainty. This can be illustrated by considering Fig. 3, where the overall engineering costs are set out against the degree of target protection required.

There is in fact an envelope of design curves for any given problem depending upon the uncertainty accepted in the final system performance and the thereby implied safety factors. The second aspect to be noted is that all engineering

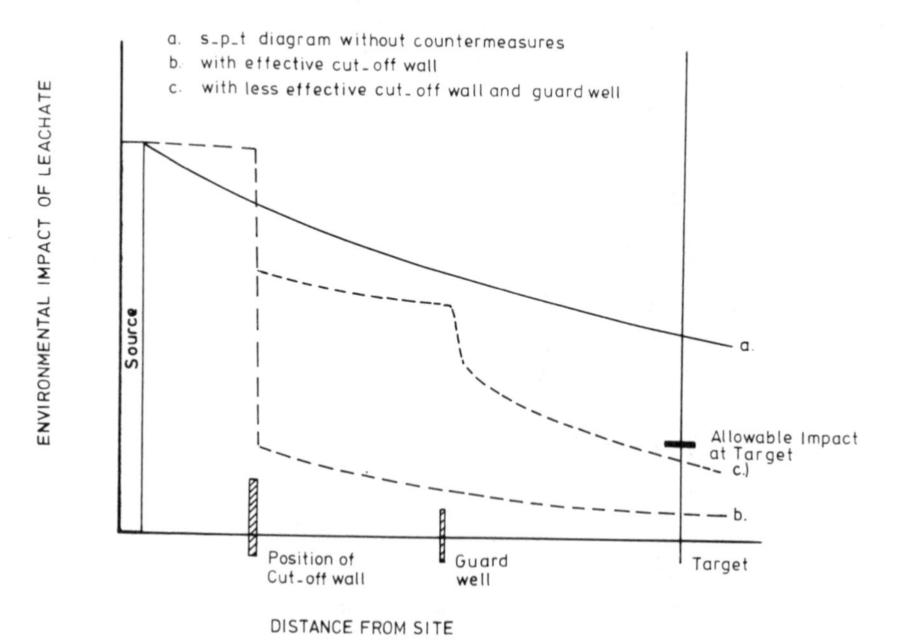

Fig. 2. Effect of countermeasures on s-p-t diagram

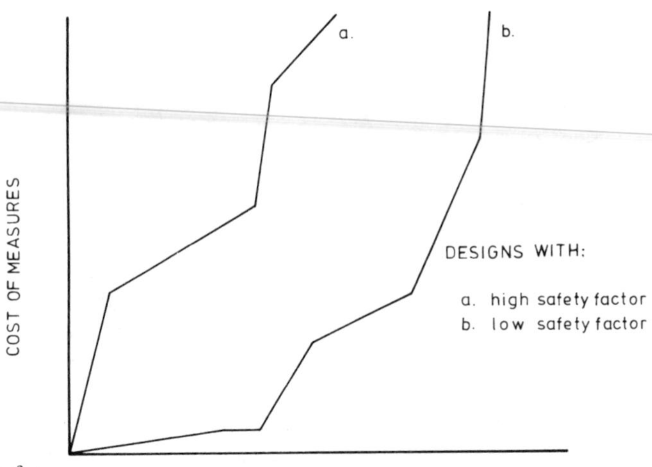

Fig. 3. Cost effect diagram for countermeasures

solutions to environmental problems must also be considered in their failure mode and must be subjected to a full risk analysis study. In the hypothetical case shown in Fig. 2 this has led to the back-up guard well being sunk in order to provide a minimum barrier that could be activated if the cut-off wall were to fail.

It is the environmental engineer's task to generate analogous information to that summarised in Figs. 1 through 3 for particular sites and scenarios.

3 Groundwater Flow and Contaminant Migration

The general methodology for contaminant migration studies is as follows:

1. Measure or calculate the regional groundwater flow;
2. Measure the regional and local groundwater qualities;
3. Measure or model the groundwater flow in the local region either at the existing site or its future location;
4. Identify the appropriate targets;
5. Specify the operation scenarios for the site;
6. Define the relevant source terms;
7. Estimate the migration parameters;
8. Model the migration patterns using the contaminant migration models;
9. Conduct a sensitivity analysis on results from the point of view of the parameter choices and levels;
10. If necessary, reiterate (7) through (10) until the desired precision is achieved.

In this section the emphasis will be placed on the techniques needed to conduct step (8) as efficiently as possible.

The groundwater flow between the source and target, usually referred to as the local flow pattern, is distinct from that of the rest of the watershed, the so-called regional pattern. In the green-field situation before the site is in place, the local system is usually controlled by the regional one and can be established by conventional geohydrological techniques (see for example Rushton and Redshaw 1975; Wen-Hsiung Li 1972).

Similar considerations apply for the chemistry of the local and regional systems. This so-called background or zero concentration profile is usually established by groundwater sampling and chemical analysis. A priori calculation is not possible.

The construction and start-up phase of a site is commonly accompanied by major disruptions of the local system. It is possible that pumping and transport water are drained from the site to the local system and surface watercourses can be radically altered. Pore water resulting from waste consolidation can also drain to the local system and there can be major infiltration changes as the balance between rainfall and evapotranspiration in the region of the site is disturbed. Finally, tailings dams can result in ponded conditions over large areas. This initial pollution impact, albeit temporary, can have serious consequences for the overall aquifer quality, introducing a slug of contaminant that slowly moves downstream, polluting wells and rivers long after site construction has finished.

For a site overlying a salt water aquifer there is an additional complication, i.e. the effective total infiltration fluxes of contaminated but comparatively fresh water can lead to the development of a freshwater lens under the site with the consequent radical changes in groundwater flow patterns.

After operation has finished, in the caretaking stages and beyond, the flow pattern settles down again to a steady state that often closely resembles the origional green-field situation.

There are two exceptions to this. Firstly, in the case where major engineering elements have been introduced into the aquifer such as cut-off walls or grout barriers and secondly, where the tailings themselves intrude into the aquifer. An example of this is where dredging spoil, which has a final permeability in the order of that of clay, is landfilled and forms a large plug in the aquifer around which the groundwater has to flow.

In all the above cases the groundwater flow pattern can be followed by computer modelling techniques. Large, three-dimensional finite element or finite difference groundwater modelling codes are widely available today and will not be discussed further except to point out that there are significant cost savings to be made by calculating the sequence of operating stages of the site as a series of hydraulic steady states rather than trying to follow the exact time dependence of the system. Typically, the hydraulic response time of the system is much shorter than the operating scenario time scales and the above simplification is usually justified. An exception is where major changes in the groundwater table are introduced.

Once the groundwater flow pattern has been established it is possible to calculate the developing contaminant distribution pattern by using the conventional convection – dispersion equation:

$$\frac{\delta c}{\delta t} + \frac{\delta s}{\delta t} = -\nabla(vc) + \nabla D \cdot \nabla C - \lambda c, \qquad (1)$$

where C and S are the dissolved and adsorbed concentrations respectively, t the time, D the hydrodynamic dispersion tensor and v the (mean local) in-pore water velocity vector. λ is the (first order) decay term for radioactive decay or biological effects.

Although in no way necessary for modern computer implementation of this model, it will be convenient for our purposes to rewrite the adsorption term in terms of the conventional retardation factor R, where:

$$R = (1 + K), \tag{2}$$

and K is the distribution coefficient given by:

$$K = s/c, \tag{3}$$

where S is expressed as mass adsorbed per unit volume associated pore fluid. This results in:

$$\frac{\delta c}{\delta t} = \frac{1}{R}\left\{-\nabla(vc) + \nabla D \cdot \nabla c - \lambda c\right\}, \tag{4}$$

which has to be solved once the spatial distribution of the parameters, the initial and the boundary conditions, are specified.

The initial conditions can be obtained from the zero situation measurements of the spatial distribution of the concentrations in the area to be studied. Often it is assumed that these are zero and that for calculation purposes the emission from the site is considered in terms of incremental load. This can be an erroneous procedure especially in areas where mine tailings are dumped near the mining activity itself and this approach can lead to large overestimates of the environmental impact of the dump.

The mathematical structure of the convection – dispersion equation requires that two boundary conditions be specified. The most common formulation is to specify the input flux as a function of time at the site and the output fluxes at the calculation domain boundaries. A less useful combination is that of concentrations at the site and domain boundaries. The input fluxes at the site are themselves usually the results of modelling studies of the hydrology and speciation chemistry applied to the waste itself.

Before discussing possible solutions to this general equation, it is worthwhile to consider it in more detail. Any solution to the equation requires input values for the water velocity (or as this equation is usually solved in combination with the Laplace formulation of Darcy's law, the hydraulic conductivities), the dispersion coefficients, the adsorption coefficients and any decay rate coefficients. These have to be specified for both space and time over the calculation domain.

These parameters can be measured in the field, in the laboratory, estimated from the nature of the ground conditions, assumed from experience or common sense, or just simply guessed. Adequate parameter estimation is the major difficulty in the effective use of predictive modelling techniques.

The parameters usually exhibit a wide variation even for rather simple geohydrological settings. Typically, the hydraulic conductivity and the adsorption coefficient can vary several orders of magnitude on a 100-m-length scale.

Fortunately, the objective of the calculation within the context of an overall safety analysis demands ranges of the results rather than specific exact values and techniques based on sensitivity analysis and stochastic approaches can be used. The major cost and time factors in the assessment study are associated with the parameter estimation. The sensitivity analysis strategy is directed to minimising these.

Usually a calculation is made using preliminary information and the results subjected to the parameter sensitivity analysis. Four possible result/parameter combinations can be expected. There are parameters which are well defined and of importance in deciding the overall outcome of the safety analysis, and parameters that are either well or poorly defined but in any case irrelevant. These three combinations do not impact on the next iteration of the parameter estimation process. However, there is a last category, i.e. the parameter is of importance but poorly defined and this has to be remeasured. The calculation is then repeated with the new estimate and the sensitivity analysis reiterated, and so on and so forth until the required precision and policy discrimination in the final overall analysis is achieved.

In practice one or at the most two iterations will be permitted and perfect answers can never be expected. This automatically implies that usually massive safety factors have to be assumed in the analysis and this alone puts a limit on the degree of effort and modelling complexity that can be usefully employed.

Examination of the equation shows that it is the velocity (or hydraulic conductivity) that controls the downstream spread of the contaminant; the dispersion coefficient controls the lateral spread and the form of the downstream contaminant front; and the adsorption coefficient the time scale within which this occurs. The adsorption could, however, delay the contaminant long enough for the decay processes to make significant inroads into the contaminant inventory on the migration paths, thereby reducing further the impact at the target.

In the following sections some consideration will be given to specific solutions of the above equation.

4 A Site in an Infinite Homogeneous Aquifer

In this section the case is considered of a site with an emission into a large aquifer of constant properties. The objective of the calculation will be to define the concentrations expected downstream from the site. Whilst the physical setting of the problem is highly simplified, it is none the less complicated enough to illustrate the major features of the problem which are to be found in more difficult cases. Furthermore, the simple analysis given here is usually the one chosen to give the first estimate of the nature of the problem and to help focus the major computer simulation, which is normally required at the final stage.

Consider the situation illustrated in Fig. 4 which shows a site above an aquifer.

The groundwater velocity in the aquifer is constant in both space and time. The aquifer is confined and has constant thickness, dispersion and adsorption coefficients. At the site the confining barrier clay has been excavated and replaced

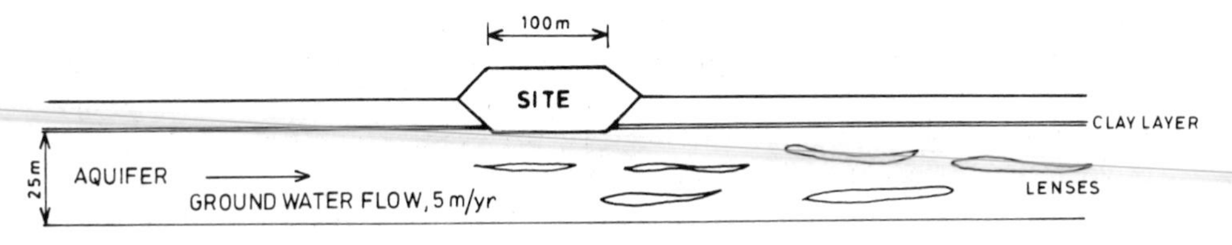

Fig. 4. Site in its surroundings

by tailings, allowing direct penetration of leachate into the aquifer (this occurs more frequently than one would hope or imagine!).

There is a flux of contaminant from the site into the aquifer. For the moment this will be considered a constant but this is in no way essential. The contaminant flux does not alter the water flow pattern under the aquifer.

These conditions have been chosen so that the convection dispersion equation has a simple analytical solution that can easily be programmed for a home or microcomputer. Thus (Wen-Hsiung Li 1972):

$$c(x,y,z,t) = \frac{M}{8\pi Dr} \exp\left(\frac{vx}{2D}\right)\left\{\cdot \exp\left(\frac{vr}{2D}\right)\left(1\text{-erf}\left[\frac{\frac{vt}{r} + r}{\sqrt{4Dt/R}}\right]\right) + \exp\left(\frac{-vr}{2D}\right)\left(1+\text{erf}\left[\frac{\frac{vt}{R} - r}{\sqrt{4Dt/R}}\right]\right)\right\} \tag{5}$$

and $r^2 = (x^2 + y^2 + z^2)$,

where M is the specific flux from the site to the aquifer (t^{-1}) and the other parameters have been defined earlier. The solution given here is for a point source, that is where the calculations are to be made for a location at some distance from the site, but can be easily extended to finite or multiple sources by summation and for time-variable sources by convolution techniques (Courant and Hilbert 1967).

For the case of a not too thick aquifer, Eq. (5) can be reduced to an even simpler form:

$$c(x,y,z) = \frac{m}{4\pi Dn} \int_0^t \frac{1}{\tau} \exp\left[\frac{-\left(x-\frac{vt}{R}\right)^2 - y^2}{4Dt/R}\right] d\tau, \tag{6}$$

where m is the specific line flux ($t\,s^{-1}m^{-1}$ aquifer thickness).

In practice Eq. (5) is chosen for cases where the bottom of the aquifer is not defined, such as for sites above fractured rock or karst formations. In this case care has to be taken in the interpretation of the definition of concentration. The concentration measured in the field (C_m) is the average over the depth of the sampling filter and is related to the point calculated concentration by:

$$c_m = \frac{1}{(h_1-h_2)} \int_{h_2}^{h_1} c(x,y,h,)dh, \tag{7}$$

and this should be taken into account where necessary. For many cases where the site is above a sedimentary aquifer, it is possible to use Eq. (6) for estimating purposes. Whilst the question as to at which point Eq. (5) has to be used rather than Eq. (6) is interesting; in practice, it is largely irrelevant. This is because the overall problem itself becomes much too complicated for this simple analysis in the immediate region of the site and recourse has to be made anyway to proper finite difference representations of the convection dispersion equation itself.

It is now possible to use these equations to illustrate some important features of the problem. Firstly, it is to be noted that they predict the contaminant plume to be ellipsoid in form with the maximum concentration on the downstream centre line. This is shown in Fig. 5.

A second general feature is that these equations have a pseudo-steady state behaviour at long times. Explicitly, Eq. (6) reduces to (Bear 1975):

$$\cdot c(x,y,\infty) = \frac{m}{2\pi D} \cdot \exp\left(\frac{vx}{2D}\right) \cdot K_o \left[\frac{v}{2D}\sqrt{x^2+y^2}\,\right]\Big), \qquad (8)$$

where K_o is the modified Bessel function of the second kind and zero order. The downstream concentrations develop to this steady state value asymtotically but in practice the value is reached at a time of about (4*R*v/x). In other words, there is a finite extent of the pollution plume for any given concentration level and parameter combination. The development towards this steady state is shown in Fig. 6 for a typical combination of parameters.

It should be noted that the retardation coefficient, R, does not occur in Eq. (8) and the maximum plume volume is not influenced by adsorption, only the time taken to reach it.

In order to gain some insight into the sort of numerical values to be expected, consider a typical case where a fly-ash site is releasing a fluoride anion at a total flux of 150 kg yr^{-1} into an aquifer of 25-m thickness with an in-pore water velocity of 5 m y^{-1} (Fig. 4). In Fig. 7 the contaminant plume extent at the drinking water concentration of 1 ppm is illustrated as a function of the dispersion coefficient.

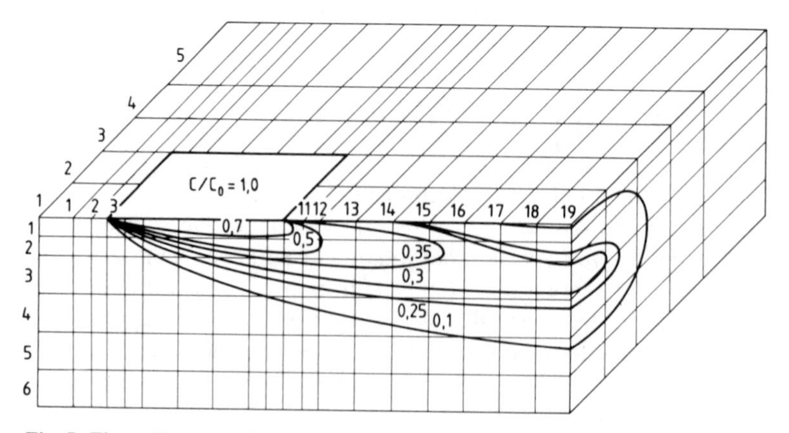

Fig. 5. Three-dimensional concentration development

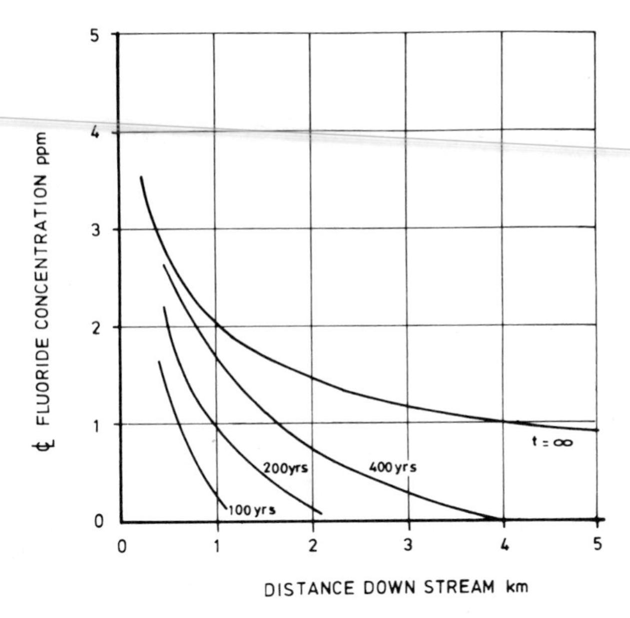

Fig. 6. Concentration buildup to steady state valve

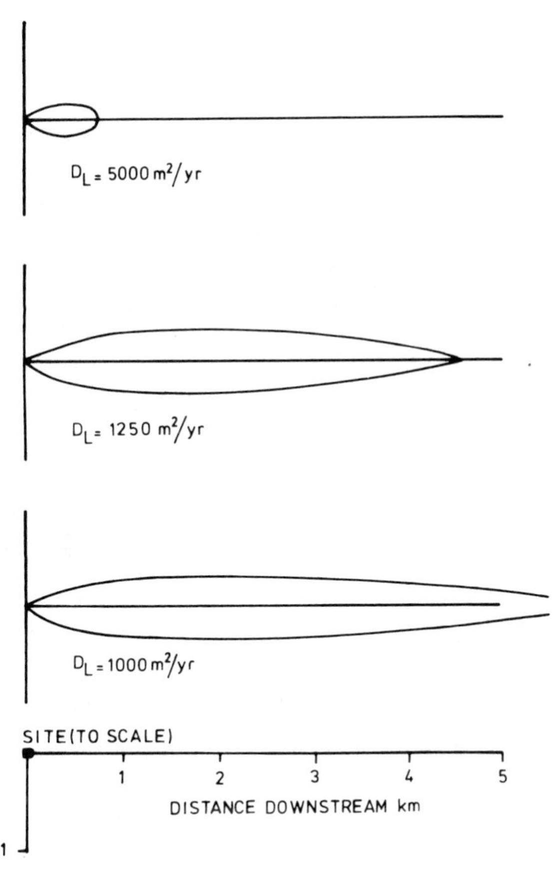

Fig. 7. The downstream plume

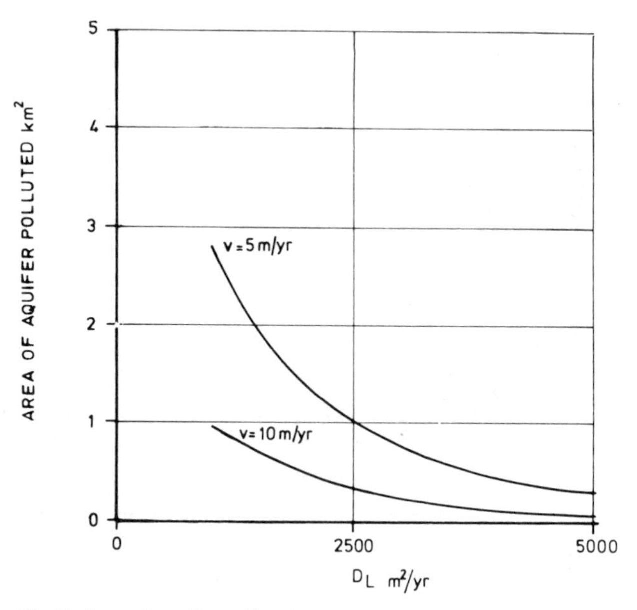

Fig. 8. Area of aquifer polluted

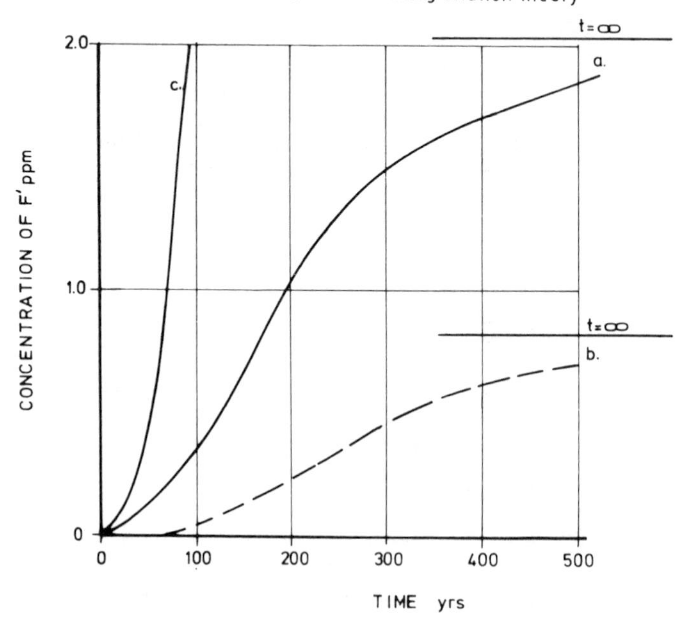

Fig. 9. Breakthrough curves 1 km downstream

As can be seen in Fig. 7, the effect of decreasing the value of the dispersion coefficient is to cause a longer plume downstream. With high dispersion behaviour the contamination is quickly washed out around the site. The area of aquifer contaminated with water that has a concentration above the MAC value is shown in Fig. 8 as a function of both velocity and the dispersion coefficient.

Finally, and again as an illustration, the concentration build-up curve at a point downstream is shown in Fig. 9.

This can be compared to that obtained from a one-dimensional simulation using the "dilution factor" source concentration (emission per unit width of a site normal to groundwater flow/groundwater flux). This is also shown in Fig. 9 and illustrates that on the flow axis the one-dimensional calculation gives a large overestimate of the concentration development. There is a considerable improvement by treating the problem as a two-dimensional one.

Figure 9 also illustrates another important feature of the breakthrough of a contaminant at the target and that is the so-called toe effect. This is the small but often environmentally significant concentration buildup ahead of the main arrival time front. The extent of this early breakthrough is largely controlled by the dispersion coefficient reflecting the heterogeneous nature of the system. In many problems it is the arrival time of the toe that is significant and not that of the main front.

5 Adsorption and the Problem of Real Systems

In the previous section we have considered the case of a non-adsorbing component and the maximum plume size that can be expected. It has been shown that there is a steady state maximum size of this plume that is reached typically in thousands of years. In this section the effect of adsorption of the migrating components on the development time of the plume will be examined in more detail.

These effects can be of dominating importance in the safety analysis. The retardation coefficients as given by Eq. (3) can vary from < 5 for sandy systems to more than 10,000 for heavy metals on clays and peats. Extensive reviews of adsorption measurements on a laboratory scale have been published (see for example Bolt and Bruggenwert 1978).

In general terms, soil organic material, clay surfaces and precipitated iron and aluminium oxides are responsible for adsorption of heavy metals and polar organics. Non-polar organics and anions are much more poorly adsorbed. Adsorption on the active surfaces of soil components should be distinguished from precipitation from the pore solution itself. However, in many operational procedures to determine the adsorption coefficient these phenomena are difficult to separate and are often lumped together in the presentation of the results.

Unfortunately, there is little or no field adsorption data available at this time. Furthermore, the actual adsorption behaviour is controlled by a wide range of physical and chemical factors such as the state of the adsorbant surface, pH, redox potential, organic matter content and component speciation. These factors cannot only change in time, but also under the influence of the pore water chemistry

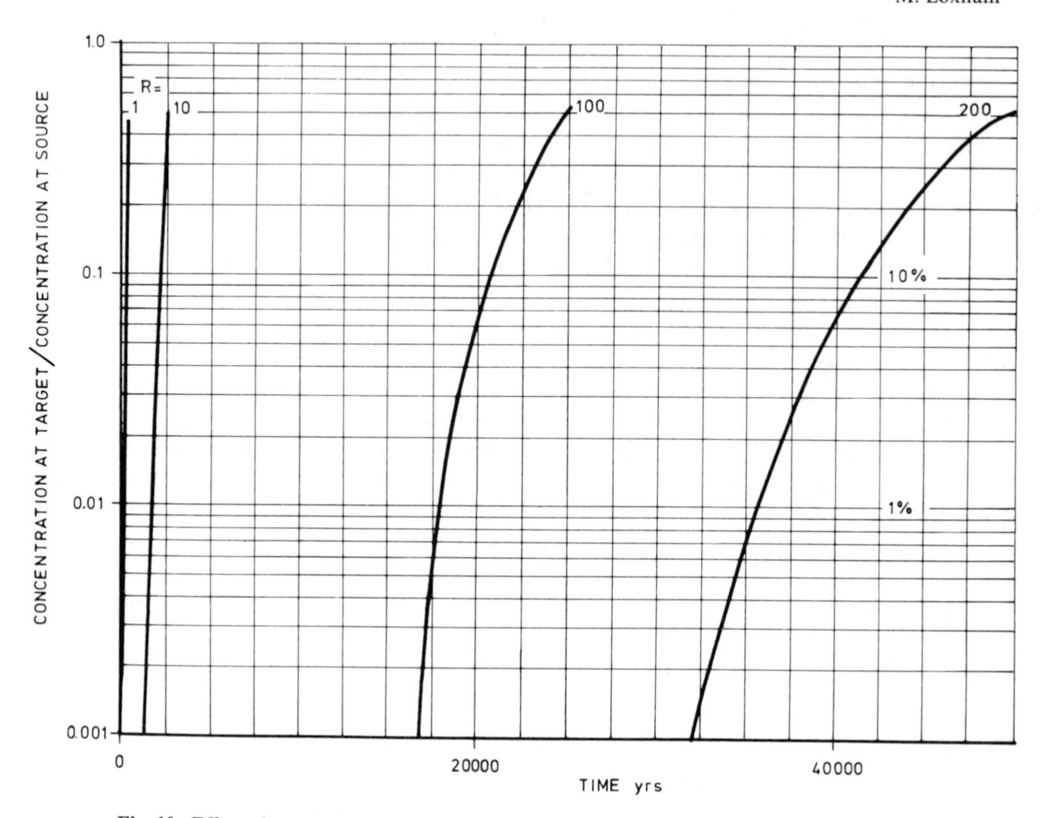

Fig. 10. Effect of retardation concentration development

changes associated with the migration front itself. Finally, it is not uncommon for different laboratories to report adsorption coefficients that differ by two or three orders of magnitude for nominally identical systems.

The influence of different adsorption coefficients on the breakthrough behaviour for the example chosen above (cf. Fig. 9) is shown in Fig. 10.

The effect of a high adsorption coefficient can be to delay the breakthrough until essentially geological time scales. This is true for both the main front and the toe of the breakthrough.

The adsorption can be seen both as a barrier slowing down the migration front and as a buffer resulting in the storage of toxic material leaching out of the site in the surrounding soils and thus preventing its convection to the target. It is, however, only rarely that the total adsorption capacity of the soils on the path to the target is so large that all the site inventory can be held back in this way.

For a component undergoing radioactive or biological decay the adsorption barrier can delay the migration effectively enough to ensure that none of the migrating component reaches the target. This is illustrated in Fig. 11. This is the basis of adsorption barrier technology in radioactive waste isolation.

Usually the quoted valves of the adsorption coefficients are obtained from laboratory experiments on disturbed samples and give maximum values.

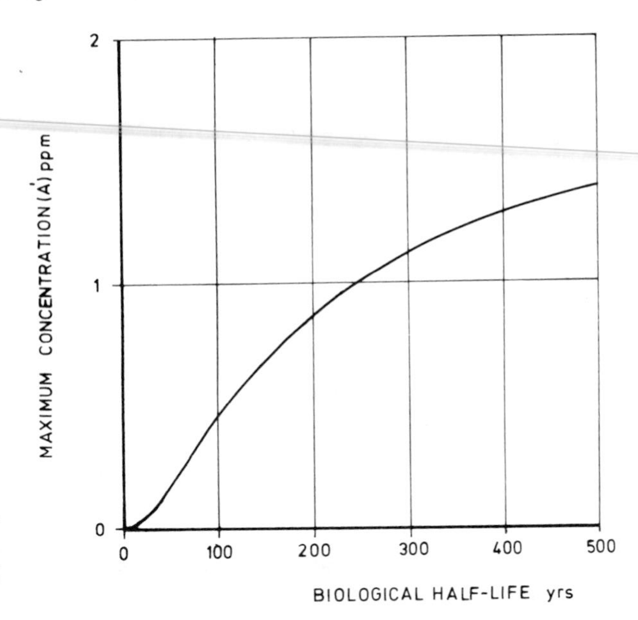

Fig. 11. Breakthrough curve for an organic component (A) at 1 km downstream assuming different half-lives

The major difficulty in assessing the field effective value of the adsorption barrier lies, however, in the heterogeneous nature of real soils and aquifers. Typically, the permeabilities and adsorption coefficients can vary over several orders of magnitude on a 100 m-length scale in a sedimentary environment.

The adsorbing soil components such as clays tend to occur as finite lenses, reflecting their sedimentary origins. These lenses have generally a low hydraulic conductivity so that the water (and contaminant) tends to flow around rather than through them. In this way it is possible that some contaminants that would otherwise have been retarded, can by-pass the adsorption sites and break through earlier than expected. This effect is seen most clearly in the toe of the breakthrough curve.

The actual mobilisation of the potential adsorption capacity is a function of several factors such as the actual geometry of the system, the lens properties and the time scales of the flow regime. This last dependency arises because for clay lenses the hydraulic conductivity is so low that the adsorption sites can only be accessed by diffusion mechanisms and these operate independently of the surrounding water velocity.

In practice two extremes can be identified, one where all the adsorption sites can be accessed and one where only molecular diffusion access is possible. The difference between these two extremes is shown schematically in Fig. 12.

In order to assess the availability of adsorption capacity in lens systems, calculations were made (Loxham 1983; van Meurs et al. 1985) on predefined heterogeneous systems using one of the large, modern, finite-difference, contaminant-migration codes (DSML 1982). The calculations showed trends similar to those given in Fig. 12, including the fast breakthrough at low initial concentrations.

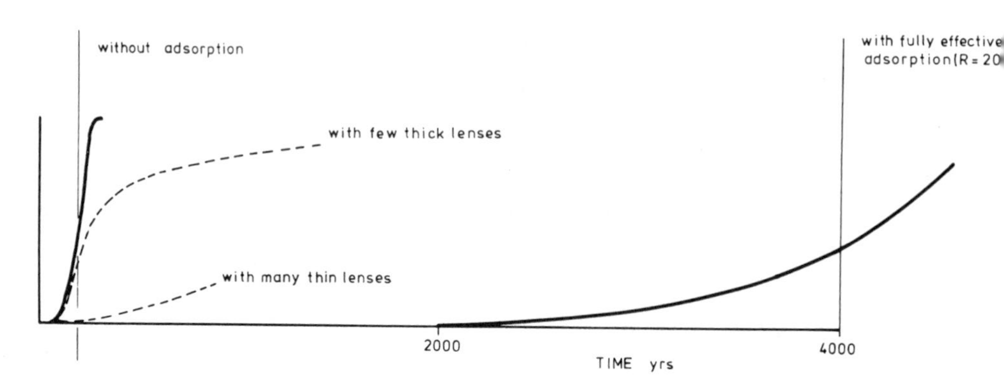

Fig. 12. Schematic effect of concentrating the adsorption capacity in fewer thicker lenses

In practice the amount of profile information required to perform such calculations will never be available. The question then arises as to what is the minimum of field information required to support a given conclusion.

The field information required concerns the actual position, or at least the thicknesses and stratagraphic successions of the lenses. These can be determined from borehole logs and geophysical techniques.

Three levels of information can be identified:

1. Overall mean soil composition without structure differentiation;
2. Stratagraphic succession with layer thickness data but without spatial distributions;
3. Complete profile information.

Studies (Loxham 1983) have been made on the utility of the information at each of these levels and the level of modelling sophistication appropriate to the field data base for landfilling low level radioactive wastes. The conclusions of these studies have general applicability to this sort of contaminant safety assessment. It was concluded that contaminant modelling with the last level of information was superior to the simpler systems, but involved a prohibitive investment in field work. Models using only first level information generally gave a large overestimate of the value of the adsorption barrier and correspondingly low (sometimes fractional!) overall safety factors in the site design. Surprisingly, calculations, using level two information, where the lens thicknesses were available but not their positions relative to each other, gave quite reasonable results.

In Fig. 13 one of the results of the study is illustrated. In this figure the apparent retardation coefficient found from the fully heterogeneous calculation divided by the maximum value is used as a measure of the mobilisation of the adsorption barrier. It can be seen that for very thin lenses all the adsorption capacity can be taken into account and for very thick lenses it is better to discount the adsorption barrier completely. Similar calculations can be performed for other sites and stratas but the results show the same trends. For lenses of intermediate thickness simple but adequate estimates can be made using dual porosity models (Skopp and Warrick 1974; Loxham and van Meurs 1984).

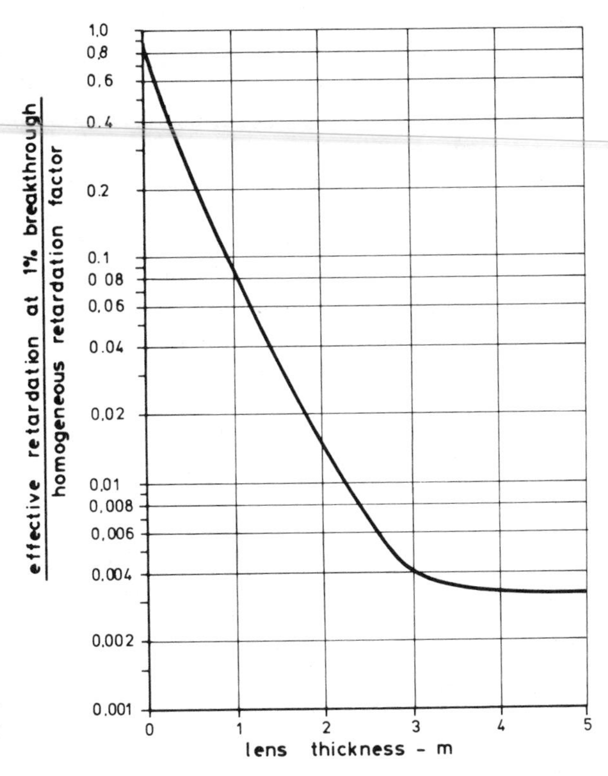

Fig. 13. Influence of lens thickness on effective retardation factor for clay lenses

This study suggests that it is possible to tailor the field work down to determining the stratagraphic data of the soils on the paths and obtaining undisturbed samples of representative strata for laboratory determination of the adsorption behaviour.

Finally, it should be noted that for fracture rock systems and karsts, a similar problem occurs in that the bulk of the rock has a volumetric holding capacity for contaminants that can only be accessed by diffussion from the hydraulically active fissures. However, the systems do differ in that in fissure systems the adsorption capacity proper tends to be concentrated at the surface of the fissure or in microcracks in its vicinity (Neretnieks 1980; Tang et al. 1981; Sudicky and Frind 1982).

6 Engineered Pollution Control Options

It was pointed out earlier that remedial actions for the case where the environmental impacts of the site on the quality of the aquifer are unacceptable can be directed towards either the source or pathway. Source-orientated measures can include prevention of leaching by excluding infiltration, stabilisation of the tailings by chemical means or as a last resort excavation and removal (which presents a problem elsewhere). These methods will not be considered here and attention will be directed to pathway control, such as cut-off walls, grout barriers, guard wells, etc.

These methods can be very effective and have found wide application. However, one general feature of all engineering options for path control is that they are relatively short-lived measures and cannot be expected to guarantee, on their own, long-term environmental safety. In other words, they should only be considered where the passive long-term behaviour of the site tends towards one of acceptable environmental impacts.

An example of this is the landfilling of PFA behind a deep cut-off wall. In the short term the wall protects the aquifers in the surroundings from the initial high leachate peak. This is washed out of the ash in a period of some 50 years by infiltrating rainfall which, in turn, is pumped up from within the site and treated. In the long term, when it is to be expected that the cut-off wall might fail, the residual impact of the remaining leachate will be acceptable for the surroundings. This strategy is illustrated in Fig. 14.

A similar, "fail-safe" strategy has been proposed for an offshore harbour, dredging-spoil dump site (van Meurs et al. 1985) in the Netherlands. In this case the initial peak leachate load is caused by consolidation water being pressed out into the underlying strata and by the initial chemistry of the spoil. This consolidation water is, however, less dense than the surrounding seawater and forms a freshwater lens under the site. The fluid dynamics of this lens are such that contaminant is substantially prevented from migrating into the surrounding aquifers as illustrated in Fig. 15 and this saves the considerable costs of a deep cut-off wall.

In the long term the spoil consolidates to give a monolith with a very low permeability and correspondingly low pore water emission flux. Whilst this flux is no longer capable of maintaining the favourable barrier function of the freshwater

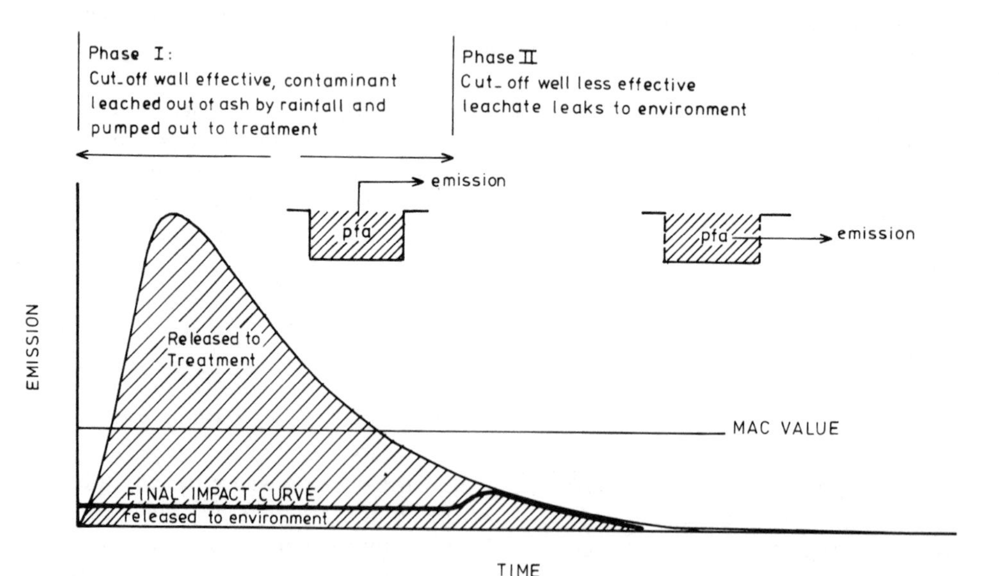

Fig. 14. PFA containment strategy

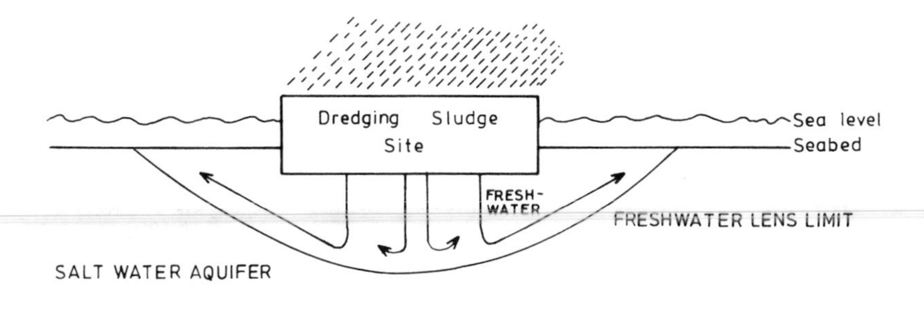

Fig. 15. The action of a fresh (ground) water lens

lens, the associated contaminant emissions are low enough for there to be no long-term pollution problem from the site.

In the past much has been made of guard well techniques to protect aquifers. In principle, the well is sunk and pumped at a rate sufficient to pull all the contaminants into the well and away from the target. Many calculation techniques have been developed to optimise the well placement and pumping rates (van den Akker 1982).

The method has, however, several serious disadvantages. The major one is that the performance of the system depends heavily on the knowledge of the hydraulic conductivity distributions of the aquifer, and these are precisely the quantities subject to the most uncertainty. A second disadvantage is that in practical terms, it would appear that a trade-off has to be made between a few wells pumping a large volume of slightly contaminated water with correspondingly high effluent treatment costs and many small "precision wells" of uncertain performance and very high operating costs. This method has achieved most success in protecting single abstraction points at the target rather than preventing significant emission from a site itself.

In principle, deep cut-off wall isolation technology has the advantage that it uncouples the hydrological regimes within the site from those in the aquifer. Furthermore, construction is possible in a wide range of soil types, also under water. The major disadvantage is that the wall has to be constructed down to an impermeable layer at the bottom of the aquifer so effectively isolating the site (and the groundwater under it) from the rest of the aquifer.

If this strata is absent within economic depths, then the method fails. Attempts to engineer such a horizontal layer by grouting techniques has usually proved either ineffective or prohibitively expensive considering the large areas involved.

A large amount of computer modelling effort has been dedicated to exploring the minimum barrier performance that such a horizontal barrier should have before it starts to significantly impact on the overall safety of the site. This is because in practice the strata actually underlying the sites are either leaky, thin, damaged or all three. Risk analysis techniques are often employed because of the options coupling the site-water management to the leak across the layer.

A typical example of cut-off wall technology is the site at Zelzate in Belgium where the highly contaminated canal dredging spoil is landfilled from the heavily

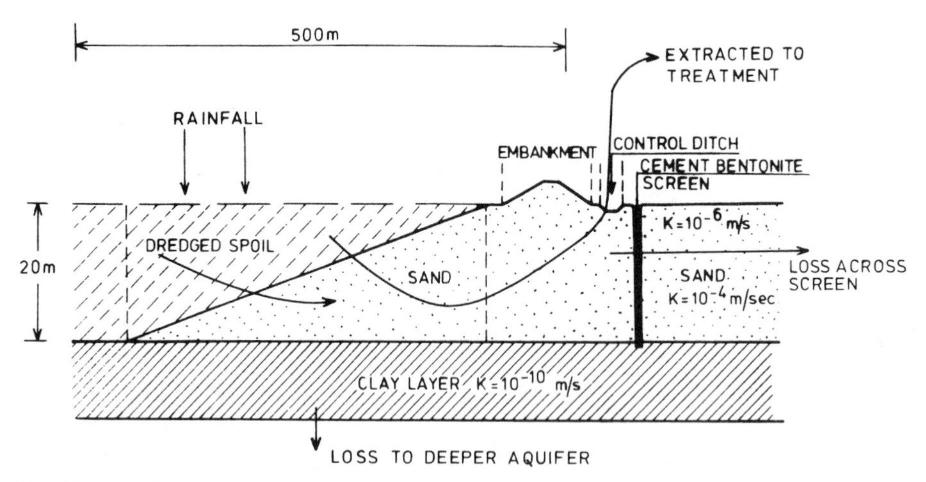

Fig. 16. Migration patterns at Zelzate

industrialised waterway between Gent and the Western Scheldt in the Netherlands (Loxham and Westrate 1985). A cross-section with the primary contaminant migration patterns is shown in Fig. 16.

The cut-off wall was constructed from cement-bentonite using the panel technique (Xanthakos 1979) down to a thick clay stratum at a depth of 24 m. The wall has a width of 0.8 m. The key to the success of the site is the ring canal shown. This is held on water level control, collecting the initial peak leachate from the tie site for the first 70 years. In this period about 2% of the mobile toxic components are lost via the wall and the underlying clay to the local aquifer and the rest via the ring canal. In due course the quality and quantity of the leachate from the site will rely on the cut-off wall to protect the aquifer, unnecessarily and the basic principle of long-term fail-safe behaviour is satisfied.

References

Akker C van den (1982) Numerical analysis of the stream function in plane groundwater flow. Thesis Univ Delft

Bear J (1972) Dynamics of flow in porous media. Elsevier, New York

Bear J (1975) Hydraulics of groundwater. Elsevier, Amsterdam

Bolt GH, Bruggenwert MGM (1978) Soil chemistry. Elsevier, New York

Courant R, Hilbert D (1967) Methoden der mathematischen Physiko Springer, Berlin Heidelberg New York

DSML (1982) The contaminant transport model VERA. Delft Soil Mech Lab, Delft, The Netherlands

Fried JJ (1975) Groundwater pollution. Elsevier, Amsterdam

Kirkham D, Powers WL (1972) Advanced soil physics. Wiley-Interscience, New York

Loxham M (1983) The influence of soil heterogeneity on the migration of radionuclides in the soil and soil-water systems Final Rep EEC Res Contract 193–81–6–WAS–NL EEC Comm, Brussels

Loxham M, Meurs GAM van (1984) Int Radioactive waste management Conf BNES, London, p 291

Loxham M, Weststrate FA (1985) The use of cement-bentonite cut-off wall to contain dredging spoils. Eng Geol 21:359–365

Meurs GAM van, Loxham M., Weststrate FA (1985) Heterogeneity and its impact on the spread of pollutants. In: Assink JW, Brink WT (eds) Contaminated soil. Nijhoff, Dortrecht, pp 79–87
Nerethieks I (1980) Diffusion in a rock matrix. J Geophys Res 85(88):4379–4397
Rushton KR, Redshaw SC (1970) Seepage and groundwater flow. John Wiley, New York
Skopp J, Warrick AM (1974) A two phase model for miscible displacement of reactive solutes in soils. Soil Soc Am Proc 38(4):545–550
Sudicky GA, Frind EO (1982) Contaminant transport in fractured porous media. Water Resour Res 18(6):1634–1642
Tang DH Frind EO, Sudicky GA (1981) Contaminant transport in fractured porous media. Water Resour Res 17(3):555–564
Wen-Hsiung Li (1972) Differential equations of dispersion and groundwater flow. Prentice Hall, New York
Xanthakos, P (1979) Slurry walls. McGraw Hill, New York

Pre-Mine Prediction of Acid Mine Drainage

K.D. FERGUSON[1] and P.M. ERICKSON[2]

1 Introduction

The uncontrolled release of acid mine drainage (AMD) is perhaps the most serious impact mining can have on the environment. The cumulative length of streams worldwide affected by acid mine drainage is not known, but over 7000 km of streams in the eastern United States are seriously affected by acid drainage from coal mines (Kim et al. 1982). In addition to low pH and high acidity, acid mine drainage from metal mines often contains dissolved heavy metals in toxic concentrations (Table 1). The processes that produce AMD are natural, but they are accelerated by mining and can produce large volumes of contaminated effluents.

Put simply, acid mine drainage arises from the rapid oxidation of sulfide minerals, and often occurs where such minerals are exposed to the atmosphere by excavation from the earth's crust. Road cuts, quarries, or other rock excavations can expose these minerals, but metal mines are a primary source since economically recoverable metals often occur as orebodies of concentrated metal sulfides (e.g., iron-pyrite, FeS_2; copper-chalcopyrite, $CuFeS_2$; zinc-sphalerite, ZnS).

Each of the basic steps in the mining process may expose metal sulfides and be a source of acid mine drainage (Fig. 1). Water flowing through the physical mine excavations, ore, mine, and mill wastes, and metal concentrates, can become contaminated. The prediction and prevention of acid mine drainage from these sources are key elements of a strategy to control pollution from mining operations. While knowledge of the acid formation process is incomplete, several factors are known to control the production of AMD.

In this chapter, we discuss these factors, review the techniques and models currently used in North America to predict mine drainage quality, and present some practical constraints, results of some verification studies, and the current approach to prediction. This chapter is intended to be a review of the science, and the art, of acid mine drainage prediction for metal mines.

While approaches and techniques have been developed solely for the prediction of AMD from metal mines, much more research has been conducted on the factors controlling AMD production and prediction in the coal mining region of the United States. Much of the knowledge gained from these coal mining studies is relevant to metal mining AMD prediction. This chapter includes extensive reference to those studies and their findings.

[1]Environmental Protection, Kapilano 100, Park Royal, West Vancouver, B.C. V7T 1A2, Canada
[2]US Bureau of Mines, Cochrans Mill Road, P.O. Box 18070, Pittsburgh, PA 15236, USA

Table 1. Examples of acid mine drainage quality

Parameter[a]	Seepage from abandoned uranium mine tailings pond in Ontario	Waste rock dump seepage from active silver mine in British Columbia	Mine water from abandoned underground copper mine in British Columbia	Runoff from abandoned coal refuse pile in Ohio
pH	2.0	2.8	3.5	2.0
Sulfate	7440	7650	1500	> 20000
Acidity	14600	430	–	21000
Iron	3200	1190	10.6	7000
Manganese	5.6	78.3	6.4	8.4
Copper	3.6	89.8	16.5	0.1
Nickel	3.2	8.0	0.06	2.2
Aluminum	588	359	–	760
Lead	0.67	< 2	0.1	–
Cadmium	0.05	0.5	0.143	–
Zinc	11.4	53.2	28.5	1.9
Arsenic	0.74	25	< 0.05	–
Reference	Hawley 1977	EP[b] 1985 unpublished data	EP 1985 unpublished data	USBM[c] 1981 unpublished data

[a]Units are mg l^{-1} except pH.
[b]EP = Environmental Protection (Canada).
[c]USBM = United States Bureau of Mines.

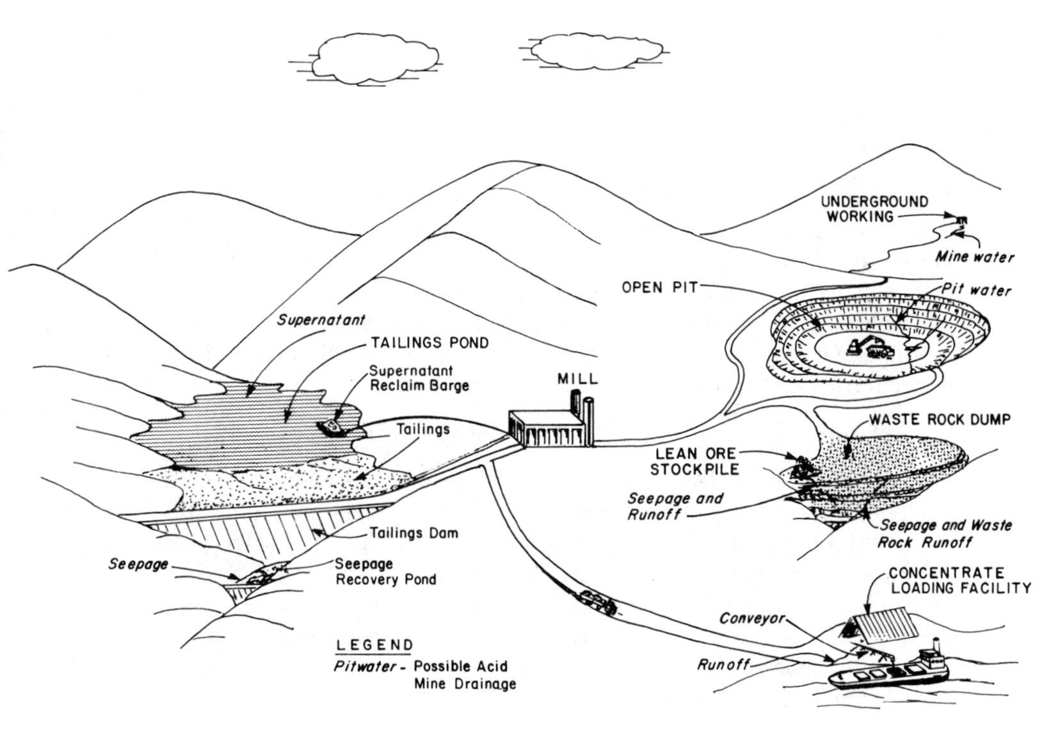

Fig. 1. Major components of a metai mine and sources of acid mine drainage

2 Factors Controlling Drainage Quality

The factors controlling AMD formation may be categorized as primary, secondary, and tertiary. The primary factors are those directly involved in the acid production process. Secondary factors control the consumption or alteration of the products from the acid generation reactions, while tertiary factors are the physical aspects of the waste materials or mine site that influence acid production, migration, and consumption. The three types of factors control the AMD reactions and the quality of water emanating from the mine waste. Other physical and chemical processes may affect the quality of the drainage shortly after it leaves the acid-producing material; these "downstream factors" are also briefly discussed.

2.1 Primary Factors.

The primary factors include pyrite and other sulfide minerals, oxygen, water, ferric iron, and iron-oxidizing bacteria which play key roles in the acid production reactions. Four stoichiometric equations describe the production of AMD from pyrite. The first reaction (Eq. 1) occurs abiotically and with catalysis by *Thiobacillus ferrooxidans*, although the importance of biological mediation is not clear:

$$FeS_2(s) + 7/2\ O_2 + H_2O \rightarrow Fe^{2+} + 2SO_4^{2-} + 2H^+. \tag{1}$$

Direct oxidation of pyrite by oxygen (Eq. 1), along with iron oxidation, hydrolysis, and precipitation (Eq. 2 and 3), predominates at high pH:

$$Fe^{2+} + 1/4O_2 + H^+ \rightarrow Fe^{3+} + 1/2\ H_2O; \tag{2}$$

$$Fe^{3+} + 3H_2O \rightarrow Fe(OH)_3(s) + 3H^+. \tag{3}$$

These three relatively slow reactions comprise the initial stage in the three-stage, AMD-formation process described by Kleinmann et al. (1981). Stage 1 persists as long as the environmental pH around the waste particles is only moderately acidic ($pH > 4.5$). The transitional stage 2 occurs as the pH declines and the rate of iron hydrolysis (Eq. 3) slows, providing ferric iron oxidant. Stage 3 consists of rapid acid production by the ferric iron oxidant pathway and becomes dominant at low pH, where ferric iron is more soluble (Eq. 4):

$$FeS_2(s) + 14Fe^{3+} + 8H_2O \rightarrow 15Fe^{2+} + 2SO_4^{2-} + 16H^+. \tag{4}$$

Without catalysis by *Thiobacillus ferrooxidans*, however, the rate of ferrous iron oxidation in acid medium would be too slow to provide a significant oxidant concentration (Singer and Stumm 1970). As such, this final stage in the AMD process only occurs when the bacteria have become established. Equations (2) and (4) combine to form the cyclic, rapid oxidation pathway primarily responsible for the high contaminant loads observed in mining environments. Regardless of reaction pathway, each mole of pyrite oxidized yields 2 mol sulfate, 2 mol hydronium ion, and 1 mol iron.

The process of pyrite oxidation is summarized in Fig. 2.

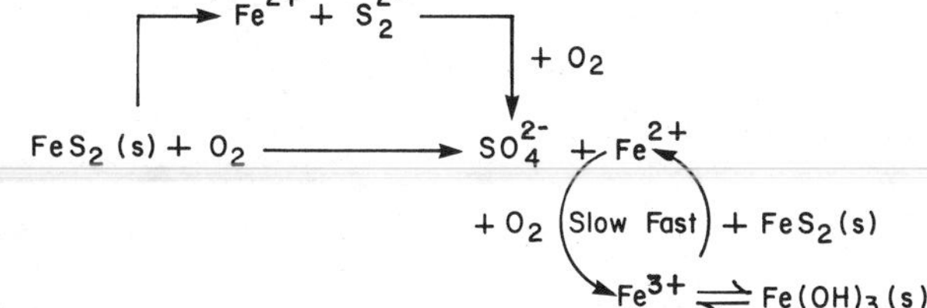

Fig. 2. Reaction pathways for the oxidation of pyrite (Stumm and Morgan 1981)

Oxygen is an essential reactant in the oxidation of sulfide minerals and is required by the aerobic autotrophic bacteria. There is confusion, however, on the dependence of overall acid production rates on oxygen availability. Various reports indicate reaction orders from zero to unity with respect to oxygen; some suggest that the reaction order changes with oxygen partial pressure (Harvard University 1970; Smith and Shumate 1970; NUS 1971; Morth et al. 1972; Nicholson 1984). Some of the conflict undoubtedly arises from the variety of experimental conditions used in the studies.

The replenishment of oxygen within a mining waste from the atmosphere is probably required to sustain the rapid, bacterially catalyzed oxidation reactions of stage 3. Significant sulfide oxidation may not occur in the mining waste where oxygen levels in the atmosphere surrounding the mineral are below 1 to 2%.

Water serves several roles in the acid production system, acting as a reactant, as a reaction medium, and as a product-transport solvent. The first two roles can be considered primary factors, as discussed by Smith and Shumate (1970) and Morth et al. (1972). Bacterial activity may be limited by moisture availability (Belly and Brock 1974; Kleinmann 1979).

Different forms of pyrite may exhibit different reactivities and, hence, affect the rates of reactions. Iron sulfide minerals present as isometric pyrite may not be as reactive as orthorombic marcasite or hexagonal pyrrhotite (Hawley 1977).

While pyrite is the dominant metal sulfide mineral in most sulfide deposits and, as such, plays the key role in the formation of AMD, other metal sulfides are also important in determining the quality of mine drainage. Metal ions can be released from the metal sulfide mineral by direct oxygen oxidation, bacterial oxidation, or acidified ferric sulfate dissolution (Dutrizac and MacDonald 1974). Autotrophic bacteria, notably *Thiobacillus ferrooxidans* and *Thiobacillus thiooxidans*, can attack most sulfide minerals under suitable conditions (Bruynesteyn 1982; Lundgren et al. 1972). Reactions for bacteria and ferric ion attack of some other metal sulfide minerals are shown in Fig. 3.

Galvanic corrosion may also occur during bacterial leaching of mixed sulfide minerals where one sulfide mineral is preferentially leached over another. For example, pyrite in direct contact with chalcopyrite does not react as vigorously as it does in isolation (Dutrizac and MacDonald 1974; Murr and Mehta 1982). The anodic reactions of chalcopyrite are illustrated in Eq. (5) (Murr and Mehta 1982):

$$CuFeS_2 \rightarrow Cu^{2+} + Fe^{2+} + 2S^\circ + 4e^- . \qquad (5)$$

CHALCOPYRITE (Dutrizac & MacDonald 1974; Murr & Mehta 1982)

$$CuFeS_2 + 4Fe^{3+} \rightarrow Cu^{2+} + 5Fe^{2+} + 2S^0$$

$$CuFeS_2 + 4Fe^{3+} + 2H_2O + 3O_2 \rightarrow Cu^{2+} + 5Fe^{2+} + 2H_2SO_4$$

$$2\,CuFeS_2 + {}^{17}\!/_2O_2 + H_2SO_4 \xrightarrow{\text{bacteria}} 2\,CuSO_4 + Fe_2(SO_4)_3 + H_2O$$

CHALCOCITE (Dutrizac & MacDonald 1974; Bruynesteyn & Duncan 1977)

$$Cu_2S + 4Fe^{3+} \rightarrow 2Cu^{2+} + 4Fe^{2+} + S^0$$

$$Cu_2S + H_2SO_4 + {}^1\!/_2O_2 \rightarrow CuS + CuSO_4 + H_2O$$

$$CuS + 2O_2 \xrightarrow{\text{bacteria}} CuSO_4$$

SPHALERITE (Dutrizac & MacDonald 1974)

$$ZnS + 2Fe^{3+} \rightarrow Zn^{2+} + 2Fe^{2+} + S^0$$

$$ZnS + 2O_2 \xrightarrow{\text{bacteria}} ZnSO_4$$

GALENA (Dutrizac & MacDonald 1974)

$$PbS + Fe_2(SO_4)_3 \rightarrow PbSO_4 + 2FeSO_4 + S^0$$

MILLERITE (Bruynesteyn & Hackl 1984)

$$NiS + 2O_2 \xrightarrow{\text{bacteria}} NiSO_4$$

Fig. 3. Reactions involving other sulfide minerals in the formation of acid mine drainage

2.2 Secondary Factors

Regardless of the presence of pyrite and other sulfide minerals, AMD may not be present at a mine site due to secondary factors. The most important secondary factor is the neutralization of acid by alkalinity released from carbonate minerals in the mine waste, such as calcite ($CaCO_3$) and dolomite ($CaMg(CO_3)_2$). The neutralization by calcite of acidity produced from pyrite oxidation may be represented as follows (Williams et al. 1982):

$$FeS_2(s) + 2CaCO_3(s) + 15/4O_2(g) + 3/2H_2O \rightarrow Fe(OH)_3(s) + 2SO_4^{2-} + 2Ca^{2+} + 2CO_2(g). \quad (6)$$

There is some evidence that carbonate minerals may also inhibit sulfide oxidation and, thus, should also be considered primary factors.

The dissolution of carbonate minerals yields alkalinity according to complex solution equilibria that depend on carbon dioxide partial pressure, temperature, mineral form, and solution composition. Geidel (1980) found that the release of alkalinity is limited by solution equilibria, while the accumulation of acidity from sulfide oxidation is controlled by kinetics. In some cases, buffering by carbonate minerals is sufficient to maintain the higher pH environment represented by Kleinmann's stage 1. Relative rates of reactions of sulfide and carbonate minerals determine the resultant drainage quality, ranging from high pH-low sulfate in carbonate-dominant materials to low pH-high sulfate in carbonate-deficient materials; rock high in both pyrite and carbonate content can produce high pH-high sulfate drainage (Caruccio 1968). Low pH-high sulfate water is the classic AMD composition.

Other secondary factors include alteration of oxidation products by further reaction, such as ion exchange on clay surfaces, gypsum precipitation, and acid-induced dissolution of other minerals. These reactions change the character of the drainage, often adding different heavy metals (Al, Mn, Cu, Pb, Zn, etc.) and replacing iron with alkaline earth cations from carbonates.

2.3 Tertiary Factors

The physical characteristics of a mining waste, the spatial relationship between wastes, and the hydrologic regime are some of the tertiary factors that affect the AMD process. The important physical characteristics include particle size, physical weathering tendency, and mine waste permeability. The rate of acid production is a function of the particle surface area since that property reflects the amount of sulfide exposed for reaction. In stage 1 of the process, the rate of acid production is linear with respect to the exposed surface area (Nicholson 1984; Ferguson 1985). However, coarse-grained mining wastes allow greater oxygen advection, and hence a greater depth of active acid generation, than fine-grained waste. In very coarse metal mine waste rock dumps, air convection is promoted by wind action, barometric pressure changes, and internal dump heating from the exothermic oxidation reactions. Under these conditions, active acid generation may occur throughout the dump rather than being limited to the surface zone, as in finer grained mining wastes such as tailings.

Water from open pits and underground workings may not be as contaminated as seepage from waste rock and tailings due to a smaller exposed surface area of sulfide minerals. Fine-grained tailings exhibit a very high exposed surface area, but acid mine drainage production may be limited by oxygen transfer and the high pH of the tailings water during the mine's operating life. Low-grade ore stockpiles are physically similar to waste rock dumps but often contain more sulfide minerals. These stockpiles can become one of the most significant sources of AMD if they are not processed through the mill within a reasonable time after mining. Concentrates have extremely high exposed surface area and sulfide mineral contents. However, water contact with concentrates is discouraged for economic reasons, therefore, they are not usually sources of AMD, except where spills occur.

The physical weathering tendency of a waste controls its permeability in the long term and affects oxygen and water transfer. As the permeability decreases, the rate of acid mine drainage production should slow. However, observations of some high sulfide mining wastes indicate that significant loadings of AMD products can continue to be released for many decades after mining ceases.

The sequence of stacking mining wastes as part of the mining plan may affect the quality of mine drainage. Calcareous materials placed above pyritic waste may reduce acid generation by providing alkalinity to infiltrating water (Infanger and Hood 1980). According to a mathematical model of stage 1 of the AMD process, $CaCO_3$ would be depleted in a 0.2-m limestone layer overlying tailings containing 5 wt.% sulfur in about 7 to 30 years for each weight percent of $CaCO_3$ present in the cover (Nicholson 1984).

Water infiltrating a mine waste participates as a primary factor in acid production reactions in the unsaturated zone. Significant acid generation below the water table may not occur, as inundation of the wastes limits oxygen transfer. A fluctuating water table, however, can periodically wet and dry acid-producing materials allowing oxidation during declines in water table elevation (Caruccio et al. 1980). Under these conditions, a surface dressing of calcareous materials would probably not improve drainage quality.

Acid may be generated to the greatest extent during the dry periods between flushing by precipitation, runoff, or groundwater. Moreover, the concentration of contaminants may be proportional to the length of time between flushings. Frequent flushings would then result in mildly acidic leachates, while accumulated products would yield high contaminant concentrations at longer flushing intervals (Geidel 1980). This is consistent with observed seasonal high concentrations in spring and early fall discharges from mines.

2.4 Downstream Factors

Where AMD is formed within a mining waste or product, natural processes some distance away from the source affect the quality of the drainage and its impact on the environment. These include both physical processes, such as dilution, and chemical processes. Once above the ground surface and separated from sulfide minerals, bacteria and oxygen convert the ferrous ions in AMD to ferric ions. This reaction consumes oxygen and 1 mol hydrogen ion per mole of iron. Subsequent hydrolysis of ferric ion produces 3 mol hydrogen ion per mole of iron. Consequently, some distance downstream from the AMD source, the oxygen content and pH of the water may decrease, and the stream bed becomes coated with iron precipitate. Further downstream, the pH of the stream will rise due to dilution and chemical processes. Between pH 5 and 6 most of the iron, silver, and lead precipitate as metal hydroxides. Reactions with carbon dioxide and carbonates raise the pH to around 7 or 8 (Wildeman 1981). Much of the copper and zinc may precipitate as hydroxides, while sulfate remains in solution. Since sulfate is produced from the primary AMD reactions and remains in solution at relatively high concentrations and pH, it is often used as a downstream water quality indicator of sulfide oxidation and possibly, AMD formation within mine wastes.

Some distance from the source, the receiving stream will fully assimilate the acid mine drainage, but much environmental damage may have occurred in the preceding stream reach.

3 Acid Mine Drainage Prediction Techniques

The prediction of acid mine drainage is needed to determine prior to mining if the quality of waters draining a mine site will exceed regulatory standards. If the oxidation processes become established, mine water contaminant levels can exceed these standards by many orders of magnitude. Thus, accurate prediction of AMD is required both to protect the environment and to ensure that resources are expended wisely to prevent or control AMD.

From the previous discussion of the many primary, secondary, and tertiary factors that control the acid mine drainage process, it may appear that the prediction of mine water quality is an impossible task. However, only a limited number of these factors may actually control the acid mine drainage process in most practical situations; therefore, models used for AMD prediction may be relatively simple.

Several predictive techniques have been developed based upon simplified AMD production models and are routinely used to estimate the quality of drainage prior to mining. Most of the methods provide only a qualitative prediction of acid mine drainage formation. A few models have also been developed to quantitatively estimate the composition of mine drainage and these are most frequently used to predict the success of measures to control AMD formation.

Methods to predict acid mine drainage can be divided into five groups:

1. Geographical mining comparisons;
2. Paleoenvironmental and geological models;
3. Geochemical static tests;
4. Geochemical kinetic tests;
5. Mathematical models.

3.1 Geographical, Paleoenvironmental, and Geological Models

Historical information on the quality of drainage from mining in geographical proximity to a proposed mine may be valuable in AMD predictions. The approach may be useful at metal mines where the orebody has been mined in the past. In those cases, abandoned tailings and waste rock dumps should be examined for evidence of acid mine drainage. This technique, however, is generally more applicable to stratigraphically continuous coal mines in sedimentary sequences than to metal mines in igneous and metamorphic occurrences.

Paleoenvironmental models have been used to predict the quality of coal mine drainage on a regional scale. Pyritic sulfur formed in marine or brackish water environments may have a greater tendency to generate AMD than that formed in freshwater environments (Williams et al. 1982). Fine-grained "framboidal" pyrite may be formed in marine environments and may be more reactive than other pyrite crystal structures (Caruccio 1975).

The paleoenvironmental approach is not applicable to metal mines. However, geologists usually develop a geological model for an orebody to estimate reserves and prepare a mining plan. Such a model is a valuable tool for interpolating results of acid potential tests for two or more cores, and for extrapolating results beyond the limits of testing.

3.2 Geochemical Tests

Both static and kinetic geochemical tests are based primarily on the assumption that geochemical reactions are the main factors that control mine drainage quality, although the kinetic tests may also consider the effect of other selected factors in AMD production.

The static geochemical test methods are characterized by analysis of the ability of a rock to produce acid and base. Acid potential can be measured in several ways. Most commonly, the total sulfur content of the sample is measured and the value is converted to the stoichiometric equivalent of acid produced on complete reaction. Alternatively, the sulfide sulfur content may be determined and used in the calculation if significant organic or sulfate sulfur is suspected. The base potential (acid consumption) is measured by the ability of the sample to neutralize strong acid, either by direct or back titration. A net base (net neutralization) potential is calculated from the results by subtracting the acid potential from the base potential: in general, a negative net value indicates a potential acid producer.

Both of the most widely used geochemical analytical techniques are static methods: the Acid-Base Account method for coal mines in the United States, and the B.C. Research Initial Test for metal mines in Canada. The procedural outlines are shown in Table 2. The costs per test shown in Tables 2 and 3 are the approximate prices of analyses by some commercial laboratories in North America. They are given to illustrate the relative costs of the various tests; prices for a particular test may vary between laboratories.

Table 2. Static acid mine drainage prediction techniques

Test	Description	Laboratory cost per test[a]	References
1. Acid/base Accounting	Potential acidity is calculated from total or pyritic sulfur analysis Neutralization potential is obtained by adding a known amount of HCl, heating sample, and titrating with standardized NaOH to pH 7 Potential acidity is subtracted from neutralization potential. A negative value below 5 t $CaCO_3$/1000 t of rock indicates a potential acid producer Paste pH is also determined	\$25 to \$50	Sobek et al. (1978) West Virginia Surface Mine Drainage Task Force (1979)
2. B.C. research Initial Test	Acid-producing potential is calculated from total sulfur analysis Acid-consuming ability is obtained by titrating with standardized sulfuric acid to pH 3.5 Acid-producing potential is subtracted from acid-consuming ability. A negative value indicates a potential acid producer	\$100 to \$175	Bruynesteyn and Hackl (1984) Bruynesteyn and Duncan (1979)

[a] 1985 United States dollars.

Table 3. Kinetic acid mine drainage prediction techniques

Test	Description	Laboratory cost per test[a]	References
1. B.C. Research Confirmation Test	Sample placed in 250 ml Erlenmeyer flask with 70 ml nutrient media, culture of *Thiobacillus ferrooxidans* at pH 2.2–2.5 Flask placed on gyratory shaker at 35°C in CO_2-enriched atmosphere pH monitored and additional sample added If pH rises substantially, then sample nonacid producer If pH remains low, then sample potential acid producer	$300	Bruynesteyn and Duncan (1979) Bruynesteyn and Hackl (1984)
2. Shake flasks	Sample placed in 1 liter Erlenmeyer flask with 600 ml water or nutrient solution Series of samples tested at various starting pH, inoculation, and temperature Samples incubated for up to 3 months Leachates analyzed weekly and bi-weekly for range of paremeters	$1500 to $3500	Davidge (1984) Halbert et al. (1983)
3. Soxhlet reactor	Standard or modified Soxhlet reactor used Water placed in reservoir, vaporized and passed into condenser. Condensed liquid drips into thimble with sample and then back into the reservoir Leachate analyzed after 64–192 h for a range of parameters	$100 to $300	Sullivan and Sobek (1982) Renton (1983)
4. Humidity cell	Sample placed in plexiglass container connected to humidified air Dry air passed over sample for 3 days, humidified air for 3 days, and 200 ml water added on seventh day Leachate removed and analyzed for range of parameters Procedure repeated for 8–10 weeks	$100 to $500	Caruccio et al. (1977) Caruccio et al. (1980)
5. Columns/ lysimeters	Sample placed in column and periodically leached by distilled water Samples of leachate analyzed for range of parameters Usually leached for 8–10 weeks minimum Several variations of setup and leaching procedures in literature	$500 to $1500	Sturey et al. (1982) Apel (1983) Ritcey and Silver (1981)
6. Test plots/ pits/piles	Run of mine or modified sample placed on impervious surface Precipitation provides leachate which is collected Samples of collected leachate analyzed for range of parameters Test usually run for at least 1 year	$100 to $25000	Eger et al. (1981) Murray and Okuhara (1980)

[a] 1985 United States dollars.

The main advantage of static methods is their simplicity. However, interpretation of the results is difficult when a sample exhibits similar acid and base potential. These procedures also ignore the disparate nature of the reactions involved. The acid-producing sulfide oxidation is under kinetic control, while the base-producing carbonate dissolution is an equilibrium process, as discussed previously.

Kinetic overburden analysis methods were designed to consider the relative rates of reaction. These methods attempt to simulate the weathering process that leads to acid and base dissolution. The main difference between static and kinetic tests is that the latter class considers the opposing acid production and consumption reactions together. A great number of protocols have been developed for kinetic tests of coal and metal mine materials (Table 3). In all cases, however, water is added to a rock sample, acid-producing and -consuming reactions are allowed to proceed, and samples of leachate are analyzed. The test conditions of water application include continuous or intermittent inundation, continuous or intermittent simulated rainfall, and alternate exposure to humid atmospheres and leaching. Some procedures require inoculation with *Thiobacillus ferrooxidans* to simulate the rapid stage of oxidation.

Kinetic methods have greater expense and time requirements than static tests. Moreover, there are no generally accepted criteria for the evaluation of test results other than relative differences in leachate quality; interpretation of the results and application to a mine proposal are left to the investigator. Investigators usually follow trends in pH, sulfate, acidity or alkalinity, and metals. The pH of the collected leachate helps to identify the stage of the AMD process. Sulfate production can be correlated with sulfide oxidation rates. The acidity or alkalinity of the leachate is an indication of the relative rates of acid production and consumption. Metal analyses may identify species that are readily released by weathering. To the degree that kinetic tests simulate field processes at an accelerated rate, and these field processes are understood, kinetic methods may be expected to provide reasonable predictions of AMD formation. However, these tests cannot model all of the factors that control AMD production in the field, nor are laboratory rates the same as field rates (Lapakko and Eger 1980; Bruynesteyn and Hackl 1984).

Of those kinetic tests gaining acceptance, the B.C. Research Confirmation Test is probably most widely used; it is currently applied frequently at base metal and gold mines in Canada. Positive results from this test are considered confirmation that the microbiologically catalyzed reactions can become self-sustaining (Bruynesteyn and Hackl 1984).

The other kinetic tests are used to some extent in Canada and the United States; some have been used in mining applications, while others were published as research methods. An advantage of kinetic techniques is that the procedure can be altered to estimate the role of a given parameter, such as particle size or oxygen availability on AMD production. In some tests, such as columns and test piles, lithologies may be stacked or mixed in an attempt to model the effect of the mining plan on AMD production.

3.3 Mathematical Models

Several mathematical approaches have been developed to model the acid mine drainage formation process and to quantitatively predict the composition of mine drainage (Table 4). None of the models consider all of the primary, secondary, and tertiary factors in AMD production. Three of the models (Caruccio and Geidel 1981b; Halbert et al. 1982; Nicholson 1984) can be used to predict whether a mining waste can generate sufficient acidity to allow the AMD process to enter the final stage 3. One of these models (Halbert et al. 1982) and the other four can be used to estimate mine water quality from bacterially catalyzed AMD processes. All models require geochemical and waste characteristic input data, and two (Caruccio and Geidel 1981b; Halbert et al. 1982) utilize results from specially designed kinetic tests.

None of the mathematical models have been used extensively to predict AMD. Moreover, while field data were used to develop the models, very few independent tests have been conducted to verify their application. The use of these and other mathematical models may increase as they become better known and refined, reflecting an improved understanding of the factors affecting AMD production. They are not discussed in subsequent sections of this chapter due to their limited use to date.

4 Limits on Prediction Accuracy

The ability to make accurate pre-mine predictions of AMD with the techniques previously discussed is hampered by the difficulty in obtaining representative geochemical samples, and in projecting the effectiveness of AMD prevention techniques to field situations.

Geochemical samples must be representative of the mine waste material. Composites of split drill core or randomly selected grab samples may be satisfactory. However, the number of samples required for testing depends on the variability of mineralization and should be decided by a geologist (Duncan and Walden 1975). This approach could lead to representative samples, in the sense of best representing important strata, if the geological factors are selected wisely and the local geology is known.

Beyond this general guideline, little work has been done to develop sampling criteria. Barth et al. (1981), in a state of the art review, reported that regulatory requirements for drill holes to characterize overburden from coal mines in the United States ranged from one per eight acres to one per 160 acres. Although not stipulated by the regulations, certain drilling methods may provide better sample integrity. The sampling pattern also influences the results. On a square grid, a high sample density gives the best representation. However, the optimum density would incur large drilling costs.

While composite samples may give a good estimate of the acid generation potential of waste rock or tailings as a whole, they do not provide information on

Table 4. Mathematical models for acid mine drainage prediction

MODEL	APPLICATION	FACTORS IN AMD PRODUCTION CONSIDERED: PRIMARY					SECONDARY			TERTIARY			MODEL OUTPUT
		Pyrite Content	Oxygen Availability	Water Limitations	Fe+3 Oxidation	Bacterial Oxidation	Carbonate Content	Alkalinity Production	Secondary Reactions	Site Hydrology	Waste Characteristics	Mining Plan	
Caruccio & Geidel 1981a and b; Caruccio 1984	Backfilled Coal Strip Mines	●	P	P	P		●	●		●	P	P	- Initial Minimum Acid Load expected from Backfilled Mine under Abiotic Conditions.
Jaynes et al. 1984a and b	Reclaimed Coal Strip Mine Overburden Dumps	●	●		●	●	P	P	●	●	●	P	- Long Term (Years) Oxidation of Pyrite and Production of Iron and Acidity - Other Parameters (Pyrite Consumption, Oxygen Profiles, etc.) may also be modelled.
Ricca & Schultz 1979	Coal Strip Mine Overburden Dumps and Refuse Piles	●	●	P	P	P				●	P	●	- Daily Fluctuations in Flow, Acidity, and Iron
Halbert et al. 1982; Halbert et al. 1983	Metal Mine Tailings	●	●	P	●	●	P	●		●	●	P	- Equilibrium Chemical Characteristics (pH, Iron, Sulfate) of Porewater
Nicholson 1984	Metal Mine Tailings	●	●	P			●	●	P	P	●	P	- Initial Acid Production and Pyrite Oxidation at Porewater pH 7-8
Cathles 1979; Cathles & Apps 1975	Copper Mine Waste Rock Dumps	●	●	P	●	●			●	●	●	P	- Short and Long Term Copper Leaching Rates - Other Parameters (Temperature, Oxygen, etc.) may also be modelled
Davis & Ritchie 1983	Metal Mine Waste Rock Dumps	●	●		P	P					●	●	- Short and Long Term Sulfate Production Rates - Other Parameters may also be modelled

● Factor considered in Model P - Factor partially considered in Model the variability of acid generation potential within an orebody. Monitoring by the US Bureau of Mines at several regraded coal mines recharged predominantly by rainfall showed that near-surface material contributed the major contaminant loads. In one case, a perennially saturated, highly pyritic zone at depth contributed virtually no oxidation products, while near-surface spoil of lower pyrite content produced disharge acidities of 1000 mg l^{-1} as $CaCO_3$ (Lusardi and Erickson 1985). At sites such as this, predictions should be based on the nature of material that will comprise the near-surface backfill after mining, rather than the entire overburden.

Tailings and waste rock dumps tend to be more homogeneous than coal mine backfills. However, even here, the nature of near-surface materials significantly affects drainage quality and, therefore, must be given special consideration in AMD predictions.

Special material handling procedures, designed to lessen the probability of acid generation, can be incorporated into a mine plan. Techniques such as segregation or blending of acid-generating and -consuming rocks, waste compaction, underwater tailings disposal, encapsulation of potentially acidic waste, rapid revegetation, and construction of infiltration galleries and low permeability caps are just some of the methods suggested to control AMD production (West Virginia Surface Mine Drainage Task Force 1979; Halbert et al. 1982; Smith and van Zyl 1983). The effectiveness of these techniques to prevent AMD formation has not been proven. However, use of these techniques in a mining operation, whether planned or unplanned, affects the accuracy of AMD predictions.

5 Verification of Prediction Techniques

Pre-mine prediction of acid mine drainage has only become widespread in North America within the last decade under the impetus of mining regulations. The number of cases, particularly for metal mines, is still small. Moreover, several years of testing the actual mine water quality after start-up are required to determine if predictions were correct. This may explain, in part, why so little work has been done to verify the relationship between AMD prediction results and post-mining drainage quality. This step is critical if predictions are to be made with confidence.

Some investigators have shown a good correlation between laboratory static and kinetic tests applied to coal and metal mine materials (Duncan and Walden 1975; Sturey et al. 1982; Hood and Oertel 1984; Ferguson 1985). Others found a poor correlation and suggested that since static tests oversimplify AMD processes, they should not be used for prediction except for obvious situations (Caruccio et al. 1980; Williams et al. 1982; SENES 1984).

A few studies have shown that AMD predictions using static and kinetic techniques correlated well with actual mine water quality. Ritcey and Silver (1981) found the results of lysimeters and shake flask tests agreed with results from large lysimeters and field test pits of tailings from an uranium mine in Ontario, Canada. Hood and Oertel (1984) found the results of column tests correlated with conductivity and total dissolved solids levels in waters draining some coal mines in the

United States, although a scaling factor was required. Column leaching tests of overburden from three coal mines in the United States correctly predicted the quality of mine drainage (Mercier 1978). Ferguson (1982) reviewed published static test results from 13 metal mines in Canada and found that AMD was accurately predicted in all but one case.

In a separate study, static and kinetic tests were performed on 22 rock samples from seven metal mines in British Columbia and Yukon, Canada (Ferguson 1985). Samples of tailings and waste rock were obtained from both active and abandoned mine sites. The total sulfur contents ranged from 0.13 to 49.2%, and neutralization potentials ranged from 0 to 258 t $CaCO_3$ equivalent per 1000 t material. The very high percent sulfur contents of some samples yielded a definite prediction for acid production according to static methods. The static and kinetic tests correctly predicted the formation of AMD in all but six cases: three were incorrect and three were inconclusive. A few results from the kinetic tests (humidity cells) were difficult to interpret due to indefinite pH trends. The results supported the use of static tests for predicting AMD, but suggested there was only a poor relationship between the magnitude of the net base value (net neutralization potential) and the amount of acidity produced during the initial stage of the AMD process. The static tests may be useful for qualitative AMD predictions but not for estimates of actual mine water quality.

Pre-mining static tests of waste rock samples from a silver mine in British Columbia correctly predicted that acid mine drainage would be generated, but low rainfall and cold temperatures were erroneously considered by the interpreter to limit formation in the field (Wilkes 1985). Within 2 years from the commencement of mining, acidic drainage was flowing from the waste rock dump, mine plant site, haul road, tailings dam, and open pit.

Although static and kinetic tests of tailings from two Canadian mines have indicated that AMD would be generated, little acidic leachate has been found after 4 to 10 years of operation (Wilkes 1985; Duncan 1975). Residual alkalinity provided by lime added in the milling process may be limiting acid production in these tailings. Whether this protection will be effective in the long term is of critical importance in predicting the environmental impact of these mines and designing the abandonment plans.

Ferguson (1985) reported a third case where static tests indicated AMD would be generated at a Canadian metal mine, but no acidic leachate was found after 3 years of mining. In that study, a kinetic test (humidity cell) was also performed, in which the leachate remained strongly alkaline during the 10-week test indicating AMD would not be generated. However, the tailings from this mine are deposited primarily under water, which may inhibit acid generation. This may be a case where tertiary factors dominate in determining acid production.

The US Bureau of Mines is currently evaluating the accuracy of one static and two kinetic techniques for 30 surface coal mines in the eastern United States. Sites were deliberately selected to represent the most difficult prediction cases; that is, where the pyrite and carbonate content approach stoichiometric balance. Field and laboratory work have been completed, and the data are being analyzed for statistically significant relationships between overburden analytical results and actual drainage quality. Preliminary results from eight mines suggest that the expected drainage acidity did not correlate well with actual mine water quality. Work is ongoing to evaluate the effect of mining and reclamation procedures, and to develop improved mathematical analyses of results.

6 Approach to Prediction

While errors have been made in AMD predictions and the relative merits of static, kinetic, and mathematical approaches are debatable, the growing body of successful predictions suggests that an investigator may approach the problem of prediction with some confidence. Moreover, the potential environmental impact of acid mine drainage and cost of remedial work necessitates that predictions be made despite the potential for error.

Investigators often favor their most familiar testing procedure, usually a static geochemical method. These relatively fast, low cost techniques allow many tests to be made of the orebody and waste material. The tests establish the relative quantity and magnitude of potentially acid-generating and -consuming materials. Also, with many tests, the spatial variability of AMD potential within a rock mass to be mined can be determined. Static tests are almost always used in an initial testing program at a mine property.

Experienced investigators compare the data from the testing program to the results from past successful and unsuccessful predictions. This comparison is a critical step that has not yet been developed into a systematic, empirical method. Accurate predictions do not follow, a priori, from the testing programs.

Fortunately, in many metal mining situations, the great abundance or general lack of potentially acid-producing sulfide minerals relative to acid-consuming carbonate minerals in the ore and waste indicates a clear and correct prediction. However, where the potential for AMD has been predicted by static geochemical testing, the investigator must design and develop measures to prevent AMD formation, a far more difficult task and one for which the simple static approaches have not been specifically designed. Here, the investigator may try the more complex kinetic and mathematical approaches. With these more complex techniques, the design of experiments and interpretation of results becomes more difficult, with large confidence limits placed on the findings.

In the end, the conclusions of the studies and recommendations of mitigation measures for a proposed mine are influenced by the sensitivity of the receiving environment to acidic discharges. The environmental sensitivity is dependent upon the physical, chemical, and biological nature of the water bodies that would receive the mine drainage. Where receiving environments are extremely sensitive, comprehensive AMD potential testing procedures with conservative assumptions must be used. Advice from mining staff, AMD researchers, and regulatory agencies will be of value in developing the testing program. Moreover, since prediction in these complex cases invariably requires consideration of many of the primary, secondary, and tertiary factors that control AMD processes, a multidisciplinary approach must be taken in developing the model of mine drainage formation.

Studies of the factors controlling mine drainage quality and methods of prediction for coal and metal mines are being conducted by several agencies, universities, and mining companies in the United States and Canada (Miller 1980; Kim et al 1982; Monenco 1984; SENES 1985; CANMET 1985). These studies should result in improved prediction and control of acid mine drainage.

7 Summary

The control of acid mine drainage represents one of the greatest environmental challenges facing the mining industry and regulatory authorities. To meet this challenge, accurate predictions of AMD must be made.

The processes that produce AMD are not completely understood, but several factors are believed to control the quality of mine drainage. Key factors, such as sulfide and carbonate content, reactive surface area, oxygen and water transfer rates, and bacterial activity have been used to develop models of the AMD process and techniques to predict the formation of acidic drainage. Geochemical static tests, which simply examine the balance between acid-producing sulfides and acid-consuming carbonates, are used most frequently. Kinetic tests and mathematical models in varying complexity attempt to simulate the AMD formation process. In general, these methods are used infrequently.

Despite the importance of AMD prediction, few verification studies have been conducted. However, these studies do indicate that prediction techniques provide an insight into the possible role of AMD processes in determining mine drainage quality. Accurate predictions can be made if the user recognizes and considers the effect of assumptions in the technique at the mine site under study. The development and use of tests and models constitutes the science of prediction. The selection of an appropriate test or model for a specific mining proposal, and translation of the test or model result into a correct prediction, is the art. Much basic and applied research on the AMD process is still required to transfer this art to a science of universal application.

Acknowledgments. The authors gratefully acknowledge helpful discussions and technical review of this chapter by Dr. R.L.P. Kleinmann, Mr. A.A. Sobek, Dr. F.T. Caruccio, Ms. P.M. Mehling, Mr. J. Scott, and Mr. J. Ingles.

References

Apel ML (1983) Leachability and revegetation of solid waste from mining. US Environ Protect Ag Rep EPA-600-52-82-013

Barth RC, Cox LG, Giardinelli A, Sutton SM, Tisdel LC (1981) State of the art and guidelines for surface coal mine overburden sampling and analysis phase I report: survey of current methods and procedures. US Office of surface mining contract J5101047, p 256

Belly RT, Brock TD (1974) Ecology of iron-oxidizing bacteria in pyritic materials associated with coal. J Bacteriol 117:726–732

Bruynesteyn A (1982) Application of microbiological methods to underground leaching of uranium ores. Worksh Microbiological leaching of ores, Moscow, Tula, June 1982

Bruynesteyn A, Duncan DW (1977) The practical aspects of biological leaching studies. Proc 12th Int Mineral processing Congr, Sao Paulo, Brazil

Bruynesteyn A, Duncan DW (1979) Determination of acid production potential of waste materials. Met Soc of AIME Pap A-79-29, p 10

Bruynesteyn A, Hackl RP (1984) Evaluation of acid production potential of mining waste materials. Miner Environ 4:5–8

CANMET (1985) National uranium tailings program – interim report 1983–1985. Can Centre for mineral and energy technology NUTP 85-6, Ottawa, Ontario

Caruccio FT (1968) An evaluation of factors affecting acid mine drainage production and the groundwater interactions in selected areas of western Pennsylvania. Preprints: 2nd Symp Coal mine drainage research, bituminous coal research, Monroeville, Pennsylvania, pp 107–152

Caruccio FT (1975) Estimating the acid potential of coal mine refuse. In: Chadwick MJ, Goodman GT (eds) The ecology of resource degradation and renewal. Blackwell, London, pp 197–205

Caruccio FT (1984) The nature, occurrence and prediction of acid mine drainage from coal strip mines. A study guide for a mini-course. 1984 Nat Symp Surface mining, hydrology, sedimentology and reclamation, Dec 1984, Lexington, Ken

Caruccio FT, Geidel G (1981a) Estimating acid loads and treatment costs of coal strip mines. Proc Symp Surface coal mining and reclamation, 27–29 Oct 1981, Louisville, Ken

Caruccio FT, Geidel G (1981b) Estimating the minimum acid load that can be expected from a coal strip mine. In: Grave DH (ed) Proc 1981 Symp Surface mining hydrology, sedimentology, and reclamation. Univ Kentucky, Lexington, pp 117–122

Caruccio FT, Ferm JC, Horne J, Geidel G, Baganz B (1977) Paleoenvironment of coal and its relation to drainage quality. US Environ Protect Ag Rep EPA-600/7-71-067, p 108

Caruccio FT, Geidel G, Pelletier A (1980) The assessment of a stratum's capability to produce acid drainage. In: Grave DH (ed) Proc Symp Surface mining, hydrology, sedimentology and reclamation. Univ Kentucky, Lexington, pp 437–443

Cathles LM (1979) Predictive capabilities of a finite difference model of copper leaching in low grade industrial sulfide waste dumps. Math Geol 11(2):175–190

Cathles LM, Apps A (1975) A model of the dump leaching process that incorporates oxygen balance, heat balance, and air convection. Metall Trans 6B:617–624

Davidge D (1984) Oxidation of Yukon mine tailings. Environ Protect Serv Reg Progr Rep 84–15, p 39

Davis GB, Ritchie AIM (1983) A model of pyrite oxidation in waste rock dumps. Meet Mining, milling, and waste treatment including rehabilitation with emphasis on uranium mining. Wastewater Assoc, 4–9 September 1983, Darwin, Aust, pp 22-1–22-13

Duncan DW (1975) Leachability of Anvil ore, waste rock and tailings. Dep Ind North Dev ALUR 74-75-37, Ottawa, Ontario

Duncan DW, Walden CC (1975) Prediction of acid generation potential. Proc Seminar Ser: Mining effluent regulations/guidelines and effluent treatment technology as applied to the base metal, iron ore, and uranium mining and milling industry. Environ Can, 17–18 Nov 1975, Banff, Alberta

Dutrizac JE, MacDonald RJ (1974) Ferric ion as a leaching medium. Miner Sci Eng 6:59–100

Eger P, Lapakko K, Weir A (1981) Heavy metals study – 1980 progress report on the field leaching and reclamation program. Minn Dep Nat Resour, p 31

Ferguson KD (1982) Acid generation potential at the Quinsam Coal project. Appendix. In: Quinsam coal task force report, environment Canada/Fisheries and oceans Canada. Vancouver, British Columbia

Ferguson KD (1985) Static and kinetic methods to predict acid mine drainage. Int Symp Biohydrometallurgy, 22–24 August 1985, Vancouver, British Columbia, p 37

Geidel G (1980) Alkaline and acid production potentials of overburden material: the rate of release: Reclam Rev 2:101–107

Halbert BE, Scharer JM, Chakravatti JL, Barnes E (1982) Modelling of the underwater disposal of uranium mine tailings in Elliot Lake. Proc Int Symp Management of wastes from uranium mining and milling, 10–4 May 1982, Abuquerque, New Mexico

Halbert BE, Scharer M, Knapp RA, Gorber DM (1983) Determination of acid generation rates in pyritic mine tailings. 56th Annu Conf Water pollution control federation, 2–7 October 1983, p 15

Harvard University (ed) (1970) Oxygenation of ferrous iron. Rep US Environ Protect Ag EPA 14010-06/69, p 220

Hawley JR (1977) The problem of acid mine drainage in the province of Ontario. Ontario Minist Environ, p 338

Hood WC, Oertel AO (1984) A leaching column method for predicting effluent quality from surface mines. In: Grave DH (ed) Proc 1984 Symp Surface mining, hydrology, sedimentology, and reclamation. Univ Kentucky, Lexington, pp 271–277

Infanger M, Hood WC (1980) Positioning acid-producing overburden for minimal pollution. In: Grave DH (ed) Proc 1980 Symp Surface mining, hydrology, sedimentology, and reclamation. Univ Kentucky, Lexington, pp 325–332

Jaynes DB, Rogowski AS, Pionke HB (1984a) Acid mine drainage from reclaimed coal strip mines – 1. model description. Water Resour Res 20:233–242

Jaynes DB, Pionke HB, Rogowski AS (1984b) Acid mine drainage from reclaimed coal strip mines – 2. simulation results of model. Water Resour Res 20:243–250

Kim AG, Heisey BC, Kleinmann RLP, Deul M (1982) Acid mine drainage: control and abatement research. US Bur Mines IC 8905, p 22

Kleinmann RLP (1979) The biogeochemistry of acid mine drainage and a method to control acid formation. PhD Diss, Princeton Univ

Kleinmann RLP, Crerar DA, Pacelli RR (1981) Biogeochemistry of acid mine drainage and a method to control acid formation. Min Eng (March 1981) V. 33 No. 3:300–305

Lapakko K, Eger P (1980) Mechanisms and rates of leaching from Duluth gabbro waste rock. Proc SME-AIME fall meeting and exhibit, 22–24 October, Minneapolis, Minn

Lundgren DG, Vestal JR, Tabita FR (1972) The microbiology of mine drainage pollution. In: Mitchell R (ed) Water pollution microbiology, Wiley Interscience, New York pp 69–88

Lusardi PJ, Erickson PM (1985) Assessment and reclamation of an abandoned acid-producing strip mine in northern Clarion County, Pennsylvania. In: Grave DH (ed) Proc 1985 Symp Surface mining, hydrology, sedimentology, and reclamation. Univ Kentucky, Lexington pp 313–321

Mercier MJ (1978) A chemical weathering study of overburden materials from three surface coal mines in southern Illinois and western Kentucky. Thesis, S Ill Univ, p 111

Miller CJ (1980) Surface mining for water quality symposium – introductory remarks. Proc Surface mining for water quality. West Virginia surface mining and reclamation association – West Virginia coal association, 14 March 1980, Bridgeport, WV

Monenco Ltd (1984) Sulphide tailings management study. Rep Reactive acidic tailings program, Dep Energ Min Resour, Ottawa, Ontario

Morth AH, Smith EE, Shumate KS (1972) Pyritic systems: a mathematic model. Rep US Environ Protect Ag EPA-R2-72-002, p 169

Murr LE, Mehta AP (1982) Characterization of leaching reactions involving metal sulfides in wastes and concentrates utilizing electron microscopy and microanalysis techniques. Resourc Conserv 9:45–57

Murray DR, Okuhara D (1980) Effect of surface treatment of tailings on effluent quality. Energy Mines Resour Can Div Rep MRP/MRL 80, Ottawa, Ontario

Nicholson RV (1984) Pyrite oxidation in carbonate-buffered systems: experimental kinetics and control of oxygen diffusion in a porous medium. PhD Diss, Univ Waterloo, Ontario

NUS (1971) The effects of various gas atmospheres on the oxidation of coal mine pyrites. Rep US Environ Protect Ag EPA-14010 ECC

Renton JJ (1983) Laboratory studies of acid generation from coal associated rocks. Proc Surface mining and water quality, West Virginia surface mine drainage task force, Charleston, WV

Ricca VT, Schultz RR (1979) Acid mine drainage modeling of surface mining. In: Brawner CO (ed) Proc 1st Int Mine drainage Symp, Miller Freeman, New York, pp 651–670

Ritcey GM, Silver M (1981) Lysimeter investigations on uranium tailings at CANMET. Energ Mines Resour Can Div Rep MRP/MSL 81-36, Ottawa, Ontario

Senes Consultants Ltd. (1984) Assessment of the mechanisms of bacterially assisted oxidation of pyritic uranium tailings. Nat Uranium tailings program, Dep Energ Mines Resour, Ottawa, Ontario

Senes Consultants Ltd. (1985) Probabilistic model development for the assessment of the long-term effects of uranium mill tailings in Canada – Phase II. Nat Uranium tailings program, Dep Energ Mines Resour, Ottawa, Ontario

Singer PC, Stumm W (1970) Acid mine drainage: the rate determining step. Science 197:1121–1123
Smith CS, Zyl D van (1983) Design criteria in acid generating mine waste disposal. Proc 7th Panamerican Conf Soil mechanics and foundation engineering, pp 597–611
Smith EE, Shumate KS (1970) The sulfide to sulfate reaction mechanism. Rep US Environ Protect Ag EPA 14016 EPA, p 115
Sobek AA, Schuller WA, Freeman JR, Smith RM (1978) Field and laboratory methods applicable to overburden and mine soils. US Environ Protect Ag Rep EPA-600/2-78-054
Stumm W, Morgan JJ (1981) Aquatic Chemistry – an introduction emphasizing chemical equilibrium in natural waters, 2nd edn. John Wiley & Sons, New York, 1981
Sturey CS, Freeman JR, Keeney TA, Sturm JW (1982) Overburden analyses by acid-base accounting and simulated weathering studies as a means of determining the probable hydrologic consequences of mining and reclamation. In: Grave DH (ed) Proc 1982 Symp Surface mining hydrology, sedimentology, and reclamation. Univ Kentucky, Lexington, pp 163–179
Sullivan PJ, Sobek AA (1982) Laboratory weathering studies of coal refuse. Miner Environ 4:9–16
West Virginia Surface Mine Drainage Task Force (1979) Suggested guidelines for method of operation in surface mining of areas with potentially acid-producing materials. WV Dep Nat Resour, Charleston, WV, p 18
Wildeman TR (1981) A water handbook for metal mining operations. Col State Univ Rep 113, Fort Collins, Col
Wilkes BD (1985) Prediction of environmental impacts at a silver-gold mine. Follow-up/audit of environmental assessment results Conf, 13–16 October, Banff, Alberta
Williams EG, Rose AW, Parizek RR, Waters SA (1982) Factors controlling the generation of acid mine drainage. Final Rep US Bur Mines, Res Grant G5105086, p 256

Methods for the Treatment of Contaminated Dredged Sediments

W.J.Th. VAN GEMERT, J. QUAKERNAAT, and H.J. VAN VEEN[1]

1 Introduction

Waterways to and from industrial and trade centres must be accessible for nautical transport of cargoes all year long. For Rotterdam, the largest harbour in the world, reaching inland to a distance of up to 50 km, this constitutes the main means of survival (see Nijssen, this Vol.).

In the Rotterdam area more than 20 million m^3 sediments are dredged annually, of which about 10 million m^3 are more or less contaminated. Moreover, the quantities of the sediment which have been severely polluted by local activities in the past exceed 1 million m^3. In about 200 locations elsewhere in the Netherlands, highly contaminated sediments have been found in channels, rivers and other water areas. The quantities may vary from 1,000 up to 100,000 m^3.

After these sediments have been dredged they must be dumped at a specific site because uncontrolled dumping in the sea or on land is unacceptable due to the potential damage to the environment. Thus, special areas for controlled storage of sediments will have to be created. In the Netherlands there is much public resistance against the realization of such areas because:

1. The Netherlands is a small country with almost the highest population density in the world. The available area is therefore in very short supply.
2. The general geological condition of Dutch soil and the generally high level of groundwater make this country very unsuitable for the creation of large areas for controlled storage of contaminated sludges.

Thus, when sediments are to be stored, it is of relative importance that the total volumes remain small and that the contaminants are concentrated as much as possible. It is therefore important to know to what extent contaminated sediments may be technically treated, in order to concentrate or detoxify the contaminants.

Suitable techniques should be economically feasible, when compared to alternatives. It is also not acceptable that the techniques produce large flows presenting higher environmental risks than that of the original sludge. The environmental merits of the techniques to be applied can be quantified by creating suitable standards. These standards should not only be based on the concentrations of the contaminants, but also on short- and long-term leachability estimates (if these can be measured) and on biological availabilities. However, as long as a general agreement on the long-term validity of leaching and an availability test do not

---

[1]T.N.O./Division of Technology for Society, P.O. Box 342, 7300 AH Apeldoorn, The Netherlands

yet exist, one has to rely, at least in the Netherlands, on concentrations of contaminants to form the basis of the standards.

If a certain standard has to be reached, it is important to know first which technological means exist or are to be developed in order to solve the problem, or at least to reduce the problem to smaller or better controllable dimensions. Furthermore, it is important to have informations on the environmental and economical consequences of applying such techniques (also compared to the situation in which no technique at all is applied).

Here, a perspective is given of techniques that may be applicable in treating sediments, whether in combination with controlled dumping of residues, or as an alternative to controlled dumping. Especially those techniques characterized as multipurpose techniques, with quite universal applicabilities with regard to several kinds of contaminated sediments, seem to be favourable at first glance. As to the techniques to be examined, the Netherlands Organization for Applied Scientific Research (TNO) has been carrying out research programmes since 1980. At the moment, results are being demonstrated on a practical scale by private companies. Simultaneously, fundamental research is being carried out on techniques which may be promising in the future.

2 Properties of Contaminated Sediments

2.1 Contamination Standards

In the Netherlands there are several standards which may be used with regards to the examination of contaminated sediments (Table 1). The standards given here concern the rates of contamination in soils. There are no standards for sediments and dredged sediments at the moment. The so-called A, B and C values mentioned

Table 1. Some standards for contamination levels in the Netherlands

	Soil Protection Act[a] (mg kg^{-1})			Chemical Waste Act[b] (mg kg^{-1})
	A	B	C	
Zinc	200	500	3000	20000
Copper	50	100	500	5000
Lead	50	150	600	5000
Chromium	100	250	800	5000 (Cr^{III})
Nickel	50	100	500	5000
Arsenic	20	30	50	50
Cadmium	1	5	20	50
Mercury	0.5	2	10	50
Oil	100	1000	5000	–
EOCl[c]	0.1	8	80	5000

[a]A = Reference value for soil; B = concentration value at which a more detailed examination is considered to be necessary; C = concentration value at which a clean-up is necessary.
[b]Reference values for chemical wastes.
[c]EOCl = Extractable organic chlorine compounds.

in Table 1 have a legislative basis; they stem from the Soil Protection Act. The chemical waste values are based on the Chemical Waste Act for handling wastes.

In the Netherlands dumping of residues including sediments is prohibited at contamination levels higher than the chemical waste level (see Table 1). Thus, these types of residue have to be decontaminated or must be transported out of the country for ultimate controlled storage at a suitable location. The contamination level after decontamination must be in the region of the A value if the sedimentary matter is to be reused in any way.

2.2 Sediment Properties

Dredged sludge generally constitutes fine matter with a dry matter content of 20–50%, depending on the dredging method. Depending on circumstances, the mineral fraction < 16 μm will vary between 10–70%. The fraction >210 μm is usually not higher than a few percent. The commonest contaminations are compounds of the heavy metals (Zn, Cd, Cr, Cu, Pb, Hg), as well as organic compounds [oil, polycyclic aromatic hydrocarbons (PAH), organic chlorine compounds (EOCl)]. The metals are probably present as sulphides and carbonates, or complexly bound in the organic matter present, or bound to the minerals (see Kersten and Förstner, this Vol.).

From analyses of sieve fractions it appears that especially the metal compounds, but organic contaminations too, are found in higher concentrations in the fines (<20 μm). This has been confirmed by Salomons and Förstner (1984). Organic contaminations are often present in increased concentration in a coarse fraction (>210 μm).

A contaminated dredged sediment is in many instances being formed in water systems with its floating components (organic matter and sludge particles). When contaminations are added to the system they will, with regards to the organic microcontaminants, the poorly soluble, lyophilic compounds such as PCB, pesticides, PAH and heavier aliphatic compounds, indicate a preference to be compounded with the solid phase. Reasonable soluble organic compounds such as benzene, toluene and chlorine methanes will remain partly in the water phase, and these compounds will, together with the floating particles, to a considerably lesser degree settle to the bottom.

With regard to further processing, the following properties of dredged sediments are important:

1. Dredged sediment contains a great deal of water (on average approx. 75%).
2. The bond of contamination to sludge is often strong.
3. Dredged sludge contains a relatively high amount of fine mineral matter (fraction < 16 μm is between 10 and 80%).
4. Fines have a higher contamination content than the coarse fraction.

3 Basic Strategies of Treatment

In essence many unit operations are available for the treatment of contaminated residues such as dredged sediments. However, evidently not all unit operations are applicable or feasible in any particular situation. For example, some operations are especially designed to treat very large quantities at low costs. Other operations, in contrast, are preferably designed for small-scale applications at relatively high costs.

Processes for the treatment of large quantities of residues will generally be built up of several unit operations combined in such a way that optimum results are realized at a minimum of operating costs. In reference to environmental technology, an "optimum result" means that the environmental merit of applying a process to a residue is as large as possible, within the accepted conditions of reasonable costs. In a generalized scheme, the scenario for the treatment of residues may be visualized as indicated in Fig. 1, where generally two categories of techniques can be distinguished:

"A" Large-scale concentration techniques. These techniques are characterized by large-scale applicabilities, low costs per unit of residue to be treated and a low

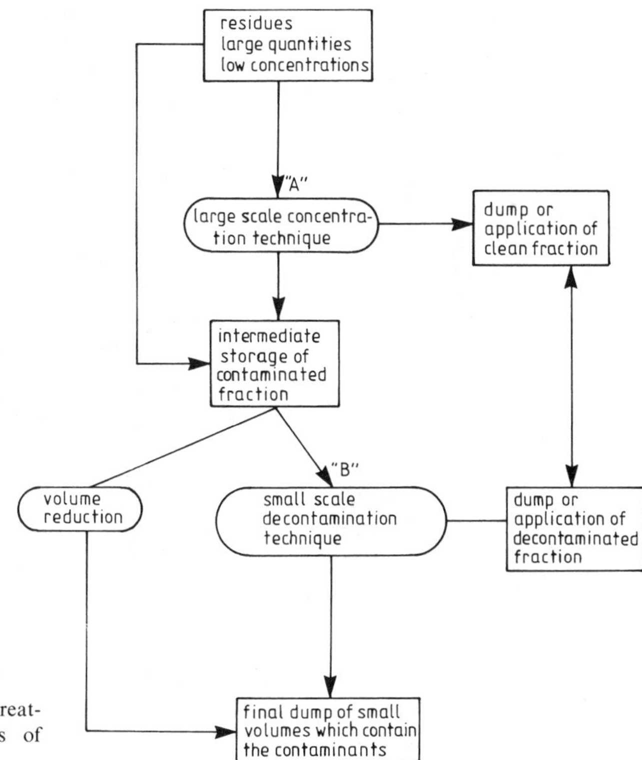

Fig. 1. Scenario for the treatment of large quantities of residue

sensitivity to variations in circumstances. It is advantageous that these techniques may be constructed in mobile or transportable plants.

"B" Decontamination or concentration techniques which are especially designed for relatively small-scale operation. These techniques are generally suited for the treatment of residues which contain higher concentrations of contaminants; they involve higher operating costs per unit of residue to be treated; furthermore, they are more complicated, require specific experiences of operators and are suitably constructed in stationary or semi-mobile plants.

It is evident that in many cases B-techniques as such are not feasible for the treatment of large quantities of residue. If, however, it is possible to preconcentrate the contaminants into concentrated flows of smaller quantities by using A techniques, it may be feasible to treat these concentrated flows by any B-technique.

Such a scenario may be set up for the specific cases of treatment of contaminated dredged sediments. This scenario is shown in Fig. 2. In reference to this scenario a few notes can be made:

1. Several capital methods are available for dredging. A definite choice has to be such that an unacceptable diffusion of contaminants is prevented, and that an additional treatment technique may be made to suit the dredging operation.

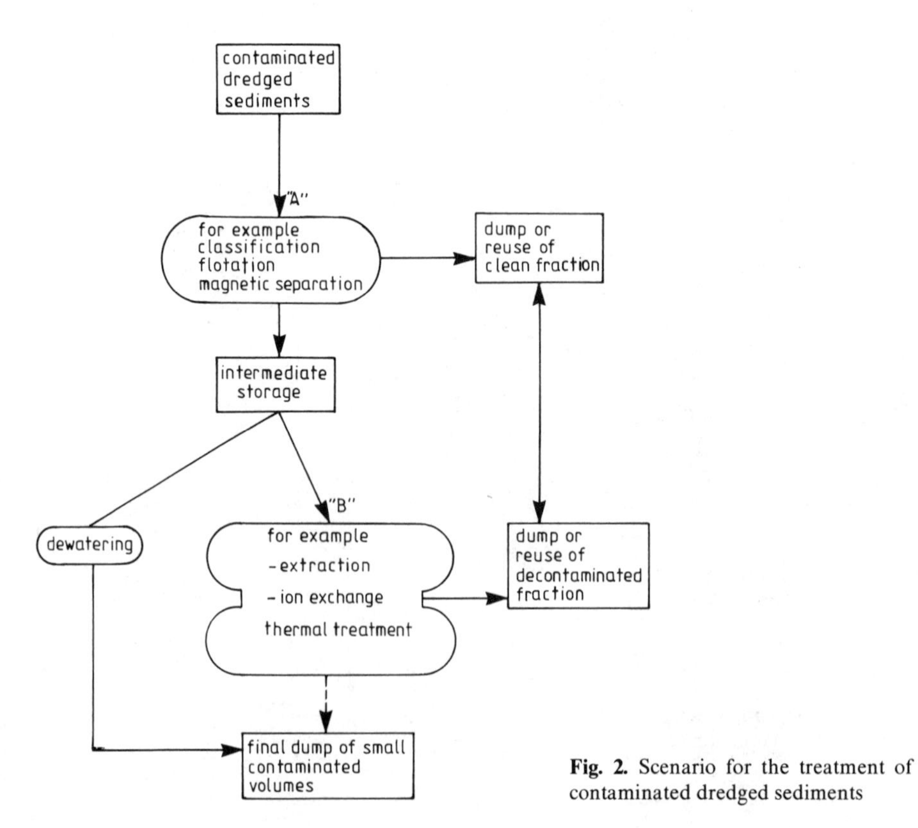

Fig. 2. Scenario for the treatment of contaminated dredged sediments

2. The A technique should be such that it may be applied in some cases to very large capacities in a mobile or transportable plant. In some cases constructing the plant on a boat is preferred, in other cases a more or less centralized treatment on land is preferred. The dredged sediments may then be transported by pipelines.
3. As to the B techniques, the choice will be defined by the kinds of contaminant in the sediments. If any cost-effective decontamination technique is not, or not yet available, one has to rely on dewatering the sediment or the concentrated flow by mechanical means or by natural consolidation, followed by additional controlled disposal. Disposal may be temporary, until feasible decontamination techniques have been developed.

   When the contaminants are of an inorganic nature, for example non-ferrometals, decontamination may be realized by techniques such as leaching, solvent extraction, ion exchange and flotation.

   When the contaminants are of an organic nature, for example oil, EOCl and/or PAH, decontamination may be based on destruction (for example combustion), solvent extraction or flotation.
4. When the environmental hazards of contaminated sediments are transferred to standardized concentration levels, it is important that the process to be applied does not result in relatively large product flows containing contaminants at concentration levels much higher that the concentration in the original sediment. This is especially true when these product flows have to be handled at high costs. It is also important that process water for sewage meets the standards for effluent quality, put forward by the local authorities.

At the moment, it seems to be an attractive proposition in the Netherlands to concentrate the contaminants into the fine fraction, for example by hydrocyclonage, followed by dewatering of the fine fraction. This fine fraction may additionally be disposed of by controlled dumping, while the coarse fraction can be disposed of without any environmental risk. However, as has already been stated, the contaminant concentration in the fine fraction must not be so high that controlled dumping of this fraction becomes much more expensive than controlled dumping of the original sediment.

In the Rotterdam harbour area the sediments have been divided into four classes (Anon 1983):

1. *Class 1.* Slightly contaminated sludge, originating from the western harbour area, specifically composed of sediment supplied by the sea. This sediment is considered to be uncontaminated. After dredging this sediment is disposed of at sea.
2. *Class 2.* Moderately contaminated sludge, originating from the middle harbour area. A mixture of class 1 and class 3 sediments.
3. *Class 3.* Contaminated sludge, originating from the eastern harbour area, specifically composed of sludge supplied by the river. The concentration of contaminants is considered to be too high for uncontrolled disposal. With regard to this sediment a large controlled dumping area is being planned. Eventually, dumping and controlling of these sediments is expected to cost several 1985-Dutch guilders per m^3 of sediment. A concentration technique may prove

attractive; however, the operating cost must be low. Moreover, the resulting concentrations of contaminants should not be so high that this stream is classified as a class 4 sediment.

4. *Class 4.* Highly contaminated sludge, locally present in a few harbours due to local waste disposal into the water. The contamination was caused by specific local activities in the past, and partly also by activities still going on. With regard to these sediments, special locations will be prepared for temporary controlled dumping. The costs are estimated to be several tens of times ten 1985-Dutch guilders per m^3 to be dumped. As far as these sediments are concerned, it seems to be attractive to develop decontamination techniques, provided the costs remain within the order of the controlled dumping costs.

For the rest of the Netherlands, it is still being discussed which procedures will have to be followed with regard to handling contaminated sediments, in relation to the concentration levels. It is, however, evident that in some cases very high costs will result. This is, for example, the case when the contamination level is high enough for the sediments to be considered as hazardous waste, as defined by Dutch law on chemical wastes (see Table). Here, the negative value of the sediments may even exceed 100 1985-Dutch guilders per m^3. With regard to these sediments it will be very interesting to develop decontamination techniques. The costs will have to be comparable with the costs of controlled dumping and will have to involve the measure of social or public acceptability, somehow expressed in a cost factor.

4 Treatment Techniques

In this Section the techniques for the A and B routes for the processing of sediments are described. For every technique concerned a general description and field for application, etc. are given.

4.1 A Techniques

4.1.1 Hydrocyclonage (Svarovsky 1984; Werther, this Vol.)

Through the centrifugal power released by the tangential flow in a hydrocyclone, particles with a higher density than the liquid will be carried to the external wall of the hydrocyclone as well as downwards alongside the wall (see Fig. 3). Because of the limited residence time in the cyclone, separation becomes increasingly poor for small particle sizes. The lower limit of the separation still attainable in practice is close to 10 μm, depending on the density of matter.

Hydrocyclonage is a relatively inexpensive technique suited for large capacities. It may be applied with regard to the purification of dredged sediment in cases in which the contaminations are concentrated in a specific fraction.

4.1.2 Flotation (de Renzo 1978)

During flotation, particles are separated from a liquid on the basis of differences in the hydrophilic character between the various particles. The principle of flotation is based on the fact that air bubbles become attached to specific particles, depending on their hydrophilic properties. The particles to which an air bubble becomes attached travel towards the surface of the water. On the surface the particles become concentrated in a layer of foam which may be skimmed. The rest of the particles remain in the water phase.

Various methods to obtain air bubbles are used. During one of these methods (dissolved air flotation), water is under pressure being saturated with air. On release, small air bubbles form. Flotation is used for example on a large scale in mining for the separation of ores. Special flotation aids are often used in this application. Dissolved air flotation is also much practiced with regard to the purification of waste water. Processing capacities of 500 $m^3\ h^{-1}$ are known in this context. Application concerning dredged sediments is in the first instance directed at eliminating hydrophobic contaminations such as oil. A possible alternative application constitutes the separation of metal sulphides through flotation after they have first been flocculated with the aid of selective flocculation agents.

Flotation may therefore be applied for the benefit of a concentrating technique A as well as a purification technique B.

4.1.3 High Gradient Magnetic Separation (de Renzo 1978)

During HGMS, magnets capable of exciting high magnetic fields (up to 20,000 G, in combination with cryogenic techniques up to 100,000 G) are used to realize high magnetic gradients in space partly filled with, e.g. steel wool. The gradients come in the order of 1,000 G μm^{-1} (compared to conventional magnetic separators, 0.0001 G μm^{-1}).

When a suspension is being led through a high gradient magnet, the ferromagnetic and probably also the paramagnetic compound, providing one or the other has a sufficiently large diameter (in practice in excess of approximately 1 μm), will be kept back by the magnet. In this way a separation between magnetic and non-magnetic compounds becomes possible.

HGMS may be used to separate ferromagnetic to weakly paramagnetic particles from a suspension. The particles may have a minimal diameter of approximately 1 μm.

The particles may even be smaller under certain circumstances. In principle the following may, among other factors, be distinguished:

1. Ferromagnetic and paramagnetic matter and compounds (Fe, Co, Ni, Cr, Mn and some Cu compounds);
2. Non-magnetic particles bonded to a magnetic floc (Fe_3O_4 or $Al_2(SO_4)_3$);
3. Metal compounds coprecipitated with ferric hydroxide (Cd, Pb).

HGMS may be applied with regard to dredged sediments when contaminations such as magnetic particles are present, or when contaminations are made magnetic by pretreatment.

4.2 B Techniques

4.2.1 Biological Treatment (de Kreuk et al. 1984)

Known embodiments of biological purification are land farming and treatment in a bioreactor.

Land farming constitutes spreading contaminated matter over land with the objective of biological degrading of the contamination. In order to allow the biological degrading to progress sufficiently smoothly, it is important for the soil to have been adequately ventilated and for the nutrients (N, P, K) to be present in their correct concentration. Through cultivation processes such as ploughing, harrowing and fertilizing these conditions may be met. Partitions will have to be installed, should the necessity arise, in order to prevent emissions to the air or to the underground.

Furthermore, the contaminated sludge may be treated in a bioreactor, during a process resembling composting, or in a slurry reactor.

Naturally, only biologically degradable contaminations may be processed by way of biological treatment. The concentration of the contaminations to be decomposed, or other components (e.g. heavy metals) in the matter to be processed may not be so high as to allow them to become toxic.

A limiting factor with regard to the application of land farming is the speed of the biological degrading. The speed in question is relatively low, so that in the case of purification a period of at least 1 year will have to be reckoned with. A second limitation is the required thinness of the ground layer in which the required ventilation may be practiced. Processing is estimated to be limited to a maximal layer thickness of 40 cm.

4.2.2 NaOCl Leaching (Kerr 1972)

According to laboratory investigations, mercury in sediment and dredged sediments may, under certain conditions, be leached with sodium hypochlorite (NaOCl).

The technique can only be applied with regard to mercury compounds present in the sludge in a form oxidizable with hypochlorite. This becomes possible when the mercury is present as a sulphide or an organic compound, and when the compounds are sufficiently accessible to the oxidizing agent. Experiments with dredged sediments with Hg contents of 6 to 500 mg kg^{-1} produce separation percentages of 95 when 1% hypochlorite is applied.

4.2.3 Acid Leaching (Müller and Riethmayer 1982)

The principle of acid leaching has been patented and has been described in the literature. It is based on the separation of metals from sludge by dissolving them in an acid environment. The technique comprises the following processing steps:

1. Extraction of metals with acid at pH = 0.5–1.0, extraction time 10 to 15 min;
2. Separation of purified sludge from the acid solution followed by rewashing with decanting centrifuges;
3. Precipitation of metals from the solution through pH increase with lime followed by carbonate precipitation with the CO_2 released with acid leaching. With this process extremely low residual concentrations of the metals are attained.

The application potential is determined by:

1. Ecological aspects; during acid leaching and neutralization, waste water with a large amount of undissolved salts will materialize. It is also uncertain whether undesirable compounds are formed when organic matter is present in the sludge;
2. Economic aspects; costs of alternative techniques or storage and the availability of a sufficient amount of HCl figure in this respect.

4.2.4 Ion Exchange

Where ion exchange is concerned, three mutually independent developments run parallel:

1. Selective cation exchange through existing, complex-forming ion exchangers (van Hoek et al. 1983);
2. Development of new, selective cation exchangers;
3. Strongly basic anion exchangers (Bolt et al. 1984; van Veen et al. 1986).

Each separate development will be briefly discussed hereafter.

a) Research on the laboratory scale was conducted with a complexing cation exchanger, which was somewhat selective with regard to heavy metals, to desorb these metals from the sludge and to adsorb them to the ion exchanger. The experiments with dredged sediments were carried out at an approximately neutral pH. With regards to selectivity and adsorption capacity, the resin TP 207 of Bauer proved to be the best. The following conclusions were drawn:

1. The process appears to involve extremely long contact times; the required times may even be in the order of 1 week. In practice, the process will consequently turn out to be difficult to realize on a continuous basis.
2. Due to the large supply of calcium, the resin's selectivity will not be high enough, thus necessitating the presence of a large excess of resin with regard to the sludge.

b) Research into the evolution of selective cation exchangers is still of a highly fundamental character. At least some years of research are expected to be required before a result, with practical applicability, may be realized.

c) The potential purification of dredged sediment through acid extraction and strongly basic anion exchange has been researched.

Metals may be mobilized through pH reduction. When the chloride concentration is sufficiently high in the liquid phase, metals such as Cd, Zn and Cu show a tendency to dissolve as negatively charged chloride complexes ($CdCl_4^{2-}$, $ZnCl_4^{2-}$, $CuCl_4^{2-}$) in solution. In this form the metals may be immobilized to a strongly basic anion exchanger, which is in chloride form. An anion exchanger is to be preferred to a cation exchanger because it is insensitive to the large supply of calcium ions in solution.

In the case of sufficiently low concentrations of chloride, the chloride complexes mentioned will be unstable and the metal ions will be converted to the cation form (Cd^{2+}, Zn^{2+}, Cu^{2+}). For this reason, the anion exchanger once charged may be regenerated through washing with water. After treatment the ion exchanger may, in principle, be used again.

Research has shown that the effect of mobilization by pH reduction is strongly influenced by ventilating the sediment. Extraction only, has minimal effect, even when pH = 1. Ventilating of the suspension greatly improves the extraction result, probably as a result of oxidation. A more or less optimal result is obtained through ventilating at pH = 4 and subsequently extracting at pH = 3.

4.2.5 Solvent Extraction (Bailes et al. 1980a,b)

The principle of organic extraction is based on the fact that organic contaminations (oil, chlorinated hydrocarbons and polycyclic aromatics) are separated from the sludge as they dissolve in an organic solvent. The solvent itself cannot dissolve in water and prefers a lower specific mass.

Applicability is limited to sediment contaminated with organic matter: oil, oily substances, chlorinated hydrocarbons and polycyclic aromatics. Furthermore, the distribution coefficient of the selected solvent has to be high enough, and in view of reuse it may be advantageous when the solvent proves to be highly volatile.

Because of limitation to capacity, inherent to an extraction arrangement, application will be found especially where processing of relatively small quantities of sediment (e.g. less than 50,000 t annually) is concerned.

4.2.6 Thermal Treatment (van Veen 1984)

The principle of thermal treatment is based on the fact that organic matter (oil, EOCl, PAH) is separated from contaminated dredged sediments by sediment treatment in an oven.

With regard to the technical realization, a number of possibilities are available, such as rotating ovens in which heat is either directly transferred by the flue gases, or indirectly by way of heat exchangers. Instead of a rotating oven, a fluid bed oven may be applied.

An alternative for direct combustion of the organic matter is the separation of the matter, in the first instance, from the sludge through evaporating. The evaporated matter be re-treated separately, e.g. by way of re-combusting.

With regard to application, temperatures in the oven or in the re-combustor will have to be sufficiently high to realize complete combustion. In the case of oily substances a temperature between 800–850°C will probably be sufficient, but in the case of combusting persistent chlorinated hydrocarbons, temperatures of up to 1400°C in the re-combustor of the processed gas will sometimes be required.

A second limitation is the large quantity of water (without preliminary dewatering, about 70% of the total amount) which has to be evaporated. Calculations show that the sludge's oil content is seldom high enough, thus the addition of external energy is not warranted.

Combustion on the spot will involve high investment. Furthermore, the capacities attainable in practice are limited to approx. 50,000 t annually, so that application appears to be limited to those volumes of sludges or those partial flows heavily contaminated with oil.

4.2.7 Wet Air Oxidation (Ely 1973)

Waste water or sediment containing organic contaminations are, in combination with air, brought into a reactor. At pressures of 100 to 200 bar and temperatures between 230° and 315°C, oxygen from the air will react with the organic fraction. The organic compounds are converted by the reaction, and when the reaction progresses fully, into carbon dioxide and water.

The technique is especially applicable in places where waste is, on the one hand, too diluted with water to merit economically feasible combustion, and on the other hand, where waste has become too toxic for biological treatment (land farming, composting) or dumping. In existing systems, suspensions up to concentrations of approx. 15% of dry matter may be processed. Maximal processing capacities in mobile plants amount to approx. 3 $m^3 h^{-1}$.

4.2.8 Dewatering (Svarovsky 1977; see Werther, this Vol.)

Dewatering is of course not a decontamination technique, but rather a decontamination-related method for further concentration of the contaminated part of the dredged sediment into a smaller volume. Dewatering a sludge at a reasonable speed and a reasonable separation efficiency usually requires conditioning of the sludge. A form of conditioning constitutes the coagulation-flocculation process, in which, through addition of chemicals, particles in the sludge coagulate to form flocs. The flocs may be separated more easily from the water phase than the actual particles. Noted conditioning agents are $FeCl_3$, $Ca(OH)_2$ and poly-electrolytes. After conditioning the sludge is dewatered in an apparatus. Noted dewatering appliances are the filter press, belt press-filter press and decanting centrifuge. Preceding the mechanical dewatering is in some instances a dewatering step, executed by way of sedimentation.

Coagulation-flocculation followed by mechanical dewatering is especially suited where dewatering of finely suspended and colloidal matter is concerned. This technique is generally applied with regard to matter "difficult" to dewater. The maximal processing capacity is approx. 30–50 $m^3 h^{-1}$ per dewatering appliance.

4.2.9 Electroosmotic Dewatering (Lockhart 1983)

The electroosmotic mobility (mu) in a colloidal or suspended system is determined by the zeta potential and the electric field intensity. In fact, the quantity of water collected by the electrokinetic flow is proportional to the zeta potential of the colloidal particles in the system, and inversely proportional to the specific conductivity. On the basis of this principle, it becomes possible, through the application of electroosmosis, to dewater systems up to at least 50% of dry matter.

When applied to dredged sediments, it becomes possible in principle to dewater fine mineral fractions, materializing as one of the product flows originating during certain processes (e.g. hydrocyclonage), to such an extent that a concentrated residual matter will form. In the case of controlled storage this entails that less storage capacity is required. The technique is an alternative to natural consolidation.

To conclude this Section, Table 2 gives a survey of the state of the art, environmental merit, research still to be carried out, as well as a survey of the global costs of the discussed A and B techniques.

5 Some Examples

5.1 Final Scenario

Through the use of the basic strategies as shown in Section 3, in combination with experimental results, an operational, final scenario for the treatment of dredged sediments can be derived. In Fig. 3 the final route is given. First of all, the sediment is classified into a coarse and fine fraction through hydrocycloning the starting material under investigation.

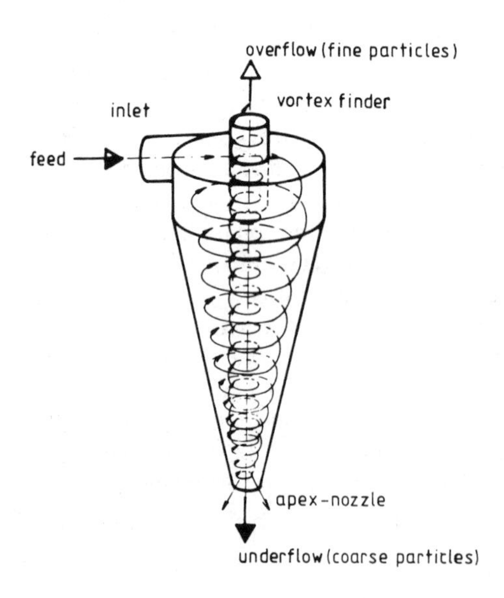

Fig. 3. The flow in a hydrocyclone

Table 2. Survey of a number of properties of treatment techniques with regard to dredged sediment

Technique/principle	State of the art	Environmental merit for sediments	Research to be executed	Global costs per m^3 of dredged sludge[a]	Remarks
Hydrocyclonage	Research is being conducted on the application and possible implementation of dredging operations of Rotterdam's sludge class 3. Laboratory investigation into sludges of different origins has been conducted	Hydrocyclonage concentrates a contamination in a partial flow. This partial flow may subsequently be dumped or further processed	Research on the applicability of hydrocyclonage with regard to various contaminated underground soils (channels, rivers, canals) and on the optimization of the process, starting from different working conditions	1–10, depending on separation limit and capacity	Research completed with regard to Rotterdam's harbour sludge with several cyclones. Demonstration on a practical scale has taken place
Flotation	Application with regard to soil treatment at demonstration level	Treatment of sediment in which contamination has been introduced as particles, or of sediment contaminated with oil	Research on the applicability of flotation with regard to sediment with many small particles (< 20 mu)	5–50, depending on consumption of chemicals and particle size.	At the moment, the possibility to separate metals from sediment by selective flocculation/flotation is being researched
High Gradient Magnetic Separation	Applicability has hardly been investigated	Purification of sediment where heavy metals are concerned	Applicability with regard to various sediments contaminated with heavy metals, or overflow water post-sedimentation or consolidation	2–20	High costs of investment. However, low processing costs due to high processing capacity. Technique still in evolutionary stage of development
Biological treatment	Application with regard to soil reorganization	Purification of sediment where biodegradable components are concerned.	Acceleration of degradation through a more process-directed approach.	Up to 40, depending on speed of biodegradation.	Bottlenecks are: required land area ($4\ m^2\ t^{-1}$), degradability of the contamination evaporation to the air

Table 2 *(continued)*

Technique/principle	State of the art	Environmental merit for sediments	Research to be executed	Global costs per m^3 of dredged sludge[a]	Remarks
NaOCl leaching	As far as it is known technique not yet evaluated to practical scale, or not yet in the process of being developed	Separation of mercury detoxifies sediment	The principle must be further investigated	Unknown	The principle has not been further developed after laboratory investigation
Acid leaching	Laboratory research on Neckar sludge completed	Purification of sediments where heavy metals are concerned	Feasibility investigation with regard to various sediments	50–100 depending on acid consumption	High lime content in sediment results in large volumes of residues
Ion exchange	Laboratory research on harbour sludge completed	Purification of sediment where heavy metals are concerned	Attainability investigation with regard to various sediments contaminated by heavy metals	20–50	Preliminary research on laboratory scale is showing favourable results with regard to cadmium
Solvent extraction	Application in combination with hydrocyclonage with regard to harbour sludge has been researched on laboratory scale	Purification of sediment where organic contamination is concerned. Finished extraction agent may be recovered in view of reuse.	Laboratory investigation on: type of solvent, separation sediment/solvent, recovery and purification of solvent	10–50	Extracting agent probably has residual value due to caloric content. Preliminary research on laboratory scale is being conducted at the moment.

Thermal treatment	Application with regard to contaminated soil on practical scale	Purification of sediment where organic contamination is concerned	Laboratory investigation: dewatering, combustion, gas washing	>100	Degrading chlorinated hydrocarbons such as PCB requires high temperatures (approx. 1400° C). The evolved gases must be recombusted. A gas washer is required when undesirable emissions occur in the atmosphere
Wet air oxidation	Not yet applied	Degradation of organic compounds Degradation may be incomplete	Fundamental potential of applicability with regard to sediment research	>100	Implementation minute due to limited capacity per plant
Mechanical dewatering	Has already been applied on practical scale for a long time	Reduction of volume with regard to (controlled) dumping	Method best suited to select various sediments	Up to 20, depending on consumption of chemicals	Limited capacity up to approx. 50 $m^3 h^{-1}$ per appliance
Electroosmotic dewatering	Experimental stage	Reduction of volume to identical degree as attained by natural consolidation	Applicability and attainability investigations	?	

[a]These costs exclusively involve the named processing step and are represented in Dutch guilders (1985). Excluded are costs of dredging, transport, pretreatment.

Classification with regard to particle size is generally a suitable technique for the concentration of contaminants, as the fine fraction has higher contamination levels than the coarse fraction. Using hydrocyclones for this purpose is logical because:

1. Hydrocyclones classify particles in a slurry;
2. Hydrocyclones classify at cut sizes up to 10 μm.

Hydrocyclones are therefore a suitable A technique.
There are two options for treatment of the fine fraction:

1. A dewatering route;
2. A decontamination route, i.e. a B-related technique.

Dewatering as a technique can be used when decontamination is not feasible or unavailable on a practical scale. As most of the decontamination techniques are not applicable as yet, dewatering may be used in short-range operations. It is also possible that dewatering will be required after decontamination.

In some cases it is required to treat the water separated in the dewatering process. This depends, among other things, on the type of dewatering technique. For instance, the filtration procedure obtains generally a better water quality than the technique of centrifugation.

Experience has shown that feasibility of decontamination of the contaminated sediment is dependent on a set of factors such as:

1. The type of contaminant;
2. The complexity of the contaminants (cocktails);
3. The environmental merit.

Investigations so far have shown that not all decontamination techniques are universally suitable. Each case has its own specialties. Thus, in each case some preliminary experiments are required in order to choose a suitable decontamination technique, and also to indicate when these techniques are operational.

5.2 Some Results

According to the final scenario a number of contaminated sediments have been treated (Fig. 4). In Table 3, for example, a description is given of four investigated types of sediment. In Figs. 5 and 6 some additional key results of these experiments are given. From these results the following appears:

1. The coarse fraction, the so-called underflow of the hydrocyclone, invariably has a much lower contamination content with respect to the original matter.
2. The dry matter partition (efficiency), which is the part of the feed dry matter represented in the underflow, is different with regard to the sediments. Sample 1 is a sediment with a relatively large clay fraction, while sample 2 is a sandy sediment.
3. The purification techniques chosen for the sediments are suitable for the decontamination of the fines.

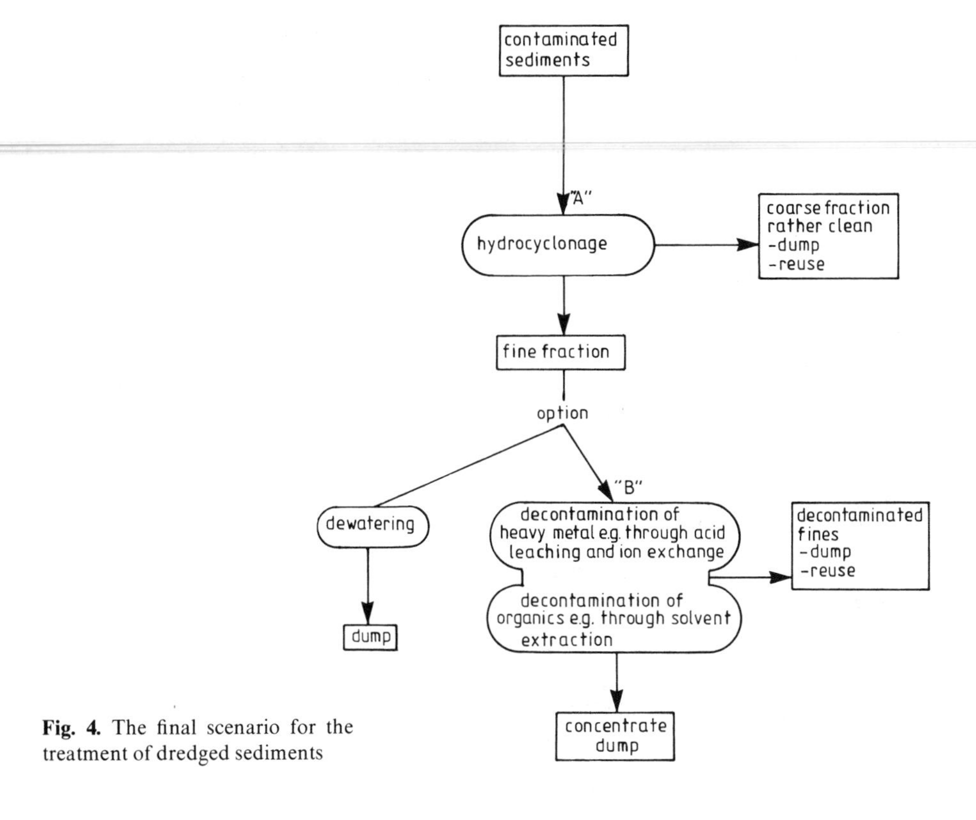

Fig. 4. The final scenario for the treatment of dredged sediments

Table 3. Some information concerning the examined sediments

Sediment	Origin	Main contaminant	Decontamination technique B[a]
1	River	Cadmium	Acid leaching/ion exchange
2	Harbour	Oil	Solvent extraction
3	River	Cadmium	Acid leaching/ion exchange
4	Canal	Chromium	Not examined

[a]Decontamination has not yet been optimized; results are mentioned in Figs. 5 and 6, which may therefore serve as indicators.

It must be emphasized that most results have been derived from relatively simple laboratory tests without any directional optimization. This has to be implemented of course when treating polluted dredgings at a specific site, operating with the special specification involved for reusing the material.

In conclusion it can be stated that hydrocyclonage may be considered as a suitable A technique in which the contaminants concentrated in the fine fraction will be recovered. Furthermore, decontamination techniques concerning the contaminated fraction also occur.

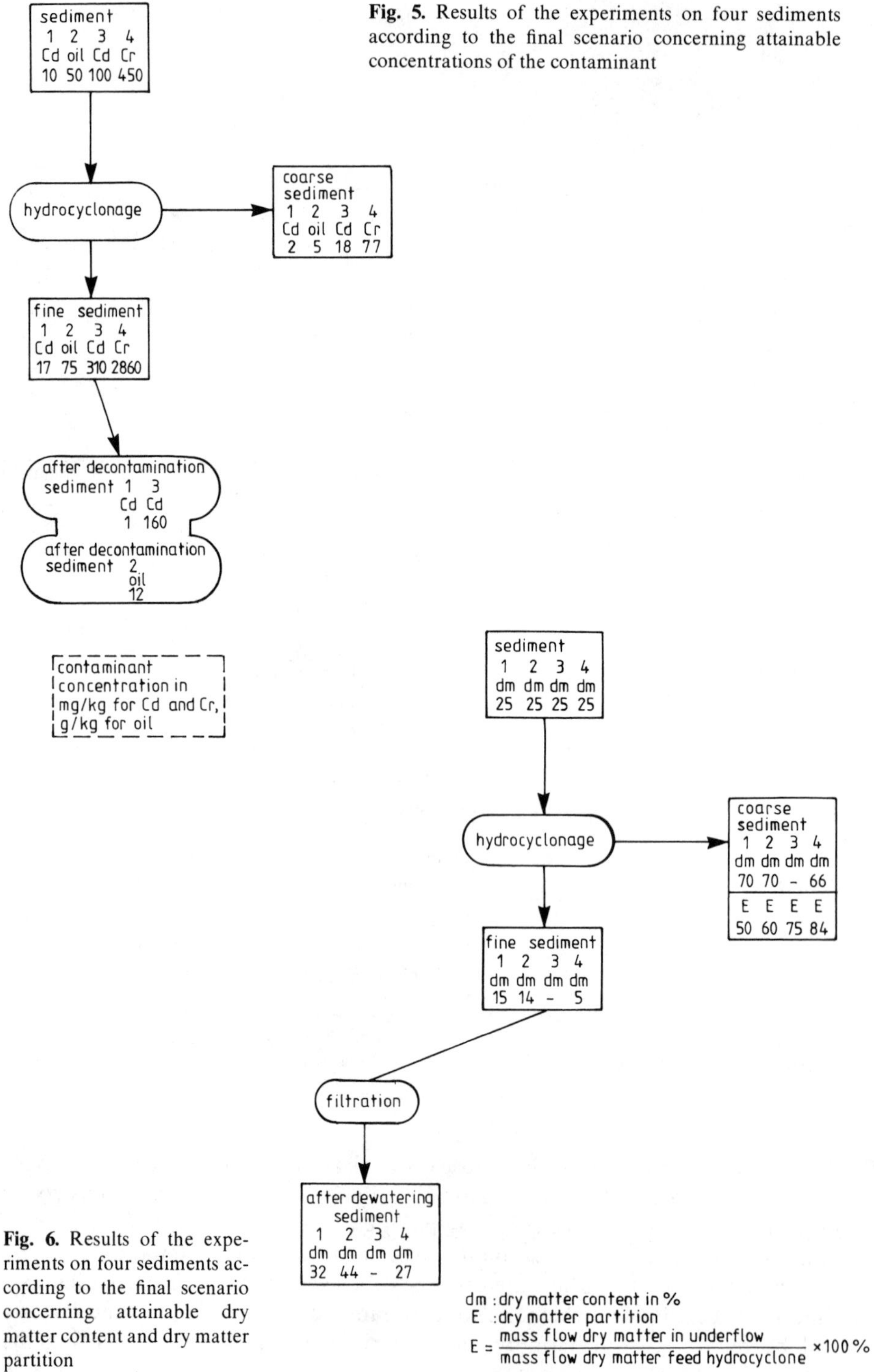

Fig. 5. Results of the experiments on four sediments according to the final scenario concerning attainable concentrations of the contaminant

Fig. 6. Results of the experiments on four sediments according to the final scenario concerning attainable dry matter content and dry matter partition

6 Final Remarks

Although in the past few years considerable progress has been made with regard to the treatment of different types of contaminated sediment, it has to be stated that there is as yet no accepted universal methodology. The main argument in this respect is the fact that recognition and subsequent acknowledgement of the problem has until now only arisen in a few places over the world.

This signifies, moreover, that research in the field of technical science is still for a large part determined by the locally specific peculiarities of the contaminated sediment. On the basis of sedimentological and geochemical insight, however, it should be possible to apply the same experiences to all contaminated sediments as far as the systems tend to be natural ones. Localities dominated by industrial sludge undoubtedly merit separate treatment especially designated to the individual case.

There are still ongoing disputes on the significance of grain-size affects. Indeed, it seems that the finest fraction (with the largest adsorbing surface) may be held responsible for the storage of contaminants, whether of an organic or inorganic character. However, it has to realized that far too little research in the field of physical science has been conducted to determine whether the medical and biological effect (health, ecological balance, etc.) of the contamination present in the sludge will indeed only be found in the fine fraction. Naturally, a concentration of the contamination in a certain fraction and its subsequent separation from the original matter will be considered to be a step in the right direction, although it should not be overlooked that perhaps some compounds in the coarse fraction may not necessarily prove positively harmless.

Furthermore, it may be noted that the question whether a certain technique or process will be applied will only partly depend on the technological state of the art. Another important incentive is the relevance to solve the problem as it is felt by the public.

Next, this public opinion will be represented by politicians. Thus, during the determination of the feasibility of a certain technique to be applied for the treatment of contaminated dredged sediments, a time factor has been built in: certain technologies will not be considered relevant at the moment they are presented by the technologists, but rather at a later time. This moment will be largely determined by external factors, such as a suddenly increased public concern for the environment after some incidents have happened.

Finally, it is important to realize that processing dredged sludges may be hindered by the increasingly high degree of complexity of the contamination. As the cocktail of contaminants in this type of solid waste material becomes more complex, the final route towards treatment may also become more complex, and more expensive. In the beginning, at least in places where there is sufficiently suitable and inexpensive space at disposal, dumping will be preferred to purification. It goes without saying that with the passing of time, the ecological bill may be presented, as has become abundantly clear in the Netherlands.

On the other hand, the economical bill is time-dependent. In the 1960s, for example, water pollution abatement was focussed on macroorganic pollutants, in the 1970s on heavy metals, in the 1980s on PCB and PAH, and presently on dioxines. Considering the pollutant concentration, the levels for macroorganics are milligrams per litre, for dioxines nanograms per litre. This means that remedial actions have to deal with decreasing concentrations, accompanied by increasingly severe standards, due to increasing toxicities of the pollutant. It is obvious that this will result in higher abatement costs.

This review is merely a state of the art. It may be expected that much experience will be gained with relatively large-scale treatment of sediments in the next few years. As the problem, from an environmental point of view, becomes an issue which cannot be avoided any longer, an increasing demand will develop with regard to practical knowledge and experience in the contemplated field. In view of the experience gained in the Netherlands thus far, it becomes natural to demand attention with regard to dialogue and cooperation as early as possible.

References

Anon (1983) Rotterdam Havenslib. Dienst van Gemeentewerken Rotterdam

Bailes PJ, Hanson C, Hughes MA (1980a) Liquid-liquid extraction: the process, the equipment. In: Ricci L (ed) Separation techniques 1: liquid-liquid systems. Chemical engineering, McGraw Hill Publications Co. New York, NY pp 217–231

Bailes PJ, Hanson C, Hughes MA (1980b) Liquid-liquid extraction: non-metallic materials. In: Ricci L (ed) Separation techniques 1: liquid-liquid systems. Chemical engineering, pp 232–237

Bolt A, Tels M, Van Gemert WJT (1984) Recovery of pure metal salts from mixed heavy metals hydroxides sludges. In: Thome-Kozmiensky KJ (ed) Recycling international. EF-Verlag für Energie- und Umwelttechnik, Berlin, pp 1025–1031

de Kreuk JF et al. Hanstreit AO, van Gemert WJT, Janssen DB, Rulkens WH, Van Veen HJ (1984) Literature study on the feasibility of microbiological decontamination of polluted soils. TNO Rep MT/R 83/223, 62 pp

de Renzo DJ (1978) Unit operations for the treatment of hazardous industrial wastes. Noeges Data Corp, pp 535–551, and 590–609

Ely RB (1973) Wet air oxidation, Pollutesne Enginaerne vol. 5, 37 pp

Hoek GL van, Gommers PJ, Overwater JA (1983) Removal of heavy metals from polluted sediments – ion exchange, a possible solution. Proc Int Conf Heavy metals in the environment, vol. 2, Heidelberg, Sept 1983. CEP Consultants Edinburgh, pp 856–859

Kerr RS (1972) Control of mercury pollution of sediments. NTIS Rep No PB-213771

Lockhart NC (1983) Electroosmotic dewatering of clays. II. Influence of salt acid and flocculants. Colloid Surf 6: 239–251

Müller G, Riethmayer S (1982) Chemische Entgiftung: das alternative Konzept zur problemlosen und endgültigen Entsorgung schwermetallbelasteter Baggerschlämme. Chem Z 106:289–292

Salomons W, Förstner U (1984) Metals in the hydrocycle. Springer, Berlin Heidelberg New York

Svarovsky L (1977) Solid-liquid separation. Butterworths, London

Svarovsky L (1984) Hydrocyclones. Holt, Rinehart & Winston, London

Veen F van (1984) Thermische grondreinigingstechnieken. In: Ontwikkelingen bodemreinigingstechnieken. Colloq, April 5, 1984. Minist VROM

Veen HJ van, Gemert WJT van, Hazewinkel JHO (1986) Reiniging van onderwaterbodems 12: Procestechniek

Classification and Dewatering of Sludges

JOACHIM WERTHER[1]

1. Introduction

The disposal of dredged sediments has become an ecological problem in Hamburg as well as in other seaports since these sediments have turned out to be contaminated with heavy metals (Table 1). The traditional solution of using the dredged sludge for filling up low marshland is no longer accepted by the public. The port authorities are therefore forced to look for alternative solutions to this problem, the dimension of which is obvious from the quantities which have to be dredged annually. In the fairways and harbour basins of the port of Hamburg about 2.5 million m^3 of mud and sand have to be dredged per year (Christiansen et al. 1982) and the corresponding figure for Rotterdam is of the order of 20 million m^3. Facing this problem Hamburg's port authorities have initiated a research program, the Hamburg Dredged Material Research Program, in the scope of which the reasons and origin of sedimentation and the contaminations in sediments as well as alternative methods for treatment and deposit of harbour mud are being investigated (Christiansen et al. 1982).

One such alternative method for treating the harbour sludge is the classification of the dredged material which has been under study by the author's Chemical Engineering group at the Technological University Hamburg-Harburg since 1980. The idea of classification is based on the finding that the heavy metals are mainly fixed to the fine particle fractions. A classification will thus separate the sludge into an uncontaminated sand fraction and a fine sludge fraction where the heavy metals are concentrated. The advantages of this process are:

Table 1. Contents of heavy metals in harbour sediments (Tent 1982)

Metals	Contents in mg kg^{-1} dry material
As	50 – 150
Pb	150 – 250
Cd	7 – 25
Cr	150 – 300
Cu	250 – 600
Hg	10 – 20
Zn	1500 – 2500

[1]Technological University Hamburg-Harburg, Eissendorferstraße 38, D-2100 Hamburg 90, FRG

1. The sand may be sold on the market for building purposes;
2. The mass of contaminated material is greatly reduced.

This latter aspect is of utmost importance with respect to the application of new technologies for the treatment of contaminated sludge. Technologies like chemical extraction of the heavy metals or microbial leaching or even the stabilization of the sludge by chemical agents are economically feasible only if the mass of contaminated material to be handled is reduced as much as possible.

In the present contribution the results obtained by the author's group will be described in detail. Since experimental experience has shown that the fine sludge suspensions produced in the classification process have very poor dewatering characteristics, attemps were made to adapt dewatering technologies available in the process industries to the special conditions found in the dewatering of fine sludge suspensions. The investigations have shown that the mechanical dewatering of sludge offers economically acceptable alternatives for the time-consuming and ecologically doubtful dewatering of sludges in polders.

It should be noted that the results and experiences reported here have been obtained exclusively with sludge from the port area of Hamburg. As is well known the sludge characteristics vary widely at different places. Experience with the mechanical processing of sludge from the Neckar area has recently been reported by Beitinger et al. (1985).

2 Characteristics of Harbour Sludges

2.1 Fractional Distribution of Heavy Metals in Harbour Sludges

Sludge samples have been classified into size fractions, the heavy metal contents of which were subsequently measured. Results of two different samples are shown in Figs. 1 and 2. Although the total concentrations in the different samples vary widely, the fractional distribution of the heavy metals is always such that most of the heavy metal contents is found in the smallest size fraction below 25 μm. Roughly, between 80 and 90% of the heavy metals are found in the particle size fraction under 25 μm. Between 95 and 98% of the heavy metals are associated with the size fraction below 60 μm. The result that the heavy metals are mainly fixed to the very fine particles of clay and organic matter confirms earlier findings of other authors (e.g. Lichtfuß and Brümmer 1977).

2.2 Particle Size Distributions of Harbour Sludges

Some examples of measured particle size distributions are shown in Fig. 3. The measurements were obtained by sedimentation analysis. Thus, the parameter characterizing the fineness of the solid material is the measured terminal velocity of the respective size fraction which has only for reasons of clarity been converted here to a particle size (i.e. the diameter of an equivalent sphere) under the

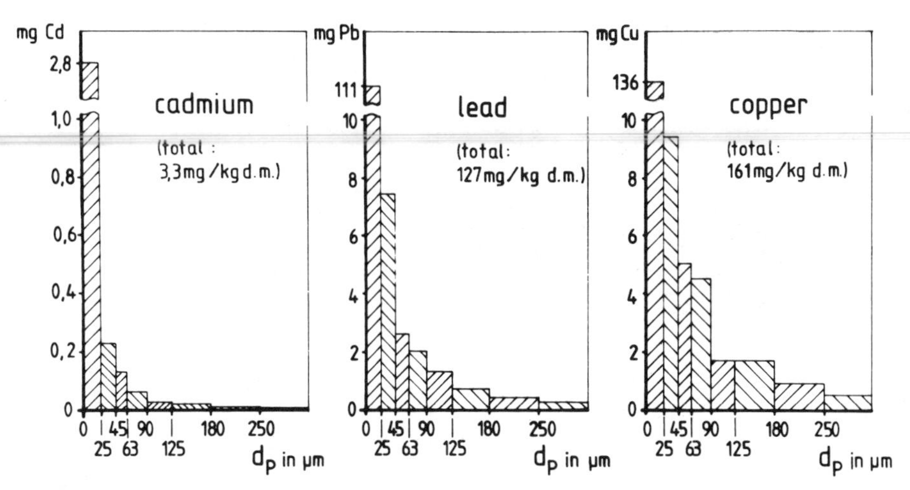

Fig. 1. Fractional distribution of heavy metals (sample taken from the Köhlfleet harbour). The figures are based on a sludge sample of 1 kg dry material (d.m.) which has been separated into size fractions; d_p = particle diameter

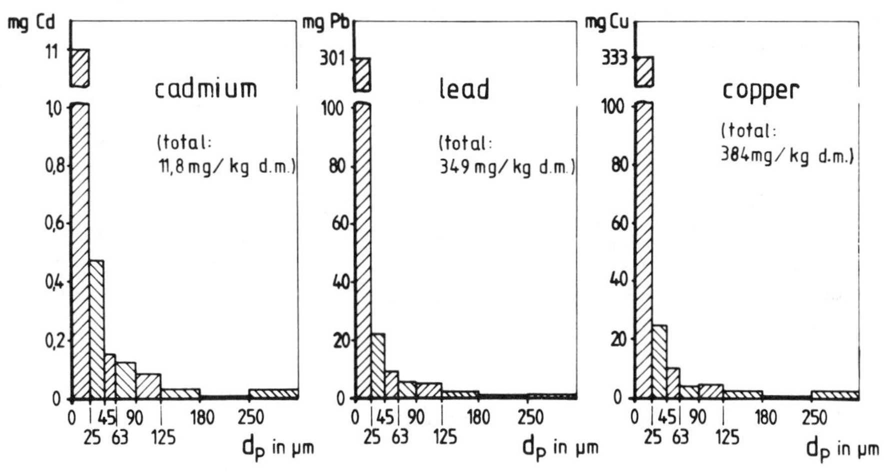

Fig. 2. Fractional distribution of heavy metals (sampling location "Alte Süderelbe"); d_p = particle diameter

additional assumption of an average solid density of 2000 kg m^{-3}. The validity of Stokes' law has been assumed. The sedimentation fluid was either water or ethylene glycol. The latter fluid was used in the case of the coarser fractions.

As is obvious from Fig. 3 the particle size of the sludge varies widely depending mainly on the location of dredging. The characteristics range from coarse to extremely fine sized distributions. Curve D in Fig. 3, for example, may simply be regarded as dirty sand, whereas curve A corresponds to an almost sand-free sludge.

The widely varying characteristics of the sludge make the development of suitable classification and dewatering processes quite difficult. Since these varia-

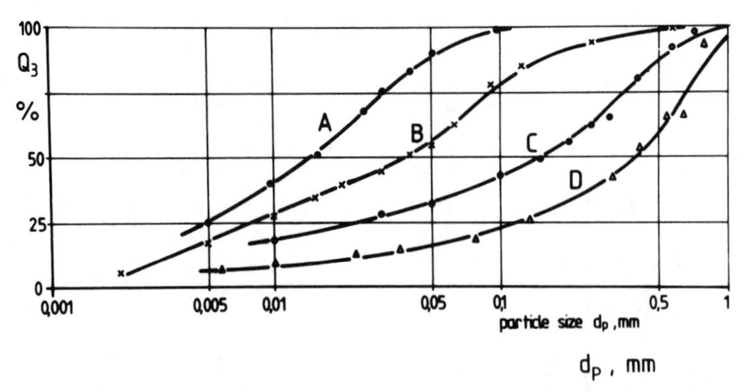

Fig. 3. Particle size distributions obtained by sedimentation analysis. Sludge samples taken from different harbour basins; Q_3 = cumulative wt%; d_p = particle diameter

tions may occur rapidly and unexpectedly during the dredging process and since furthermore the transport and unloading of sludge-carrying barges by pumping is accompanied by segregation effects, the control and adaption of the classification process to varying feed conditions have posed severe problems in the development and design of this process.

One general conclusion may be drawn from Fig. 3 when compared with Figs. 1 and 2. Since 95 and 99% of the heavy metals are associated with the size fraction under 60 μm and since furthermore this same size fraction occupies between 20 and 90% of the total sludge mass (50% on the average), this means that by separating the fraction below 60 μm from the sludge nearly all of the heavy metals will be concentrated in approximately half of the sludge mass. The sludge disposal problem will thus be reduced to the handling and disposal of half of the original sludge mass.

2.3 Contents of Organic Matter in Harbour Sludges

Sediments taken from the harbour basins generally contain between 10 and 20 wt% organic matter. In order to investigate the possible interrelation between the particle size distribution in the sludge and its contents of organic material, sludge samples were classified by sieving and the fractional contents of organic matter determined by burning at 600 °C. The results indicate a strong relationship between the contents of organic matter in the sludge and the mass fraction of fines in the sludge. Figure 4 presents this conclusion in the form of a plot of the contents of organics against the mass fraction of particles with sedimentation velocities under 2 cm min^{-1}, which corresponds to particle sizes under 25 μm when an average solid density of 2000 kg m^{-3} is assumed. The finer the sludge material, the higher thus the contents of organic material. A classification of the sludge material will thus yield a fines fraction, the organic contents of which is much higher than that of the original sludge which may be advantageous for further thermal treatment of the fines fraction.

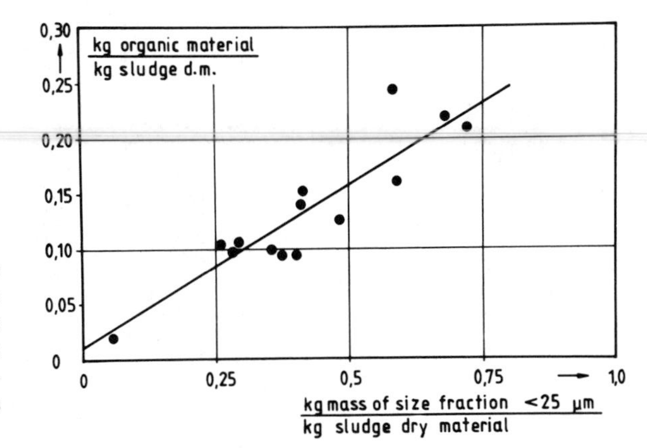

Fig. 4. Relationship between the contents of organic matter and the contents of fines in sludge samples taken from various parts of the port of Hamburg. Results obtained by regression analysis (Hilligardt 1986)

3 Classification of Sludges

3.1 Rheological Characteristics of Sludge Suspensions

In the port of Hamburg the sediment is dredged by bucket-chain dredgers and is transported in barges to pumping stations. After pumping the sludge suspension has a solids concentration between 5 and 20 wt%. Since the viscosity of this suspension is of great importance for the separation behaviour of classifying devices, the rheological behaviour of sludge suspensions was studied in a Couette-type rheometer. The example of Fig. 5 reveals a typical structural viscosity. When compared to the results of viscosity measurements and calculations of other authors (Einstein 1906; Clarke 1967; Krieger 1972; Riquarts 1981; Hanel 1984), it was found that sludge suspensions exhibited a far greater viscosity than suspensions of pure inorganic materials of the same solids concentrations, as is illustrated by Fig. 6. Included in the plot of Figure 6 are also some results of measurements on activated sludge suspensions carried out by Hanel (1984). It is remarkable that Hanel's viscosity expression fits the sludge data obtained in the present work, but only if the contents of organic matter in the sludge is inserted in his equation.

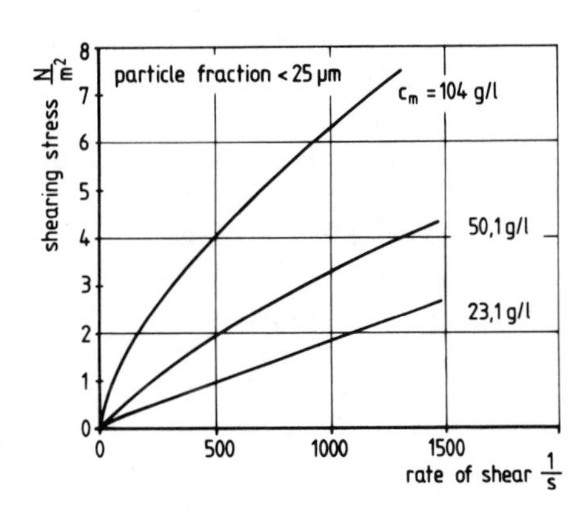

Fig. 5. Shear diagram of the particle size fraction under 25 μm obtained by sieving. c_m = Solids concentration: mass of solids dry matter relative to volume of suspension (Hilligardt et al. 1986)

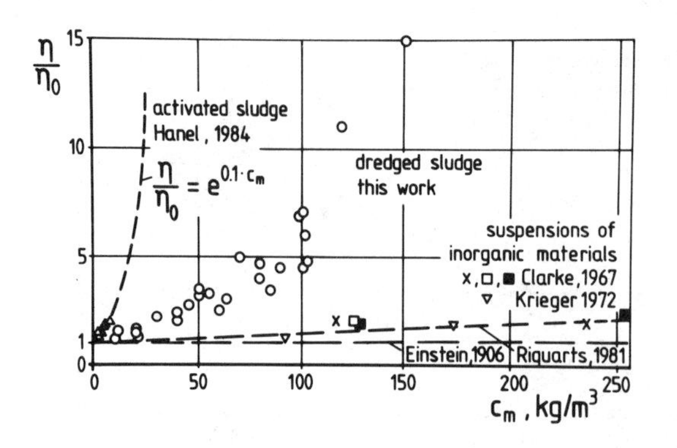

Fig. 6. Ratio of suspension viscosity to pure water viscosity for different suspensions. c_m (See legend Fig. 5, here in kg m^{-3}); η = dynamic viscosity, Pa s; η_0 = dynamic viscosity of pure water, Pa s (Hilligardt et al. 1986)

3.2 Classification Tests on the Laboratory Scale

The investigation of the sludge properties leads to the conclusion that the classification behaviour cannot be calculated from first principles. The wide particle size distribution, the distribution of densities and the complicated rheological behaviour of the sludge suspension make it necessary to study the classification process experimentally.

Different installations are available for the classification of particulate suspensions according to their sizes and/or densities in the process industries. In the course of the present investigation both hydrocyclones and elutriators have been tested.

The hydrocyclone is schematically shown in Fig. 7. The sludge suspension is tangentially introduced into the cyclone chamber where the separation of the coarse fraction from the fines is effected by the action of centrifugal forces. The coarse fraction leaves the cyclone in the underflow, while the fines are contained in the overflow. The advantage of the hydrocyclone is its simplicity and its ability to handle large throughputs compared to the cyclones' size. A disadvantage is that the sharpness of the separation is fairly low.

The elutriator shown in Fig. 8 effects a much better sharpness of separation. The basic principle of separation is here separation according to the settling velocity of the particles in an upflowing water stream. The fluidized bed of coarse particles forming at the bottom part of the apparatus aids in washing the underflow by particle-to-particle interaction in the bed.

The classification of harbour sludge was first tested on the laboratory and semi-technical scale (Dreuscher et al. 1984). Some results are shown in Fig. 9, where the particle size distributions of feed (F), underflow (U) and overflow (O) are plotted as cumulative mass distributions together with separation efficiency curves (E) for a laboratory hydrocyclone (diameter 72 mm). The separation efficiency is defined here as the ratio of mass in a certain size fraction appearing in the underflow to the mass contained in the same size fraction of the feed. As it is widely known the separation efficiency curve of a hydrocyclone is comparatively flat, thus indicating a separation which is not very sharp (Svarovsky 1984, Bradley 1965).

The underflow of the hydrocyclone still contains a considerable mass fraction in the size range below 60 μm, i.e. this underflow material will still be contaminated

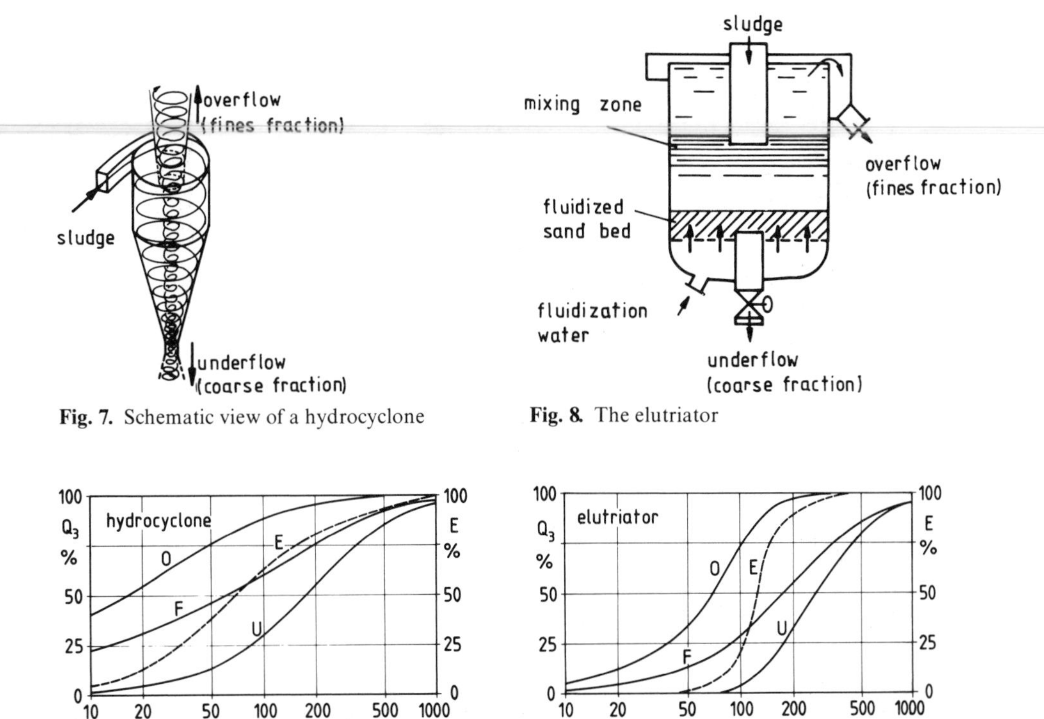

Fig. 7. Schematic view of a hydrocyclone

Fig. 8. The elutriator

Fig. 9. Hydrocyclonage of harbour sludge; laboratory hydrocyclone 72 mm diam, feed concentrations $c_{m,f} = 110\,kg\,m^{-3}$; Q_3 = cumulative wt%; E = separation efficiency, %; d_p = particle diameter (Hilligardt et al. 1986)

Fig. 10. Washing of hydrocyclone underflow in the 200-mm diameter laboratory elutriator; for definitions of Q_3, E, d_p, see legend of Fig. 9 (Hilligardt et al. 1986)

with heavy metals. The classification may either be improved by suitable combination of two or more stages of hydrocyclones or by washing the underflow of the hydrocyclone in an elutriator where the liquid-fluidized bed forming at the bottom section of the apparatus provides a very effective means to clean the coarse fraction from adhering fines. As is shown in Fig. 10 virtually no fines are found in the underflow of the elutriator. The separation procedure is further illustrated in Fig. 11 which demonstrates that the sand thus separated from the sludge has a heavy metal content which is of the same order of magnitude found in naturally occurring sandstones.

3.3 Large-Scale Testing of Sludge Classification

Since investigations on the laboratory and semi-technical scale respectively, yielded promising results it was decided to test this classification concept on the technical scale.

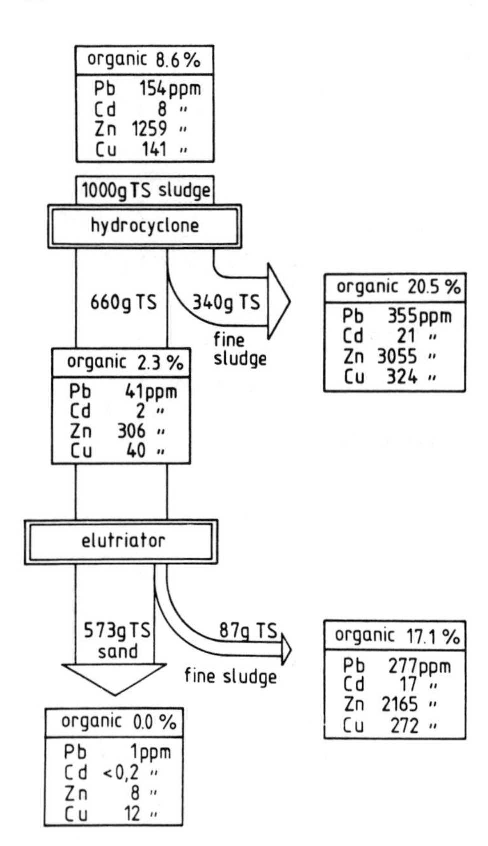

Fig. 11. Mass balance and distribution of heavy metals. The percentage organic material was determined by oxidation at 600°C; *TS* = dry material (Hilligardt et al. 1986)

Supported by the German Ministry of Research and Development, a pilot plant has been erected by Hamburg's *Amt für Strom- und Hafenbau* (Kröning 1985). The flow diagram of this plant, which has a capacity of 1000 m^3 h^{-1} suspension, is shown in Fig. 12. Figure 13 presents a view of the elutriator and hydrocyclone, which gives an impression of its dimensions. As is indicated in Fig. 12 a further purpose of this plant is to test novel methods for the dewatering of the classified sludge. Decanting centrifuges as well as filter presses are being investigated in order to find financially acceptable alternatives for the time-consuming and ecologically doubtful dewatering in polders. The pilot plant has been in operation since May, 1985 and will be in use for approximately 1 year. First results have largely confirmed the laboratory data, thus indicating the basic feasibility of the classification concept.

Figure 14 shows separation efficiency curves measured in the pilot plant with two different hydrocyclones. As illustrated by this plot the solids concentration in the feed suspension turns out to be decisive for the cut size as well as for the sharpness of separation. The influence of the feed concentration is even greater than the influence of the hydrocyclone diameter. This is an effect which is not experienced in common dressing processes in the sand industry. The physical

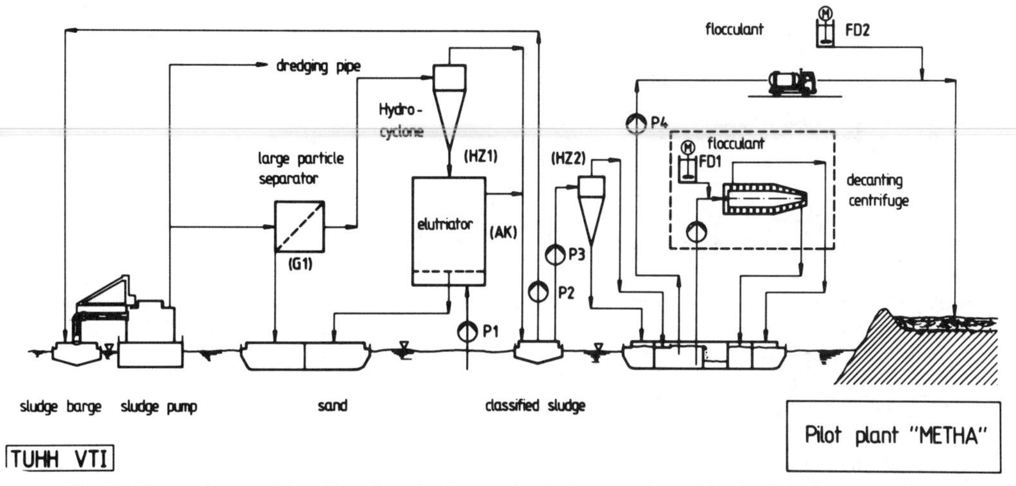

Fig. 12. Flow scheme of the pilot plant for the mechanical processing of dredged sludges

Fig. 13. Hydrocyclone and elutriator section of the pilot plant

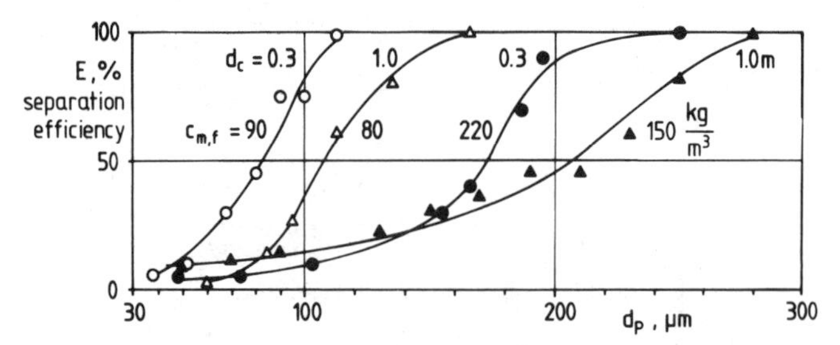

Fig. 14. Separation efficiency curves measured in the pilot plant with two different hydrocyclones for the same type of sludge; d_p = particle diameter; d_c = hydrocyclone diameter, m; $c_{m,f}$ = solids concentration at the feed suspension, kg m^{-3} (particle size distribution described by curve B of Fig. 3)

reason of this behaviour lies mainly in the strong influence of feed concentration on the rheological behaviour of the suspension (cf. Fig. 6). The technical processing plant thus will require special means for controlling the feed concentration.

4 Dewatering of Fine Sludge Suspensions

4.1 Preliminary Laboratory Experiments

Basic characteristics of the fine sludge suspension, produced in the classifying process step, are given in Table 2. The technical aims in the dewatering of suspensions are the achievement of as high a solids content as possible, on the one hand, and the clarification of the remaining water, on the other. In order to obtain some basic insight into the suitability of the respective dewatering techniques, separation by gravitational sedimentation, filtration and centrifugation respectively, were tested on the laboratory scale. A summary of many experiments is given in Table 3.

Preliminary measurements on the laboratory scale showed that dewatering by filtration was not feasible without addition of flocculants since the fine particles stopped up the filter cloth rapidly. An investigation of the settling behaviour of these suspensions yielded the result that at concentrations above 2 wt% zone settling prevails. This means that a thickening process is feasible even without addition of flocculants.

If lime or polyelectrolytes are added in considerable quantities, filtration under pressure yields excellent results as far as the solids contents in the filter cake is concerned. The costs of investment as well as of operation, however, turned out to be fairly high for the present application. Centrifugation offers the possibility of dewatering without adding flocculants, although in this case complete clarification of the overflow is not possible. On the basis of these laboratory results and on the basis of cost evaluations, it was decided to test the dewatering of the fine sludge suspension in a decanting centrifuge on the pilot scale.

Table 2. Characteristics of fine sludge prior to dewatering

Solids concentration	2–8 wt% d.m.
Size distribution	0–60 10^{-6}m
Average particle size $d_{p\ 50,3}$	5–15 10^{-6}m
Mass fraction of organic matter	15–30%

Table 3. Dewatering of fine sludge suspensions on the laboratory scale

Method	Achievable solids content (wt% d.m.)	Solids retention (%)	Flocculant requirement in kg kg^{-1} sludge d.m.
Gravitational sedimentation (thickening)			
with flocculation	8–12	80	–
without flocculation	8–12	100	1
Gravitational filtration	15–19	100	1
Filtration under pressure	35–40	100	4
Centrifugation			
without flocculation	25–33	95	–
with flocculation	25–33	100	1

4.2 Pilot-Scale Testing of the Dewatering of Fine Sludge Suspensions in Decanting Centrifuges

At the METHA (Mechanical Treatment of Harbour Sludge) plant a commercially available decanting centrifuge unit with a throughput of 3 $m^3\ h^{-1}$ suspension has been installed. A special feature of this unit is that it offers the possibility of variation of operating parameters like throughput, rotational speed and differential rotational speed over a wide range of conditions. Some typical results are shown in Figs. 15 and 16.

The measurements leading to the results depicted in Fig. 15 were obtained by centrifugation of the fine sludge suspension without adding flocculants. The centrifuge is able to remove a significant portion of the solid particles, typically between 40 and 80%, from the feed suspension. The solid product is obtained with a fairly low moisture content, typically between 50 and 60%. The moisture content is reduced by increasing the feed solids concentration as well as by increasing suspension throughput. The decreasing moisture content, however, is accompanied by a reduction in the solid separation yield, thus requiring a careful optimization of the centrifuge operation.

A complete clarification of the centrifuge overflow is not possible without addition of flocculants. Figure 16 presents some results on the specific flocculant

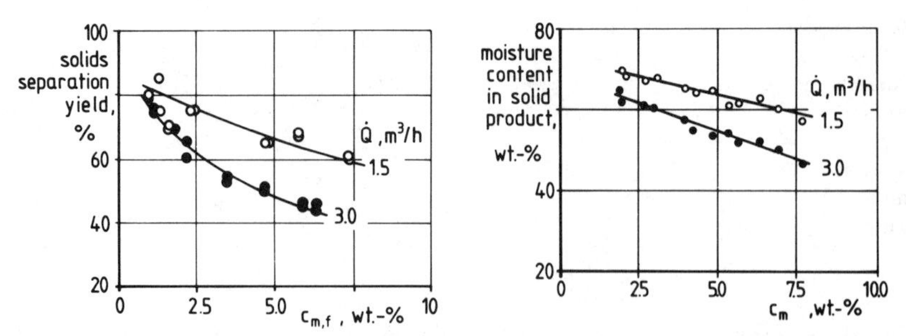

Fig. 15. Dependence of solids separation yield and moisture content of the solid product on feed concentration $c_{m,f}$ and feed throughput Q (co-current decanting centrifuge, centrifugating value = 400, differential rotational speed of the screw = 1.5 rpm); c_m = solids concentration: mass of solids dry matter relative to volume of suspension, kg m^{-3}

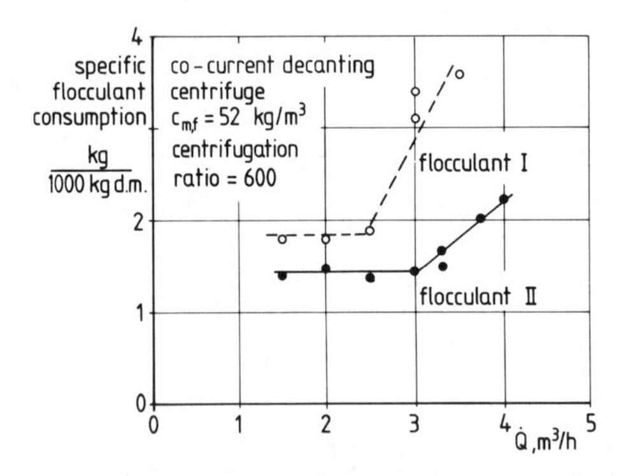

Fig. 16. Specific flocculant consumption at various feed throughputs; $c_{m,f}$ = Solids concentration of the feed suspension, kg m^{-3}; Q = Throughput of feed suspension, m^3 h^{-1}

consumption in such clarification tests. As long as the throughput is kept below a certain amount, the specific flocculant consumption is constant, whereby the level of the specific consumption is dependent on the nature of the flocculant. The rapid increase of specific flocculant consumption when this limiting throughput value is exceeded may be due to increasing shearing stress of the solids. Another point that should be mentioned is that the moisture content of the solid product is fairly high under these conditions. The moisture content typically ranges between 68 and 70%. Cationic flocculants with molecular weights of approximately 5,000,000 yielded optimum results.

5 Conceptual Design of a Technical Sludge Processing Plant

Based on the above mentioned results the flow scheme of a technical sludge processing plant has been developed (Fig. 17). The classification consists basically of a hydrocyclone separator HZ 1 the underflow of which is washed in the elutriator AK. The overflows of both hydrocyclone HZ 1 and elutriator AK are fed to the hydrocyclone HZ 2 which produces a fine sand fraction. The classification part of

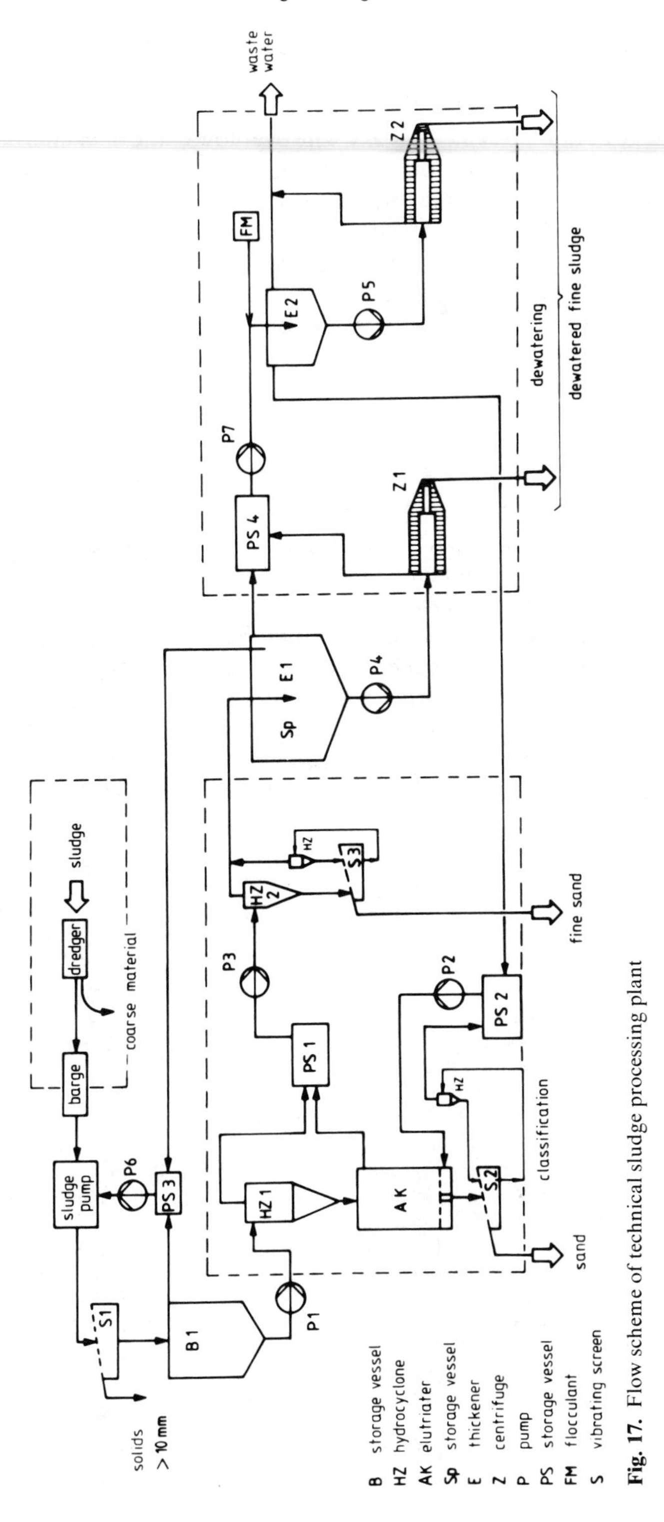

Fig. 17. Flow scheme of technical sludge processing plant

the processing plant operates batchwise following the barge-emptying sequence at the sludge pump. The overflow of the hydrocyclone HZ 2 is discharged into the storage vessel SPE1 which acts also as a thickener. The thickened sludge is dewatered in the first-stage decanting centrifuge Z 1. The overflow of the thickener SPE 1 and the overflow of Z 1 is collected in the thickener E 2, where after addition of flocculants the suspension is prepared for further dewatering and clarification respectively, in the centrifuge Z 2. The waste water will have to be processed in a separate biological waste-water treatment plant. Respective calculations have shown that the two-stage dewatering procedure not only requires less flocculant, but also yields a lower average moisture content in the solid product.

6 Conclusions

In the present review first results of the development of a new sludge treatment process are reported. The basic idea is that a classification of the dredged river sediments leads to a separation of an uncontaminated sand fraction from the contaminated harbour sludge. The separation process reduces the amount of contaminated material and thus reduces the costs of further treatment or disposal. The classification is most effectively carried out by a hydrocyclone with subsequent washing of the underflow in an elutriator. For the dewatering of the remaining fine sludge suspension, a two-stage centrifugation in decanting centrifuges has turned out to be most advantageous. The process steps have been tested on the technical scale in a pilot plant. Based on these results, the flow scheme of a technical sludge processing plant has been developed.

References

Beitinger E, Merklin-Lempp I, Meyer-Hübner V. (1985) Pilotentwicklung eines Verfahrens zur Herstellung eines Leichtbaustoffes aus feinkörnigen Fluβ- und Seesedimenten. Forschungsber T 85-012 Bundesminist Forsch Technol Bonn/FRG

Bradley W (1965) The hydrocyclone. Pergamon, Oxford

Christiansen H, Öhlmann G, Tent L (1982) Probleme im Zusammenhang mit dem Anfall von Baggergut im Hamburger Hafen. Wasserwirtschaft 72:385-389

Clarke B (1967) Rheology of coarse settling suspensions. Trans Inst Chem Eng 45:251-256

Dreuscher H, Egerer B, Bauer W, Werther J (1984) Entwicklung eines Verfahrens zur Abtrennung einer Sandfraktion aus Baggerschlamm. TIZ-Fachber 108:576-582

Einstein A (1906) Eine neue Bestimmung der Molekulardimensionen. Ann Phys IV 19:289-308

Hanel R (1984) Die Viskosität von Abwasser und Belebtschlamm. GWF-Wasser/Abwasser 125:390-393

Hilligardt R (1986) Der Einsatz von Hydrozyklonen für die mechanische Aufbereitung organikhaltiger Baggerschlämme. Diss, Tech Univ Hamburg-Harburg/FRG

Hilligardt R, Kalck U, Kröning H, Weber J, Werther J (1986) Classification and dewatering of dredged river sediments contaminated with heavy metals. 4th World Filtration Congr, April 22-25 1986, Ostend/Belgium

Kröning H (1985) Versuchsanlage zur mechanischen Trennung von Hafenschlick "METHA" nimmt den Forschungsbetrieb auf. Schiff Hafen Kommandobrücke 7:68-69

Krüger J (1972) Mechanism for Non-Newtonian Flow in Suspensions of Rigid Spheres. Adv Colloid Interface Sci 3:111–117

Lichtfuß R, Brümmer G (1977) Schwermetallbelastung von Elbe-Sedimenten. Naturwissenschaften 64:122–125

Riquarts HP (1981) Berechnung der effektiven Viskosität von Suspensionen sedimentierender Teilchen in Newtonschen Flüssigkeiten bei hohen Scherraten. Verfahrenstechnik 15:238–242

Svarovsky L (1984) Hydrocyclones. Holt, Rinehart & Winston, London

Tent L (1982) Auswirkungen der Schwermetallbelastung von Tidegewässern am Beispiel der Elbe. Wasserwirtschaft 72:60–62

Stabilization of Dredged Mud

WOLFGANG CALMANO[1]

1 Introduction

Traditional disposal alternatives of dredged material are subaqueous (open-water) disposal, application to intertidal sites and upland deposition. More recently, coastal marine disposal in capped mound deposits above the prevailing seafloor, disposal in subaqueous depressions and capping deposits in depressions have been proposed for contaminated sediments (Kester et al. 1983). These categories differ primarily in the biological population exposed to the contaminated sediments, physico-chemical conditions and transport processes potentially capable of removing contaminants from dredged material at the disposal site (Gambrell et al. 1978).

The ecological effects of heavy metals in contaminated sediments are determined more by the chemical form and reactivity than by the level of accumulation. Land deposition of dredged mud is mainly concerned with the following factors:

1. Change of pH;
2. Change of redox conditions;
3. Formation of soluble, complexing organic compounds;
4. Microbial interactions.

These factors affect the chemical forms and bonding strength of heavy metals in sediment particles and thus mobility and availability. For the deposition of freshwater sediments in intertidal and coastal marine environments, the effect of salinity is of great importance. Remobilization of certain metals, e.g. Cd, is due to increased ion concentrations and formation of soluble chloro-complexes (Ahlf 1983, Calmano et al. 1985).

In order to minimize potential adverse effects upon disposal of dredged material, chemical and mechanical stabilization techniques, especially for upland disposal alternatives, are proposed and discussed in the following sections.

2 Factors Affecting Chemical Stability of Sludges

The most effective physico-chemical environment for immobilizing the most potentially toxic metals in dredged materials is:

[1]Technical University of Hamburg-Harburg, Eissendorferstr. 40, D-2100 Hamburg 90, FRG

1. Near-neutral pH;
2. Strongly reduced;
3. Non-saline; and
4. Especially, where sulphides are present (Gambrell et al. 1978).

For many metal examples a clear relationship has been found between pH values and dissolved metal concentrations. With decreasing pH, the metal concentration in solution increases. Figure 1 shows the dissolution of Cd in soil as a function of pH (Alloway and Morgan 1986).

Acidity not only imposes problems in increased availability and toxicity of metals, but also in other aspects of pollutant enrichment, ranging from the toxification of groundwater to problems concerning growth and reproduction of organisms, increased leaching of nutrients, the ensuing reduction of soil fertility and finally to the undesirable acceleration of mercury methylation in the sediments (Fagerström and Jernelöv 1972).

Therefore, the content of buffer substances in the sediments is of prime importance. Calcium carbonate and aluminium silicates are the predominant factors affecting buffer capacity which controls the magnitude of shifts in the pH after formation of acid in or addition to the system. Sediments with carbonate contents >10% are protected against acidification over a wide range. This is shown in Fig. 2 with titration curves for dredged material suspensions from the Neckar River and Elbe River (Hamburg harbour).

While the Neckar River sediment belongs to this category, the dredged mud from Hamburg harbour exhibits a very low buffer capacity as evidenced by the immediate decrease of pH after addition of acid. Such acidification effects became evident, when harbour sludge was used in agriculture and led to elevated concentrations of heavy metals in crops and vegetables (Herms and Tent 1982).

The buffer intensity of a sludge can be tested by a simple procedure. Ten percent sludge suspensions in distilled water (pH_0) and in 0.1 N acid (pH_X) are shaken for 1 h and the difference for the obtained pH values is calculated:

$$\Delta pH = pH_0 - pH_X .$$

Three categories of pH values can be established, ranging from $\Delta pH < 2$ (strongly buffered), $\Delta pH = 2$–4 (intermediate) to $\Delta pH > 4$ (poorly buffered). These

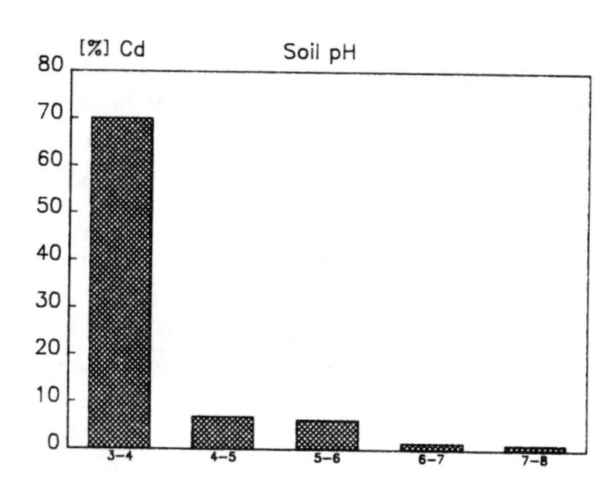

Fig. 1. Relationship between cadmium in the soil solution and pH (After Alloway and Morgan 1986)

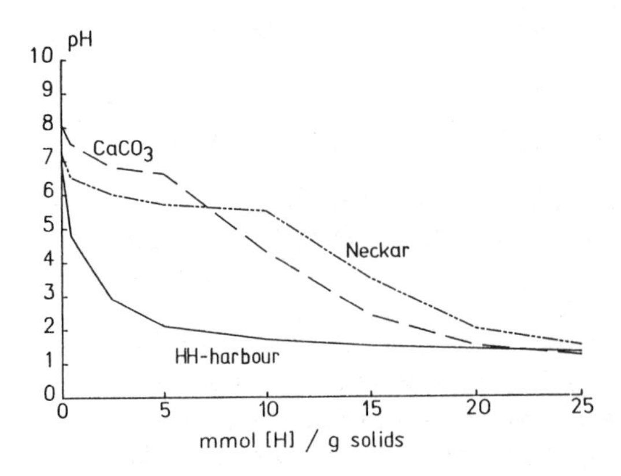

Fig. 2. pH changes in suspensions of calcium carbonate and river sediments after addition of acid (After Calmano et al. 1986)

criteria can help to decide if a dredged mud has to be stabilized, e.g. by addition of lime or limestone, before final deposition, although this procedure does not consider slowly reacting buffer substances.

3 Stabilization Techniques

In order to maintain ports and channels accessible to shipping, and also if they are polluted with contaminants, sediments have to be removed regularly. Once the potential impacts of contaminants are determined, a management plan can be formulated in order to minimize the environmental pollution upon disposal of dredged material.

Stabilization techniques proposed for solid waste materials and contaminated soils (Rulkens et al. 1985) may also be considered for strongly polluted sediments:

1. Construction of mechanical barriers (encapsulation);
2. Fixation of water by addition of chemicals;
3. Formation of sparingly soluble compounds by chemical reactions (chemical immobilization);
4. Reduction of mobility of critical compounds by changes of pH/redox conditions.

In general, solidification/stabilization technology is considered a last approach to the management of hazardous wastes. The aim of these techniques is a stronger encapsulation of contaminants to reduce the emission rate to the biosphere and to retard exchange processes respectively (Wiedemann 1982). Two steps can be differentiated:

1. At best, a material is produced which can be used for earth or landscape building like road construction, dike construction or building of walls for noise reduction, so that disposal sites are not necessary;
2. In other cases the material could be improved by a stabilization process so that it could be deposited safely and cheaply.

Most techniques used for the immobilization of metal-containing wastes are based on solidification/stabilization with cement, water glass (alkali silicate), coal fly ash, lime or gypsum. A detailed representation and discussion of stabilization techniques of hazardous wastes and remedial actions for redevelopment of contaminated soils is given by Wiedemann (1982). Some typical characteristics, compatibilities, disadvantages and advantages of these methods are given below (Wiedemann 1982; Rulkens et al. 1985) and in Table 1 (Malone et al. 1982).

Cement is mainly a mixture of oxidic calcium, aluminium, silicon and iron compounds which were activated in the kiln and set at contact with water by formation of hydrous aluminates and silicates. After corroboration these compounds are water resistant. It is known that many wastes can be stabilized by cement. For sludges with solid contents of 25%–60%, dewatering is not necessary since cement needs water for consolidation. Other absorbing materials like clays may be added. Bentonite, for example, has a large specific surface and is negatively charged with a corresponding cation exchange capacity. In addition, many inorganic and organic compounds are bound by sorption.

The solidity and permeability of the product may be controlled by the amount of added cement. The immobilization of metals by the formation of less soluble hydroxides or basic carbonates is advanced by the high pH value of cement mixtures. On the other hand, it can be expected that high pH values will have unfavourable effects on the mobility of certain toxic metals, like lead, chromium, copper and nickel which form hydroxy-complexes and show an increased solubility at pH>12. Also, the dissolution of natural organic substances (humic acids), together with complexed or adsorbed contaminants, may be enhanced.

Water glass (alkali silicate) is another useful solidification reagent for sludges. The reaction with multivalent metal ions leads to a hydrous gel, e.g.:

$$Na_2SiO_3 + CaCl_2 + H_2O = CaSiO_3 + H_2O + 2\,Na^+ + 2\,Cl^-.$$

Most transition metals form practically unsoluble silicates. The metals are immobilized in the polymerized structure, just as organic molecules and other pollutants.

Chemically unfixed are, e.g. chloride ions and monovalent cations. This follows from the nature of the water glass reaction which leads to a release of alkali ions and anions as shown above, so that leaching of additional amounts of these ions is only reduced by a decrease in the water permeability of the solidified product.

The American CHEMFIX process is based on the immobilization of hazardous wastes with water glass and is applied above all for inorganic contaminants; the Belgian SOLIROC process uses as yet unknown additives and is also recommended for solidification of organic wastes.

The *pozzolane* effect is based on the settling behaviour of coal fly ashes, cement or kiln dusts and on the reaction of these materials with lime. Similar to cement or water glass, the solidification and immobilization of contaminants with pozzolanes mainly consist in the encapsulation and low water permeability of the product. Also, the high pH values contribute to a certain stabilization of heavy metals by the formation of less soluble compounds. Pozzolanes normally show a good long-term stability, but a slow solidification.

Table 1. Compatibility of selected waste categories with waste solidification/stabilization techniques. (After Malone et al. 1982)

Waste component	S/S treatment type			
	Cement-based	Lime-based	Thermoplastic solidification	Organic polymer (UF)[a]
Organics				
Organic solvents and oils	Many impede setting, may escape as vapor	Many impede setting, may escape as vapor	Organics may vaporize on heating	May retard set of polymers
Solid organics (e.g., plastics, resins, tars)	Good – often increases durability	Good – often increases durability	Possible use as binding agent	May retard set of polymers
Inorganics				
Acid wastes	Cement will neutralize acids	Compatible	Can be neutralized before incorporation	Compatible
Oxidizers	Compatible	Compatible	May cause matrix breakdown, fire	May cause matrix breakdown
Sulfates	May retard setting and cause spalling unless special cement is used	Compatible	May dehydrate and rehydrate causing splitting	Compatible
Halides	Easily leached from cement, may retard setting	May retard set, most are easily leached	May dehydrate	Compatible
Heavy metals	Compatible	Compatible	Compatible	Acid pH solubilize metal hydroxides
Radioactive materials	Compatible	Compatible	Compatible	Compatible

[a]Urea – Formaldehyde resin.

Note: "Compatible" indicates that the S/S process can generally be successfully applied to the indicated waste component. Exceptions to this may arise dependant upon regulatory and specific situation factors.

The advantage of common handling of both sludges and pozzolanic ashes or dusts is that disposal capacity is not reduced if both wastes have to be removed by deposition.

Lime, in the form of calcium oxide or calcium hydroxide, is often used for chemical stabilization of soils. Most natural soils contain larger amounts of silicic acid or colloidal aluminium silicates in a strongly hydrous state. If such soils are mixed with lime, pozzolanic reactions occur and calcium silicates and aluminates are formed which show the known effects for heavy metals. Also, in the absence of these reactive substances a certain solidification can be obtained by a mixture of sludges with calcium oxide. It has been used especially for the treatment of oil-contaminated sludges.

Calcium oxide reacts with water to calcium hydroxide which has an increased surface:

$$CaO + H_2O = Ca(OH)_2.$$

Calcium hydroxide shows a relatively good water solubility and does not represent a stable end product. On the other hand, at contact with carbon dioxide in the air, the more stable calcium carbonate is formed under natural conditions:

$$Ca(OH)_2 + CO_2 = CaCO_3 + H_2O.$$

This reaction may be accelerated by artificial treatment with carbon dioxide gas.

The solidification of *gypsum* by addition of water is a known process in which water is included in the crystal structure and calcium sulphate dihydrate results. This process can be used to solidify sludges because the strong resorption of water by the gypsum dries up the waste, and the crystals form a strong and stable frame in which other pollutants and contaminants may be included.

In contrast to calcium hydroxide, gypsum is a stable end product, although the solubility in water is relatively high.

At present, it does not seem possible for the gypsum and cement industries to absorb the expected quantities of flue gas gypsum arising from coal-fired and lignite-fired power stations. New applications for these quantities must be developed, and one possibility may be the stabilization of heavy metal contaminated sludges.

4 Implementation Strategies

Francingues (1985) indicates three basic on-site implementation strategies according to the manner in which the chemical reagents are added to and mixed with the materials being treated:

1. In situ mixing;
2. Plant mixing;
3. Area mixing.

In situ mixing is suitable primarily for dredged slurries that have been initially dewatered. It is best applicable for the addition of large volumes of low reactivity solid chemicals. This strategy incorporates the use of construction machinery, typically a backhoe, to accomplish the mixing process. Where large containment areas are being treated, clamshells and/or draglines may be used. Major modifications to the in situ mixing strategy include the development of the reagent addition and mixing equipment that allows better control of the process.

Plant mixing strategies can be adapted for applications to slurries and solids. It is most suitable for application at sites with relatively large quantities of contaminated materials to be treated. In the plant mixing process the materials are physically removed from their location, mechanically mixed with the immobilization reagents and redeposited in a prepared disposal site. Special equipment adaptions have been utilized to handle sludges with high solid contents and contaminated soils.

Area mixing is applicable to those confined disposal sites, where high solids content slurries must be treated. The term "area mixing" is used to denote those strategies that use horizontal construction techniques to add and mix the stabilization reagents with the dredged material. Area mixing is land intensive, requiring a relatively large land area to carry out the process. Area mixing strategies present the greatest possibility for fugitive dust, organic vapour and odour generation. The typical area mixing strategy will require that the dredged material be sufficiently dewatered to support construction equipment.

The three basic implementation strategies can be incorporated into confined disposal operations in a variety of ways to improve contaminant containment. In most cases, the same immobilization reagents and equipment can be employed.

Figure 3 (Francingues 1985) shows three concepts in the application of immobilization technology at confined disposal sites.

Concept A involves alternating layers (thin lifts) of relatively clean dredged material and contaminated dredged material that is stabilized. The initial lift of clean, fine-grained sediments would be dewatered to promote densification and consolidation to provide a low permeable soil layer or natural liner for the containment area. Once this layer has achieved the desired degree of consolidation and permeability, the contaminated material would be placed on top, dewatered and stabilized in situ. This layering process would continue until the containment area was filled. A final soil cover would be added, shaped and vegetated.

The second disposal method shown in Fig. 3 incorporates soil stabilization as a treatment to produce a low permeable liner and is designated concept B. The stabilized soil layer is used to contain any leachate generated from the dewatering and long-term disposal of the contaminated dredged material. Appropriate chemical reagents are added and mixed with the disposal site soil using either the in situ or area mixing methods. Then a layer of coarse material is added above the stabilized layer to facilitate dewatering and collection of leachate. The contaminated dredged material is disposed next and dewatered. A clean layer of dredged material is used as final cover. One modification to this concept would be the additional step of stabilizing the contaminated dredged material to further protect against contaminant escape.

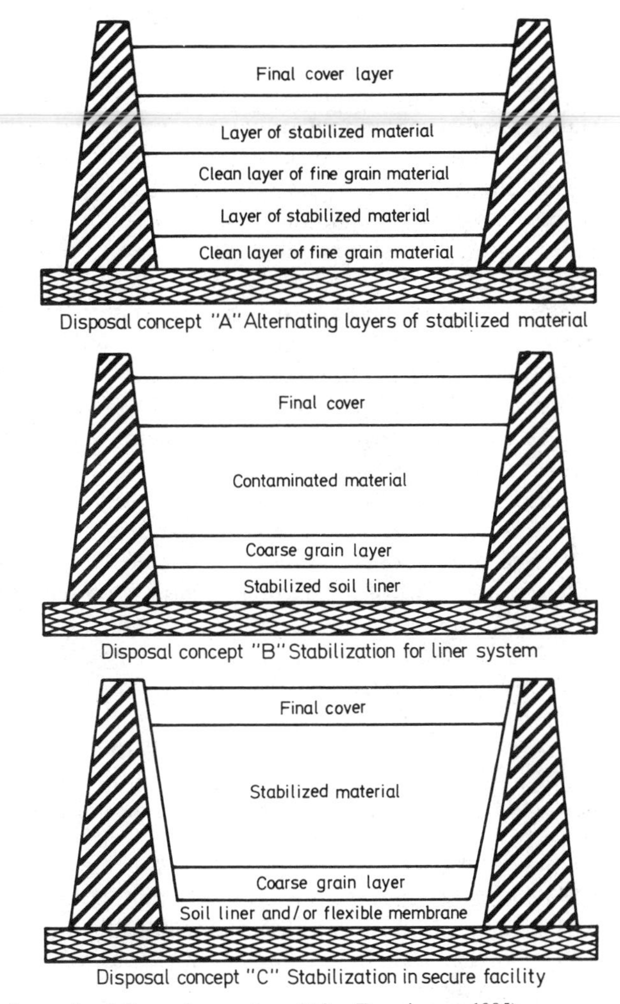

Fig. 3. Immobilization concepts for confined disposal operations (After Francingues 1985)

The final concept illustrated in Fig. 3 provides the highest degree of environmental protection. A soil liner, or flexible membrane, or both, are used to line the bottom and sides of the disposal sites. Then a coarse-grained layer is used to facilitate dewatering and leachate collection. The contaminated sediment is disposed into the lined site, dewatered and stabilized. An alternative would be first to dewater the contaminated sediments in a temporary processing facility, and then to apply the stabilization process prior to placement of the treated material in the confined disposal site. This stabilization operation would be accomplished using the plant mixing strategy (either mobile or field erected) or the area-wide mixing strategy, stockpiling the stabilized material prior to rehandling and disposal in the prepared site. The last operation in concept C would be to finally cover, shape and vegetate the site.

5 Evaluation and Efficiency of Stabilization Processes

Solidification and immobilization processes for chemical and mechanical stabilization of contaminated sludges are designed to prevent hazardous effects for the environment. These processes must include certain qualities of the material, e.g. a better resistance against erosion, a better carrying capacity or a low water permeability. Other qualities like decomposition of organic pollutants or immobilization of persistent pollutants to reduce the leachability and bioavailability are desired.

Processes on the basis of silicates, aluminates, lime and limestone will be preferred, because the chemical reactions and the formed mineral phases are similar to natural conditions. Special attention should be directed to mechanisms which are able to fix pollutants chemically (Wiedemann 1982).

To estimate short- and long-term chemical transformations the interrelations between solid phases and water have been increasingly subjected to laboratory experimentation. From country to country different procedures and tests are used and thus the results are hardly comparable. Unique criteria are rare and often criticized because of their simplicity, e.g. the German standard leaching test "S4", in which one part of a sample is shaken with ten parts distilled water for 24 h (Fachgruppe Wasserchemie 1983).

The chemistry of leaching tests was reviewed by Lowenbach (1978) and Grant et al. (1981). Comparisons of different leaching tests were conducted by Jackson et al. (1981), Ham et al. (1979) and Theis and Padgett (1983).

A critical overview of the most important leaching tests is given by Stegmann (1979, unpubl.), who compares the procedures with the German S4 test. Anderson et al. (1979) presented four factors which should be predicted by an ideal leaching test:

1. The highest concentration of a pollutant in the effluent;
2. The factors which control this concentration;
3. The total amount of a pollutant which may be mobilized from a fixed quantity of the material; and
4. The type of mobilization in the course of time.

Single reagent extraction tests are useful for defining the release characteristics of contaminants under specific conditions. Liquid/solid ratio, pH value, contact time and complexing capacity of the leachate must be considered a new for any case. The selectivity of extraction reagents can sometimes be enhanced through use of serial extraction procedures, whereby the solid material is sequentially contacted with one extractant at a time. A major advantage of this approach is that the type and sequence of extracting reagents can be altered for application to a specific material (Theis and Padgett 1983).

Column leachate tests on polluted harbour sediments were conducted to determine the types and concentrations of pollutants which might be leached in an upland disposal site (Seger and Leonard 1984). For these tests a set of three replicate columns was used. Each column was loaded to a depth of 10 in. with sediments. The system was designed to maintain anaerobic conditions which would be expected to persist at an upland disposal site. An artificial rainfall buffered at pH 4.5 was periodically added to each column to maintain a head of 1.5 to 2.0 ft. above the sediments.

In the Netherlands a working group with participants of research institutions and industry has proposed a standard leaching test for combustion residues consisting of a column leaching test and a series of cascade extractions with renewal of the contact solution up to an ultimate liquid/solid ratio of 100 (van der Sloot et al. 1984). The procedure was operated in an upflow using a peristaltic pump to maintain a constant percolation rate and a series of five subsequent extractions at a liquid/solid ratio of 20 using the same portion of residue repeatedly. The contact time for each extraction was 23 h. Agitation was effectuated by rolling tightly closed polyethylene bottles on a roller table. The obtained cumulative liquid/solid ratio was 100. A material can be solidified in such a way (monolith) that the convectional transport of contaminants due to movement of the interstitial water becomes negligible, and the diffusive flux is decisive for emission of pollutants. The diffusion process depends on the temperature and composition of the surrounding water.

For simulation of the dynamics of these diffusion processes a stand test has been designed with which it is possible to determine the effective diffusion factor for a number of materials. The test must be carried out over a long period, and for the translation of the data to practical situations very sophisticated descriptions are needed (van der Sloot et al. 1985b).

The US Army Corps of Engineers and the US Environmental Protection Agency have developed an elutriate test which is designed to detect any significant release of contaminants from dredged material (Lee and Plumb 1974). This test involves the mixing of 1 vol of the sediment with 4 vol of the disposal site water for a 30-min shaking period.

A study conducted on the factors influencing the results of the elutriate test has shown that the test as originalwy developed cannot yield a reliable estimate of the potential release of contaminants at the dredged material disposal site, since it did not define the conditions of mixing to enable a well-defined, reproducible oxygen status existing during the test period (Lee et al. 1976). As a result a modified elutriate test has been proposed in which compressed air agitation is utilized during the mixing period.

Recently, a sequential leaching study conducted on originally anoxic sediments, which had been treated with the modified elutriate test, showed that sediment components controlling metal solubilities change drastically during the short aeration period (Kersten and Förstner 1986).

Such sequential leaching techniques, which include the successive leaching of metals from ion exchangeable, carbonatic, reducible, sulphidic/organic and residual sediment fractions, can provide reliable information on the behaviour of metal pollutants before and after the application of stabilizing additives and characterize typical environmental conditions.

Most realistic, but very time-consuming and expensive, are experiments and measures in field studies on disposal sites or on comparable and equivalent models (e.g. lysimeter studies). Such disposal experiments have to be carried out as a rule over several years, must produce an enormous amount of data and can only be justified in cases in which the behaviour of a hazardous waste is mostly uncertain. Even then it is difficult to make long-term predictions over geological periods (Wiedemann 1982).

The majority of stabilization processes leads to an improvement of the physical and mechanical properties of the material. Thus, a decrease in the mobilization of hazardous substances can be obtained, but for all known and studied processes the immobilization effect is nearly exclusively based on physical encapsulation of the contaminants in a solid and low permeable mass. Chemical fixation is insignificant. It can be expected that the reaction with some additives increases the chemical reactivity and water solubility of certain pollutants. Amphoteric metals and some organic contaminants show an enhanced dissolution in an acid and alkaline milieu, thus the best alternative for upland disposal of dredged material seems to be a strongly buffered system with pH values in the range of 7 and 9.

6 Application of Stabilization Techniques for Metal-Polluted Dredged Sediments

In 1985 Battelle carried out a feasibility study on the solidification of dredged mud from Hamburg harbour (West Germany) (Battelle 1985). This study evaluated experimental works done by several firms and institutions and summarized the fundamental results. Three important questions had to be answered:

1. To what degree are the mechanistic properties of the material improved by solidification (improved possibilities to form a hill by upland disposal)?
2. How far is it possible to improve fixation of pollutants in the mud by solidification (improved disposal abilities)?
3. Are there other experiences with solidification of dredged mud, and what about the cost factor?

The investigations showed that about 14 firms and institutions were occupied with solidification of river sediments and harbour sludges. Although there have been some experiences with technical solidification of wastes for years, only results from laboratory experiments could be obtained for harbour sludges. These studies clearly showed an improvement of soil-mechanistic data by solidification. Examination of heavy metal leachability on the solidified and original harbour sludge led to the following results:

1. Solidification of harbour sludges from Hamburg and Bremen (West Germany) is practicable in the laboratory with hydraulic additives on the basis of cement and lime.
2. From the results of comparable leaching tests no significant advantage or disadvantage of the studied solidification techniques could be detected.
3. Heavy metal concentrations in the leachates in the solidified and original harbour sludge are comparable, from which can be seen, that solidification did not lead to immobilization in the studied cases.

At present, cost estimations of solidification are accompanied by many uncertainties. Battelle expects for solidification with cement, depending on the use of additives, 10 and 15% cement respectively, between 30 and 55 DM m^{-3} (deutsche marks) dewatered dredged mud. Other estimations for upland disposal of solidified sludge (15% cement) are in the order of 100 DM m^{-3} (ARGE Schlick-technik 1985, unpubl.).

At the Technical University of Hamburg experiments are being carried out to stabilize dredged mud with lime, limestone, cement and coal fly ash in different mixtures (Calmano et al. 1986). Table 2 shows the mixtures of dredged mud from Hamburg harbour and additives; as can be seen pH values differed strongly.

Acid titration curves of two examples are shown in Fig. 4. At high pH values after addition of lime, cement or coal fly ash, soluble metal bases can be formed and problems with development of ammonia result. More favourable is the addition of limestone, since pH conditions are not changed significantly and maintenance of pH at neutrality or slightly beyond favours adsorption and precipitation of most toxic metals.

The stabilized but not solidified sediments were subjected to a leaching procedure using distilled water. In this test the samples of the fine-grained and stabilized sediments were shaken for 15 h in a 1:10 solid/water ratio. The results are shown in Table 3.

Leachates show lower retention capacity for higher pH mixtures introduced by lime and cement/coal fly ash additives, especially for copper and nickel, although these mixtures exhibit a relatively high buffer capacity. The high metal mobilization correlates with high TOC, BOD and COD values in the leachates. Since natural humic acids become increasingly soluble with higher pH values, metals strongly complexed by organic matter are most likely mobilized at these pH values from the stabilized dredged materials.

Table 2. Mixtures of dredged mud from Hamburg harbour and additives

Sample	Dredged mud (g)	Lime (g)	Limestone (g)	Cement (g)	Coal fly ash (g)	Initial pH
1	100					7.74
2	80	10	10			11.76
3	80		20			8.68
4	80	10		10		11.43
5	80			10	10	11.44

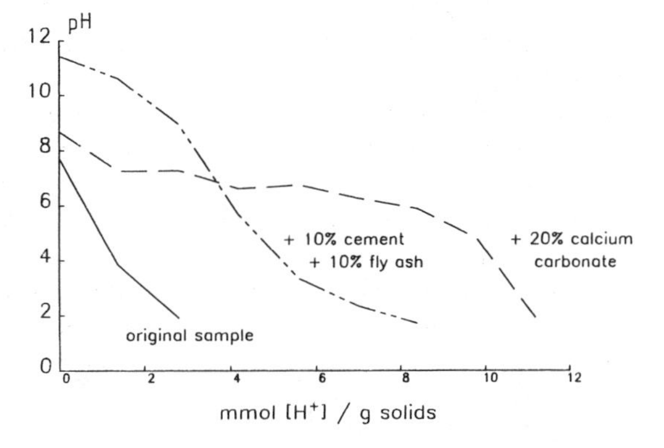

Fig. 4. Effect of calcium carbonate and cement/fly ash additives on chemical stabilization of fine-grained sediment from Hamburg harbour (After Calmano et al. 1986)

Table 3. Leachable metal portions with dist. water (% of total metal content in the dry sample) and sulphate (mg l^{-1}), total organic carbon (TOC) (mg l^{-1}), biological oxygen demand (BOD) (mg O_2 l^{-1}) and chemical oxygen demand (COD) (mg O_2 l^{-1}) in the leachates

Element	Dredged mud	Limestone	Lime/ limestone	Lime/ cement	Cement/ coal fly ash
As	1.3	5.1	0.6	0.5	0.4
Cd	–	–	–	–	–
Co	–	0.2	2.1	2.1	1.4
Cr	–	0.2	0.2	0.3	0.2
Cu	–	0.2	20.8	15.6	16.5
Hg	–	–	–	–	–
Ni	0.3	0.7	17.5	14.6	10.4
Pb	–	–	–	–	–
Zn	–	–	–	–	–
Final pH	7.5	7.8	12.3	12.0	12.0
SO_4^{2-}	750	400	n.d.	n.d.	n.d.
TOC	44	46	259	307	277
BOD	24	46	260	362	321
COD	132	177	650	781	733

n.d. = not determined

The relatively high sulphate concentrations in the leachates (the Hamburg harbour basins are situated in the freshwater part of the Elbe estuary) indicate initiation of oxidation processes in the dredged mud. As mentioned above, during disposal of anoxic sediments acidification can result from the oxidation of sulphidic compounds.

In order to trace the metal remobilization behaviour of the sediment, a dried original sludge sample was titrated with nitric acid to pH values of 5 and 3 respectively. Limestone stabilized samples titrated with the same amount of acid exhibited higher pH values of 6.8 and 5.9 respectively. The results of the retention capacity for metals are presented in Table 4.

Distinct mobilization after acidification was shown by As, Ni, Zn and especially Cd. Leachates of the limestone stabilized sample show improved immobilization. This property is not only the consequence of the increased buffer intensity, but also other mechanisms seem to retain the metals as evidenced by comparison of the original sample at pH 5 and the limestone stabilized sample at pH 5.9. Somewhat problematic remains the behaviour of As, since it was not immobilized by this chemical stabilization technique. Further studies are necessary.

Kamon (1985) investigated the possibility of using a stabilization treatment by lime and cement mixture for very soft freshwater clay.

Lime alone or cement alone did not cause any hardening in preliminary tests. Hence, a special stabilizing agent which aims at the pozzolanic reaction and can reduce the free water by enclosing it into ettringite crystals was introduced. A lime/cement mixture with aluminium sulphate was used with varying relative proportions of their contents. The proportion of lime to cement was mixed in a weight ratio of 2:1. The combination ratio of the lime/cement mixture to liquid

Table 4. Mobilization of metals after addition of acid (% of total metal content in the dry sample)

Element	Original pH 5	Sample pH 3	Limestone stabilized sample pH 6.8	Limestone stabilized sample pH 5.9
As	2.1	12.2	7.6	25.9
Cd	7.4	62.5	<0.1	0.4
Co	1.8	4.6	1.1	3.0
Cr	<0.1	0.5	<0.1	<0.1
Cu	0.2	3.8	<0.1	<0.1
Hg	<0.1	<0.1	<0.1	<0.1
Ni	10.7	37.6	1.8	5.2
Pb	<0.1	0.3	<0.1	<0.1
Zn	13.4	56.2	0.3	2.0

aluminium sulphate was varied in a weight ratio of 9:1, 8:2, 7:3 and 6:4. These stabilizing agents were added to the soil samples at 5, 10, 15, 20, 30 and 40% of the soil dry weight. The varied water contents of the soil samples were 150 and 200% which corresponded to the in situ water content. The effect of the hardening treatment was examined by the unconfined compressive strength q_u of the treated samples.

Figure 5 shows the unconfined compressive strength of a sample with a water content of 150% after a curing time of 360 days in relation to the combination ratio and added volume of stabilizers. The increasing strength in samples of 9:1 and 8:2 combination ratios and at more than 15% added volume is clearly illustrated in the slope of the figure. The hardening strength increases with the increase in the added volume of stabilizing agents. The optimum combination ratio is 9:1 to 8:2. Beyond these ratios the effect of the ettringite crystals reduces the hardening strength.

By X-ray diffraction analysis and scanning electron microscope study it was clearly seen that the pozzolanic reaction products are directly related to strength increase.

In order to trace the gas production properties of chemically stabilized sludge from Hamburg harbour, we conducted a batch test (Garbers and Krause, pers. commun.). Original and stabilized samples were sealed airtight in 120-l plastic containers equipped with an automatic leachate recycling unit. The containers were kept in temperature stabilized rooms at predefined conditions (30°C). The gas produced within the containers was sampled in aluminium gas balloons. Preliminary results on the gas and leachate production indicate that microbial metabolic reactions in chemically stabilized samples differ from those of the original dredged mud. The initial microbial degradation of the organic matter in the limestone stabilized samples seems to be inhibited. No methane was detected after 4 weeks of incubation in the gas of the stabilized mixture, whereas in the original sample a methane proportion of 25% was found after the same time. Long-term incubation experiments with lime, limestone and gypsum stabilized sediments are in progress.

Another possible solution for the solidification of dredged mud is the use as building materials. A feasibility study by Battelle showed the following possibilities (Battelle-Institut 1982):

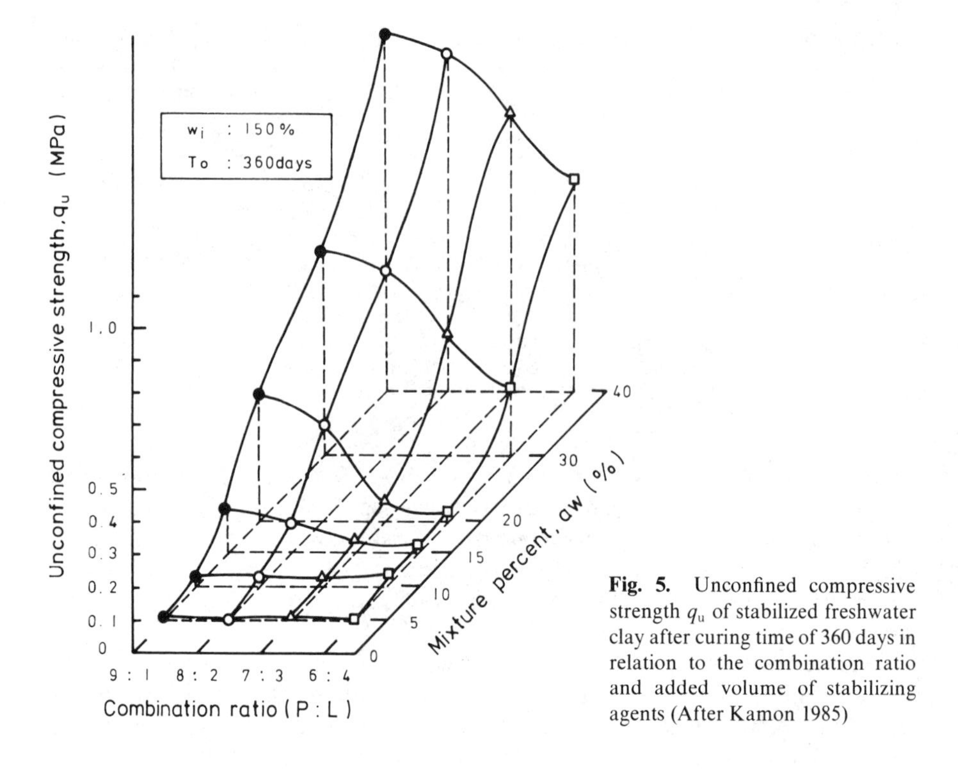

Fig. 5. Unconfined compressive strength q_u of stabilized freshwater clay after curing time of 360 days in relation to the combination ratio and added volume of stabilizing agents (After Kamon 1985)

1. Admixture for production of building stones;
2. Surrogate of raw materials for production of quartz fibre;
3. Admixture for production of cement;
4. Solidification to mudstone.

All these procedures are objected in view of recycling. Brickwork products, like tiles and clinkers, can be produced by the use of fine-grained dredged materials with high clay contents. Indeed, there are more scruples to use strong polluted sediment, especially if the material is contaminated with mercury, arsenic, cadmium and volatile organic pollutants, because a very sophisticated foul-air purification is required. If these problems are solved technically, parts of solidified dredged mud could be used, e.g. as building materials in disposal sites (in drainage layers), since the leachability of these substances is relatively favourable.

7 Conclusions

The chemistry and microstructure of stabilized dredged material have to be studied more extensively in the future. Previous work on the mechanisms of fixation by the use of scanning electron microscopy (SEM), X-ray diffraction (XRD), polarization microscopy, porosimetry and chemical leaching tests demonstrated the importance of microstructure in determining the degree of fixation of toxic substances (Kamon 1985, Poon et al. 1985, Khorasani, pers. commun.). Also, the ecological effects of the stabilized material and corresponding bioassays must be included in further investigations.

From a geochemical point of view, another natural stabilization alternative for dredged sediments could be of interest. The deposition in an anoxic and sulphidic milieu seems to be a very efficient method to immobilize most heavy metals because of formation of very stabile metal sulphides. Such a milieu is found in the marine environment, and it must be, of course, strictly isolated from the surrounding hydrosphere and biosphere. Especially for mercury-contaminated sediments a typical marine process is of interest (Berman and Bartha 1986; Blum and Bartha 1980): in the anoxic, sulphidic milieu monomethyl mercury, one of the most toxic compounds, is transformed into readily volatile dimethyl mercury and low soluble mercury sulphide (Craig and Moreton 1984).

Besides the example of the harbour sludge depot at Rotterdam, several projects are in progress, in which dredged material is deposited into submarine borrow pits and covered with clean sediments (Bokuniewicz 1982, Morton 1980). On this occasion organic pollutants are increasingly accounted for, the behaviour of which as yet has been little studied in comparison to heavy metals (Brannon et al. 1984; Förstner et al. 1985; Sumeri 1984).

In Hamburg a subsediment deposition for contaminated harbour sludge is planned (Tamminga et al. 1986). The present studies show the following technological concepts (Führböter 1985).

1. The optimal location is an area with a water depth between 2–5 m and a ground consisting of sand at least 25 m thick.
2. The site should be located in an area which is morphologically as stabile as possible with a low erosion potential by current and wave forces.
3. The most suitable technological concept seems to be the deposition into a previously dug hole which is sealed with uncontaminated material after completion of disposal.
4. The site should be enclosed by an annular dam to prevent the transport of contaminants.
5. The dredged material can be placed by spoiling or in partially dewatered form (see Fig. 6). The transport water must be recirculated and cleared off.

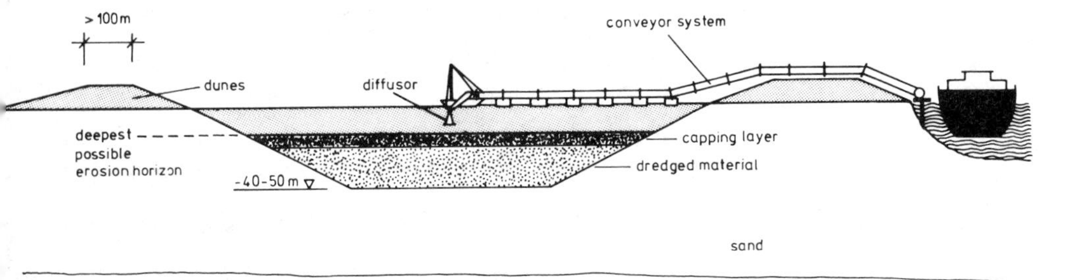

Fig. 6. Model of a planned subsediment deposition for contaminated dredged mud from Hamburg harbour (After Tamminga et al. 1986)

The disposal site is outlined for a period of 20 years. The yearly dredging volume is 600,000 m^3 partially dewatered material and 2,500,000 m^3 untreated dredged sludge respectively.

From an ecological point of view there are some open questions. Apart from long-term stability, resistance of the capping material and contaminant exchange processes through ground, dam and capping layer, the decomposition of organic material combined with gas development (H_2S) and the formation of metal-complexing compounds have to be studied. The answers to these questions require interdisciplinary efforts in research.

References

Ahlf W (1983) The river Elbe: behaviour of Cd and Zn during estuarine mixing. Environ Technol Lett 4:405–410

Alloway BJ, Morgan H (1986) The behaviour and availability of Cd, Ni and Pb in polluted soils. In: Assink JW, Brink WJ van den (eds) Contaminated Soil. Martinus Nijhoff, Dordrecht, pp 101-113

Anderson MA, Ham RK, Stegmann R, Stanforth R (1979) Test factors affecting the release of materials from industrial wastes in leaching tests. In: Pojasek RB (ed) Toxic and hazardous waste disposal, vol 2: Options for stabilization/solidification. Ann Arbor, 259 pp

Battelle-Institut e.V. Frankfurt (ed) (1982) Neue Technologien zur Behandlung von Baggerschlick – Durchführbarkeitsstudie Stufe I: Sammlung, Vorauswahl und Beschreibung der aussichtsreichsten Lösungsmöglichkeiten, 184 pp

Battelle-Institut e.V. Frankfurt (ed) (1985) Neue Technologien zur Behandlung von Baggerschlick – Durchführbarkeitsstudie II: Detaillierung. Endberich Teil I: Verfestigung von Hafenschlick, 35 pp

Berman M, Bartha R (1986) Control of the methylation process in a mercury polluted aquatic sediment. Environ Pollut Ser B 11:41–53

Blum JE, Bartha R (1980) Effect of salinity on the methylation of mercury. Bull Environ Contamin Toxicol 25:404–408

Bokuniewicz HJ (1982) Submarine borrow pits as containment sites for dredged sediments. In: Kester DR, Ketchum BH, Duedall IW, Park PK (eds) Dredged material disposal in the ocean (Wastes in the ocean, vol 2). John Wiley & Sons, New York, pp 215–227

Brannon JM, Hoeppel RE, Gunnison D (1984) Efficiency of capping contaminated dredged material. In: Dredging and dredged material disposal, vol 2. Proc Conf Dredging '84, Clearwater Beach, Fl, pp 664–673

Calmano W, Ahlf W, Förstner U (1983) Heavy metal removal from contaminated sludges with dissolved sulfur dioxide in combination with bacterial leaching. Proc Int Conf Heavy metals in the environment, Heidelberg, CEP Consultants, Edinburgh, pp 952–955

Calmano W, Wellershaus S, Liebsch H (1985) The Weser estuary: a study on heavy metal behaviour under hydrographic and water quality conditions, Veröff Inst Meeresforsch Bremerhaven 20:151–182

Calmano W, Förstner U, Kersten M, Krause D (1986) Behaviour of dredged mud after stabilization with different additives. In: Assink JW, Brink WJ van den (eds) Contaminated soil. Martinus Nijhoff, Dordrecht, pp 737–746

Craig PJ, Moreton PA (1984) The role of sulfide in the formation of dimethyl mercury in river and estuary sediments. Mar Pollut Bull 15: 406–408

Fachgruppe Wasserchemie in der GdCh (1983) Deutsche Einheitsverfahren zur Wasser-, Abwasser- und Schlammuntersuchung, 12. Lfg. Chemie, Weinheim

Fagerström T, Jernelöv A (1972) Aspects of the quantitative ecology of mercury. Water Res 6:1193–1202

Förstner U, Ahlf W, Calmano W, Lohse J (1985) Untersuchungen zum Verhalten von Hamburger Baggerschlick beim Einbringen in eine Deponie im Küstenvorfeld (orientierende Laborexperimente). Intern Stud Amt Strom- und Hafenbau, Hamburg, 62 pp

Francingues NR (1985) Identification of promising concepts for treatment of contaminated sediments. In: Patin TR (ed) Management of bottom sediments containing toxic substances. Proc 10th U S/Jpn Experts Meet, 30–31 Oct, 1984, Kyoto, Jpn, pp 162–185

Führböter A (1985) Optimierungsbetrachtungen für eine Tiefdeponie auf einer künstlichen Sandinsel in der Nordsee (Einbau von kontaminiertem Hafenschlamm im Verklappverfahren). Stud Behalf Amt Strom- und Hafenbau, Hamburg

Gambrell RP, Khalid RA, Patrick WH, Jr (1978) Disposal alternatives for contaminated dredged material as a management tool to minimize environmental effects. DMRP-Rep DS-78-8. US Army Corps Eng Waterways Exp Stn, Vicksburg (Miss)

Grant CL, et al. (1981) Comparison of laboratory methods for studying leaching of retortet oil shale. Symp Surface mining hydrology, sedimentology and reclamation, Univ Kentucky, Lexington, p 451

Ham RK et al. (1979) Comparison of three waste leaching tests. EPA 600/2–79–071, Cincinnati, Oh

Herms U, Tent L (1982) Schwermetallgehalte im Hafenschlick sowie in landwirtschaftlich genutzten Hafenschlick-Spülfeldern im Raum Hamburg. Geol Fahrb F12:3–11

Jackson K et al. (1981) Comparison of three solid waste batch leach testing methods and a column leach testing method. First Conf Hazardous solid waste testing, ASTM STP 760, Philadelphia, Pa, p 186

Kamon M (1985) Lime-cement hardening of very soft freshwater clay. In: Patin TR (ed) Management of bottom sediments containing toxic substances. Proc 10th US/Jpn Experts Meet, 30–31 Oct, 1984, Kyoto, Jpn, pp 237–248

Kersten M, Förstner U (1986) Chemical fractionation of heavy metals in anoxic estuarine and coastal sediments. Water Sci Technol 18:121–130

Kester DR, Ketchum BH, Duedall IW, Park PK (1983) Wastes in the ocean, vol 2: Dredged material disposal in the ocean. John Wiley, New York, 299 pp

Lee GF, Plumb RH (1974) Literature review on research study for the development of dredged material disposal criteria. DMRP-Rep D-74-1. US Army Corps Eng Waterways Exp Stn, Vicksburg (Miss)

Lee GF, Lopez JM, Piwoni MD (1976) Evaluation of the factors influencing the results of the elutriate test for dredged material disposal criteria. In: Krenkel PA et al. (eds) Dredging and its environmental effects. Am Soc Civil Eng, New York, pp 253–258

Lowenbach, WA (1978) Compilation and evaluation of leaching test methods. EPA 600/2–79–095, Cincinnati, Oh

Malone PG, Jones LW, Larson RJ (1982) Guide to the disposal of chemically stabilized and solidified waste, SW-872, Office of Water and Waste Management, U.S. Environmental Protection Agency, Washington DC

Morton RW (1980) Capping procedures as an alternative technique to isolate contaminated dredged material in the marine environment. In: Dredge spoil disposal and PCB contamination: Hearings before the Committee on Merchant Marine and Fisheries. House of Representatives, 96. Congr, 2nd Sess Exploring the various aspects of dumping of dredged spoil material in the ocean and the PCB contamination issue, March 14, May 21, 1980. USGPO Ser 96–43, Washington, DC, pp 623–652

Poon CS, Clark AI, Peters CJ, Perry R (1985) Mechanisms of metal fixation and leaching by cement based fixation processes. Waste Manag Res 3:127–142

Rulkens WH, Assink JW, Gemert WJT van (1985) On-site processing of contaminated soil. In: Smith MA (ed) Contaminated land – reclamation and treatment. Plenum, New York London

Seger ES, Leonard RP (1984) Column settling and column leachate test on polluted harbour sediments. 27th Annu Conf Great lakes research. US Army Corps Eng, Buffalo District, 1776 Niagara Street, Buffalo, NY 14207

Sloot HA van der, Piepers O, Kok A (1984) A standard leaching test for combustion residues. BEOP-31. Bur Energ Res Proj, Netherlands Energ Res Found, Petten, The Netherlands, 55 pp

Sloot HA van der, van der Wijkstra Jivan Stigt, CA, Hoede, D (1985a) Leaching of trace elements from coal ash and coal ash products. In: Duedall IW, Ketchum BH, Kester DR, Park PK (eds) Wastes in the oceans series, vol 4. John Wiley & Sons, New York

Sloot HA van der, Weyers EG, Hoede D, Wijkstra J (1985b) Physical and chemical characterization of pulverized coal ash with respect to cement based applications. Netherlands Energ Res Found ECN, ECN-178, 247 pp

Stumm W, Morgan JJ (1970) Aquatic chemistry. Wiley Interscience, New York London Sydney Toronto

Sumeri A (1984) Capped in-water disposal of contaminated dredged material. In: Dredging and dredged material disposal, vol 2. Proc Conf Dredging '84, Clearwater Beach, Fl, pp 644–653

Tamminga PG, Tent L, Tietjen G (1986) Neuere Konzepte zur Unterbringung von Baggergut im Küstenvorfeld in einer Untersedimentdeponie. Amt Strom- und Hafenbau, Hamburg. BUP Stud 17

Theis TL, Padgett LE (1983) Factors affecting the release of trace metals from municipal sludge ashes. J water Pollut Contrib Fed 55:1271–1279

Wiedemann HU (1982) Verfahren zur Verfestigung von Sonderabfällen und Stabilisierung von verunreinigten Böden. Ber Umweltbundesamtes 1/82. Schmidt, Berlin

Integrated Biological Systems for Effluent Treatment from Mine and Mill Tailings

HANS VON MICHAELIS[1]

1 Introduction

Biological processes are of course routinely used in the treatment of municipal waste water. Randol (1985) has just completed a comprehensive encyclopedia on *Waste Management in Mining and Metallurgical Operations* and one of the interesting discoveries is that if treated municipal waste water is used for mineral processing, e.g., flotation, then the type of biological process applied in the municipal treatment plant is all important. Waste treated by conventional biofilters is more often than not unsuitable for use in the flotation process. However, these are more sophisticated multistage activated sludge processes that seem to render this type of water suitable for this purpose without reducing recoveries. The important point is that biological processes can be developed for routine use in the treatment of waste and effluents.

Another important potential application of biological processes is in the treatment of effluents from mines and metallurgical operations either for the removal of metal contaminants, organic components, and other toxic species such as ammonia, cyanide, or for removal of undesirable anions such as sulfate, phosphate, etc.

Townsley et al. (1985) discuss the biorecovery of metals from industrial effluents by filamentous fungi; Imai (1985) discusses the utilization of sulfate-reducing and photolithotropic bacteria; Brierley et al. (1985) report on their new AMT-Bioclaim waste water treatment and metal recovery technology employing material derived from bacteral cell biomass, while Whitlock and Mudder (1985) describe the use of bacteria to successfully recover cyanide, ammonia, and low level metals from mine and mill effluents in two stages.

The aims of this chapter are twofold. The first aim is to list several additional examples of biological processes or phenomena which are either in use or under evaluation to treat or polish effluents. This list should offer encouragement that biological processes can actually work. The second aim is to encourage further investigation and development of simple integrated biological systems for effluent treatment. It is conceivably possible that integrated bioprocesses employed in an"engineered ecology" may prove useful and hopefully can be developed into systems that make maximum use of natural attributes of a site and require less attention than conventional systems.

[1]Randol International Ltd., 21578 Mountsfield Drive, Golden CO 80401, USA

2 Effluent Treatment Needs in the Mining and Mineral Processing Industry

Randol Int Ltd. (1985) concluded that:

1. Effluent treatment requirements for mining and mineral processing operations are highly site-specific and depend on climate and the extent to which a mine requires dewatering as well as the nature of the ore, tailings, and waste rock.
2. There are important unresolved environmental problems associated with effluents from some mining and mineral processing operations, mill tailings, and waste rock. These occur more often in wet, i.e., net precipitation climates, or where the mines themselves generate excess water resulting in the unavoidable need to discharge an effluent.
3. Acid mine drainage associated with mining and treatment of massive sulfide ores, coal, and tailings containing sulfide minerals may require treatment. Complexed metals associated with cyanide leaching effluents or acid mine drainage sometimes make it difficult or impossible by conventional means to meet discharge water quality permit requirements. Not only is it sometimes impossible to achieve low levels of soluble metals in solution, but finely divided suspended precipitate particles may also be difficult to remove from effluents.
4. Conventional water treatment processes, e.g., lime-settle-filtration are often expensive both in terms of capital and operating costs.
5. Effluent treatment needs can be reduced through careful water balance management with emphasis on runoff control and recycling.

Regulations require that acid mine drainage or water that has been in contact with waste rock and/or tailings must be collected and treated before it can be discharged. Randol Int Ltd. (1985) further concluded that:

6. Sulfide-bearing mines, waste rock, and tailings may continue to generate contaminated acid mine drainage for several centuries after the closure of the operation. Wixson and Davies (1985) pointed out that ancient rock and chat tips connected with old Roman lead-zinc mines in Wales are still causing contamination problems that must be corrected at considerable cost to today's taxpayers. If contaminated, runoff and seepage must be collected and treated "in perpetuity" from today's very much larger sites; the cost of doing so may become so high that it eventually may present an untenable financial drain on the company responsible for maintaining the site, or it may contribute excessively to an accumulating burden on the taxpayer.
7. In some cases "conventional" effluent treatment processes may not achieve sufficiently low levels of metals or other contaminants to avoid environmental damage. This is particularly so if complexing agents are present.

It would be highly desirable, therefore, to develop biological processes which can be employed in "passive" systems which require a minimum of ongoing maintenance. Such systems are not yet well enough defined that they can simply be pulled off the shelf and plugged in, but some components of such systems seem to be emerging and give encouragement for the continuation of further development efforts.

Serious environmental problems associated with effluents from tailings and waste rock are also caused by natural biological processes which need no help from mankind to do their job. It seems logical to try and harness nature to work for us in helping to solve these problems.

3 Integrated Biological Process Concept

The concept proposed here for further attention employs more than one biological process simultaneously or in sequence to treat or polish effluents. Sterrit and Lester (1979) reviewed biological process applications for effluent treatment and identified the possibilities of employing more than one biological process synergistically. They concluded that . . .

"The use of algal blooms in artificial meander systems for the removal of metals from mine waste streams appears to be an effective method of pollution control. If this system were combined with a bacterial system capable of raising the pH, it may possible to produce metal-free effluents with low acidity. Both systems may require amendment with nutrients, principally nitrogen for the algae and a suitable organic carbon source such as sewage sludge for the sulphate reducing bacteria."

They concluded further that "the nature of mine waste drainage, in particular its acidity, probably precludes the use of a heterotrophic bacterial system treating an organic waste to remove heavy metals and organic compounds. However, such a system operated in conjunction with a mine waste treatment scheme could provide nutrients for algae and sulphate reducing bacteria, and provide pregrown cell flocs for the removal of heavy metals."

Biological systems offer a variety of potentially useful processes. The large surface area of aquatic plants and algae present a substrate on which to support bacteria as well as a filtering and settling mechanism which proves effective in removing suspended solids that normally would not settle. Several plants, algae, and bacteria are found to take up and metabolize metals from solution. In addition, decaying biomass presents anaerobic conditions and some of the nutrients required for sulfate-reducing bacteria to function. This offers a further and powerful means of polishing solutions by sulfide precipitation without introducing excess amounts of H_2S. Finally, the large surface area of leaves above the water level in a cattail bog results in enhanced evapotranspiration which can prove useful in disposing of excess water. In addition, the biological metal removal system may be integrated with natural geochemical adsorption effects in the soil substrate. Kickuth (cited in ENDS 1985) has found that marsh plants have the ability to transfer oxygen at a remarkably high rate from the atmosphere into their rhizomes. In turn, oxygen stimulates the growth of bacteria and protozoa which are similar to those which do the work in aerobic sewage treatment processes.

Even the everyday marshland that is to be seen near most mine-mill sites wherever there is water offers both aerobic and anaerobic ecological environments within which plants, algae, bacteria, and enzymes function simultaneously. Duplication of such a range of processes in man-made systems would indeed be too complex to contemplate.

This chapter proposes that in addition to examining the subject through the eyes of specialist botanists, phytologists, microbiologists, etc., that we adopt a system's approach and work on developing an integrated biological system with enhanced effluent treatment capacity through engineering its ecology. To achieve this will require an interdisciplinary approach with emphasis perhaps more on empirical trial and error experimentation rather than theory. A similar proposal is made by Spisak (1985) who also reviews mechanisms of a few of the more interesting processes that can be involved in such systems.

Randol Int Ltd. (1985) identifies several examples where natural biological systems are having a beneficial effect on the quality of effluents from both active mill tailings and abandoned mine sites. Features of some of the observed examples are summarized in the next paragraphs. One challenge that presents itself is that in not one single case is there adequate knowledge on exactly how the natural biological systems achieve their beneficial effects and on their sensitivities or potential performance parameters in practice.

Two criticisms of natural biological systems, i.e., marshland bogs, are often raised. First, it has been stated that the area required to be effective must be very large. This may be so, but quite frequently there is no shortage of land around mining and milling operations. Furthermore, through improved engineering it is very possible that the surface area required to achieve a given objective may be greatly reduced compared with a natural marsh. Aquatic plants tend to grow in clusters and when water enters a marsh, it tends to flow around the reed clusters rather than through them. This was described by Black (undated) in an evaluation of test marshes for the treatment of domestic sewage. Root zone biotechnology systems in engineered marsh ecologies are in commercial operation for sewage treatment for small communities in West Germany according to ENDS Report 125 (1985). The same reference suggests their potential applicability in treating industrial wastes.

The biological activity of such marshes may be enhanced by engineering and maintaining them properly. The paper by Wile et al. (undated) describes how artificial marshland cells were designed and installed which had enhanced capacity for the treatment of municipal sewage. Similar principles could likely also be applied in the design of marshland cells for implementation of integrated biological process systems for mine-mill effluent polishing, even though the processes involved are different.

The second area of concern is whether accumulated metals removed from the effluent will be in a stable form or whether it is likely that they will find their way back into the environment at some point in the future. In response to this question it is of interest to use the Black Sea as an illustration.

The Black Sea is black because it is underlain by about 100 m of sediments that are black in color due to their iron sulfide content. This iron sulfide is the result of a highly efficient natural bioreduction of sulfate from the water with a resultant precipitation of iron as a sulfide. Similar effects are observed in some Dutch canal bottoms. Sulfides of most metals are significantly less soluble than hydroxides. If we are satisfied to leave metals in tailings and decantation ponds in the hydroxide form, should we not be even more satisfied to deposit them in reducing marshland conditions as sulfides?

The ultimate restoration of the marshland site may be an important factor, but remember that in some acid mine drainage treatment systems that may be needed to perform their effluent cleaning function for hundreds of years before the source of AMD is finally exhausted.

4 Applicability of Integrated Biosystems

Biological processes are not a universal solution to all effluent treatment needs. For example, in the case of a tailings impoundment from a massive sulfide mine-mill complex containing, e.g., 50 million tons of 40–50% iron sulfides, the volume of sulfuric acid and iron salts that will be generated as the material oxidizes is so large that the size of marsh to physically contain the iron hydroxides and sulfides that would be formed would be correspondingly large. In such cases it is likely to be unavoidable that a treatment plant be installed and here the emphasis would likely be on selecting a process that would generate the most compact precipitate sludge, i.e., a High Density Sludge (HDS) process. Better methods of disposing such tailings are needed which avoid generation of acid mine drainage in the first place.

However, even in such a case, a biological system may prove useful as insurance against occasional upsets in the treatment plant, or it may be used as a polishing step in conjunction with a simpler treatment plant which raises the pH and removes a precipitate by a simple settling process.

During the Randol Int Ltd. (1985) investigation, numerous smaller effluent treatment needs were encountered where 100–1000 m^3 day^{-1} of water was treated and where efficient metal removal was not achieved primarily because the clarification step failed to adequately remove suspended precipitates. In such cases the metal loading of the effluent is small and an engineered marsh with plants and algae offers a possible way to polish them rather than installing more expensive and sophisticated filtering devices.

Some older mining districts are characterized by a very large number of smaller sites scattered quite widely. These may generate mine drainage in volumes of the order of tens to hundreds of m^3 day^{-1} which nonetheless cause serious environmental damage.

According to Holm et al. (1984), there are at least 10,000 inactive metal mines in Colorado alone, only some of which are significant contributors of pollutants to the stream network. Enough to render 25 watersheds and 450 stream miles barren of aquatic life and unfit for other uses.

Clearly, the installation of a conventional effluent treatment plant at each site would be neither economically practical nor add to the aesthetics of the areas involved. In such cases, engineered integrated biological systems, as proposed by Holm et al. (1984), may develop into the most attractive answer.

Mudder and Whitlock (1984) and Mudder and Whitlock (1985) describe an innovative yet capital-intensive system employing rotating biological contactors

(RBC's) to support cyanide and ammonia-oxidizing bacteria. Operating costs of this system are low. Could there be a way in which a similar process could be made to work reliably year-round in more moderate climates in engineered artificial marsh cells in which the plants themselves serve as the substrate for useful bacteria?

Simovic et al. (1984) provided interesting insight into the "natural degradation" of cyanide in tailings ponds. This process works well during the summer months at some Canadian sites where there is abundant sunshine, plenty of wind, and adequate land to build large, shallow impoundments. Natural degradation of cyanide does not work as well during the colder months, and some mills end up storing their baren bleed through the winter. Cyanide degradation appears to occur more rapidly if the effluent is aerated, and still faster if it is mixed with sewage before aeration. This raises the possibility of using an integrated biological system to enhance natural degradation. It is known that addition of small amounts of ferrous sulfide cause detoxification of cyanide effluents. It is possible that passing effluents containing dilute cyanide and metals after mixing with chopped up sewage through a mature marshland would result in detoxification of the effluent due to naturally generated sulfide precipitation of metals. Another alternative is that the effluent-containing metal cyanide complexes might prove so toxic that they kill the aquatic biosystem. In such an event, probably the intelligent approach would be to employ the integrated biological system after natural degradation has done the bulk of the work. It is clear that industry still lacks knowledge of the extent to which such biological systems offer useful opportunities.

5 Biological Effluent Treatment – Examples

Possible applications of the biological systems are:

1. Heavy metals polishing in mine drainage and tailings decant solution.
2. Passive primary treatment in conjunction with limestone/lime addition.
3. Detoxification of cyanide, heavy metals, and ammonia-containing effluents, before or after natural degradation.

5.1 Example 1: Metals Polished by a Natural Marsh

The first example of an integrated effluent treatment system occurs naturally in upstate New York at a site where an effluent decanting from a zinc-lead mill-tailings impoundment is discharged to a swamp. The so-called swamp is a low-lying area vegetated by trees, clumps of grass, mosses, ferns, etc. Water passing through the area is caused to fan out because of minor ground relief due to tree roots, decaying logs, and frost pressure ridges. Water appears to cover about half of the surface and is several inches deep.

Water was sampled over a 5-day period at the entry point and at a culvert located about 1500 ft. away from the tailings impoundment overflow and each day's samples were analyzed for both "total" and "dissolved" heavy metals.

Table 1. Effect of "swamp" treatment on suspended solid and metal concentrations (mg l^{-1})

	Tails	Culvert	Change (%)
Suspended solids	4.4	2.7	-39
Zn	0.23	0.09	-61
Pb	0.04	0.028	-30
Cu	0.025	0.007	-72
Fe	0.14	0.05	-64
Mn	0.19	0.008	-91

The results of the five day average analysis are shown in Table 1.

It is clear that this "passive" effluent treatment system does an excellent job of polishing the tailings decant effluent. The tailings effluent at a pH of around 7.5 has relatively low metal levels in the first place largely due to the fact that mine water containing dissolved and suspended metal values is used as mill process water and much of the metals are precipitated in their high pH process.

The precise mechanism of metal removal in the swamp area was not determined. The samples were collected in May and the possibility of dilution should not be altogether ignored, although the test was conducted by a responsible person working for a major company.

5.2 Example 2: Artificial Meanders Treat Mine Water

St. Joe Minerals Company at their Pierrepont Mine in upstate New York has installed a series of artificial meandering lagoons in which water depth is controlled at about 3 ft. The purpose of the meandering lagoon is to polish dissolved and suspended metals from mine water effluent. At the time this operation was visited, the underground workings in the new mine were still fresh and metal values were low in the mine effluent. In a year or two as the passive marshland matures it will be an excellent site at which to assess the performance of this type of system. Because the meander system is shallow, water temperatures tend to rise in summer and initially, before the marshland vegetation established itself, a problem was experienced in keeping the discharge temperature within New York State permit limits for trout streams. Also, in establishing this marsh, initial difficulty was encountered in establishing a dense plant growth until the water depth in the meander system was reduced from over 4 ft. Cattails and other marsh plants are sensitive to water depth.

5.3 Example 3: Passive Mine Drainage Treatment in Colorado

Holm, Bishop, and Tempo (1984) of the Colorado Dept. of Natural Resources, Division of Mined Land Reclamation, describe work in Colorado to develop passive mine drainage treatment systems. Holm and Guertin (1984) describe the theoretical design of a passive system for AMD abatement employing a high energy cascaded flow of AMD over limestone rocks in such a way as to provide a high degree of attrition and aeration. This was followed at the Marshall No. 5 abandoned coal mine site by a settling area and with polishing being provided by aquatic plants in an ecologically engineered bog.

Holm et al. (1984) derived their concept of passive drainage treatment mainly from the observation of several natural systems in Colorado, where polluted mine drainage flows, in one case, into a small network of beaver ponds (Shuster Mine) and, in another case, into an artificial bog (Delaware Mine). In both cases a significant reduction in heavy metal concentrations and acidity was realized within a very short drainage segment. These observations were duplicated in bench-scale experiments and field trials by Lapako and Eger (1981). Field-scale investigations feasible of using a passive mine drainage treatment system were demonstrated at the Marshall No. 5 coal mine and reported by Holm (1984).

5.3.1 The Shuster Mine Drain: Natural Bog/Pond Treatment

The Shuster Mine drains polluted mine water into Oak Creek, a tributary of the Yampa River in Colorado. The mine opening is very near the town of Oak Creek. The town is located along Oak Creek about 20 miles southwest of Steamboat Springs, Colorado. Mining activity in the Coal Creek Study has been connected primarily with the extraction of bituminous coal.

The Shuster Mine drainage contains concentrations of iron (8 ppm) and manganese (1.7 ppm), cadmium (15 ppb), lead (200 ppb), and zinc (40 ppb). However, these elevated levels are not observed according to Holm et al. (1984) in Oak Creek even though the Shuster Mine drains voluminous quantities of water (0.1–9.0 cfs) relative to the flow of Oak Creek.

Oak Creek, below the Shuster Mine drainage, exceeded water quality standards once for zinc during the spring of 1980, an event that appeared to be caused by a background source. Zinc concentrations gradually increased downstream, independent of the Shuster Mine drainage. Cadmium exceeded the water quality standards for Oak Creek in the spring and the fall, but in both cases exceeded the standards above as well as below the Shuster Mine. Cadmium levels were only slightly influenced, if at all, by mine drainage. Mine drainage also had no apparent effect on lead values in Oak Creek. The pH values for the Shuster drainage ranged from 5.9–7.7. Iron and manganese were only slightly elevated in Oak Creek below the mine drain.

The impact of the Shuster Mine on Oak Creek is relatively minor even though the mine drainage is spectacular to look at, showing graphic evidence of red staining. A remarkable natural phenomenon seems to be occurring at this site. Mine water is cascaded down the face of the exposed sandstone outcrop below the adit resulting in oxygenation to saturation. Beavers have constructed several ponds in an elaborate system of canals and diversions below the Shuster adit. The result is that the heavily oxidized Shuster drainage filters through the boglike beaver dams and the effluent is gradually blended with stream water until it finally joins the main stream. According to local Oak Creek residents, fishing in Oak Creek is as good below the Shuster drain as it is above.

5.3.2 Delaware Mine – An Artificial Bog/Pond Treatment (Experiment)

One of the first experiments in Colorado with metal drainage treatment using an artificial pond and bog was done for the Delaware Mine in the Peru Creek drainage as reported by Holm et al. (1979). The Peru Creek area has been mined extensively for silver and other base metals over the last 110 years. The sources of acid mine drainage and heavy metals loading are numerous and widespread. One of the significant point sources of acidity and heavy metal loading to Peru Creek is the Delaware Mine.

The mine is located on the east side of Cinnamon Gulch, which drains into Peru Creek. The portal is blocked but effluent flows out through the rubble. There is a yellowish-orange (Fe) precipitate coating the drainage channels. In August 1978, the effluent from the Delaware Mine was rerouted into a small, heavily sodded pond that had been built for the purpose. The effluent flowed across a large area of mine waste and then through this pond. Samples were taken above and below this pond to measure the effectiveness of the pond in improving water quality.

Test results from the bog/pond experiment (Table 2) indicate that over 75% reduction in total metals loading concentration was achieved just by settling out the suspended solids and through adsorption of dissolved metals onto the bog species and soil materials in the pond.

Table 2. Delaware mine metal loading (g day^{-1}) above and below the pond (Holm et al. 1979)

Mine drainage	Flow cfs	pH	Cd	Cu	Pb	Zn	Fe	Mn
Above pond	0.33	3.9	1.2	9.0	2.0	343	525	227
Below pond	0.31	4.5	0.3	2.0	0.1	72	6	49
Reduction in metals loading			75%	78%	95%	80%	99%	83%

5.3.3 Marshall No. 5 Coal Mine, Boulder, Colorado: Passive Treatment by Design

The Marshall No. 5 Coal Mine is located 4 miles south of Boulder, Colorado. It is one of the few inactive coal mines in Colorado which discharges a significant volume of metalliferous acidic mine water which drains across Boulder City open space land.

Water treatment at the site was separated into two stages. The first stage involves oxidation and precipitation. Oxidation is enhanced by creating very turbulent flows. Turbulence is induced by placing rock obstructions in the channel. Fe and Mn are readily precipitated under oxidizing conditions according to Holm et al. (1984). Other metals coprecipitate with the oxides of Fe and Mn. The pH of the drainage tends to increase under highly oxidizing conditions. This further enhances precipitation and coprecipitation.

The first stage of the process is further designed to encourage precipitation as the drainage passes through short sections of limestone cobbles which are placed in trenches periodically between the turbulence groins.

The second stage of the process involves small rock-check dams composed of durable limestone which are designed to temporarily retain the mine drainage and allow it to filter through heavy organic peat and living bog materials. The initial concern was to select a suitable structure that would hold a deposit of peat materials and yet allow percolation of mine effluent through the peat materials. The US Forest Service design for check dams is ideally suited to the construction of an artificial peat bog because of the ease of construction, durability, readily available materials, and ability of the structure to freely pass water. The rock-check dams serve as the structure to hold the peat moss, but also tend to further aerate the acid waters. It was considered important to provide rock aprons and bank protection below each structure to prevent undercutting, as well as a properly designed spillway over each of the rock-check dams.

The bog community species that were introduced in the simulated bog at Marshall No. 5 Coal Mine included:

Aquatic species
Sphagnum moss
Water hyacinth
Pondweed
Horned pondweed
Frogbit

Riparian species
Broad-leaved cattail
Field Horsetail

Forbes
Strawberry clover
White sweetclover

Terrestrial species with high water tolerance
Grapes, Red-top
Tall wheat grass
Western wheat grass
Bann wild rye
Tall fescue
Weeping lovegrass
Birdsfoot trefoil

Trees
European black alder
Gray birch
Black locust

5.4 Example 4: *Sphagnum* Moss Oxidizes Ferrous Iron

Kleinmann (1983, pers. commun.) and co-workers observed that in a natural marsh in West Virginia, which receives water of acid mine drainage from a coal spoil bank, ferric iron is precipitated as a gelatinous hydroxide within 10 ft. of the point at which the acid mine drainage entered the bog. Ferrous iron is oxidized to ferric hydroxide when the contaminated solution comes in contact with *Sphagnum* moss in the bog. This observation prompted the US Bureau of Mines to experiment with a mobile "portabog" made of plexiglass to try to duplicate these results in a portable pilot plant. In the natural marsh ferric hydroxide is coagulated and flocculated by the plants at pH 2.5. Adsorption effects on the plant surfaces appear to be important, and there was evidence that as the flocculated precipitate settled into the deeper levels of the bog, the hydroxide was being converted to a sulfide through reaction with sulfide ions generated by in situ anaerobic biological conversion of sulfates. Subsequent experiments have shown that dead *Sphagnum* moss biomass works as well or better than the live moss, and that *Sphagnum* moss is by no means the only species that shows this phenomenon.

The US Bureau of Mines has subsequently installed several test wetland sites in West Virginia to treat acid mine drainage from coal mine spoil banks. One of the more significant aspects is the efficient removal of manganese from effluents at pH 7 in the wetland which is remarkable because in conventional precipitation systems a pH of 10 or more is needed to remove manganese.

The USBM finds that the method used to prepare an artificial cattail marsh is important if it is to work as desired in the treatment of acid mine drainage. Shallow trenches are excavated. Six inches of agricultural limestone is first laid down as a base, followed by 15 in. of compost. Cattails are planted in a layer of coarse soil spread over the compost. The cattail wetland starts working within about 2 weeks of being planted. Preliminary indications are that 200 $ft.^2$ of wetland surface area is needed per gpm of flow to treat coal spoil acid drainage. Interestingly, the agricultural lime does not become coated with ferric hydroxide and therefore does not become passivated. Possibly this is due to the reducing environment beneath the layer of compost. One hypothesis now being tested for the manganese removal mechanism is that it may be precipitated as a carbonate due to carbon dioxide generated through the reaction between the acid and the agricultural limestone. Another hypothesis is that manganese is oxidized by bacteria.

Two *Sphagnum* moss wetlands and seven *Typha* (cattail)-dominated wetlands have been installed in conjunction with the USBM in West Virginia. Another observation is that cattail wetlands outperform *Sphagnum* moss wetlands in iron removal.

The iron bacteria *Gallionella* is known to accumulate dissolved iron from solution which in turn is secreted by the organism in long strands of hydrated iron oxides. The bacterial cell lives at one end of the group of strands. The secreted oxides form a spongy, rustlike crust on the effluent substrate. *Gallionella* may also accumulate manganese in a similar manner according to Gordon (1983) and Manahan (1984). Manganese-oxidizing bacteria are employed in West Germany as a means of reducing manganese levels of effluents.

5.5 Example 5: Removal of Organics and Metals in Missouri

Jennett and Wixson (1977) pointed out that effective biological treatment the lead-zinc flotation plant tailings decant water were effectively removed in an artificial meander system in which algae were encouraged to grow. The algae also removed residual metal values from the decanted tailing effluent before it was discharged into the receiving stream.

Because algae tend to detach themselves periodically during their life cycle it was necessary to construct a final polishing lagoon at the end of the meander system. The small lagoon provided additional retention time for the effluent to settle accumulated and dislodged algae and aquatic plants and their heavy metal contents. A detailed study was completed on this system by Ernst (1976). Tests indicated that the meander system and settling lagoon removed 84% of the residual organics in the raw effluent. This passive water treatment system also resulted in a significant reduction in lead values in the effluent to the receiving stream. Whether this was due to removal of dissolved lead or not is not clear, however, it is expected that the meander system with aquatic plants and algae were very effective in removing lead carried in minute suspended particles which would normally settle inefficiently.

Jennett and Wixson (1977) pointed out that effective biological treatment through the use of meanders for metal and organics removal would require a timely plan for ultimate closure of the ponds in order to fix the metals, if the long-term benefits of such a system were not to come to naught.

5.6 Example 6: Wood Chips and Agriwaste in AMD Treatment

Several workers have experimented with the use of wood chips and other agricultural waste as a carbon source and substrate by which to treat acid mine drainage. The wood chips seem to provide a substrate and a source of carbon for a bacterial culture that is capable of removing sulfate from water. One of these workers is Lapako and Eger (1981).

CANMET has also carried out promising experiments along similar lines which resulted in a pH increase and partial removal of sulfate.

In Japanese work the effects of the addition of various cometabolites was tested. Securing large sources of such cometabolite organic chemicals for addition to acid mine drainage to support biological treatment in this way would appear to present an economic limitation.

More recently, CANMET has found that the pH of solutions draining through acid-generating tailings can be increased from pH 2 to pH 6 if the surface of the sulfide-containing tailings is covered with a layer of material (biomass and sludge) dredged from a cattail marsh.

5.7 Example 7: Cyanide Treatment in Effluents

Considerable research has been carried out in Canada into the so-called natural degradation of cyanide in effluents. The conclusion of Simovic et al. (1984) is that atmospheric CO_2 and aeration play a major role in reducing the pH of the effluent and resulting in volatilization of HCN over a period of time to the atmosphere.

A gold mine-mill complex located just south of the arctic circle has demonstrated extremely efficient natural degradation of cyanide in their tailings pond decant solution which is discharged as an effluent. Undoubtedly the large surface area of the shallow ponds, abundant sunshine in summer, and the unusually windy climate play a major role. Metals are removed with somewhat greater difficulty, especially arsenic. In temperate climates this would present an excellent opportunity to test the use of a marshland ecology for removal of metals and ammonia downstream from the cyanide degradation ponds. However, because of the short growing season a marshland would not be practical, although algae and bacteria possibly using the operation's sewage as a nutrient source could be worth further consideration as a means of enhancing metal and ammonia removal.

A second Canadian mining company (in a less extreme climate) has found that by mixing a cyanide effluent with sewage and applying vigorous aeration, the cyanide level reductions are accelerated. Presumably, the sewage provides a substrate for cyanide-oxidizing bacteria in this case. It is proposed that an engineered artificial marsh could possibly provide further cyanide and metal polishing in conjunction with a system which preaerates sewage with the cyanide effluent. The pretreated sewage would provide nutrients for both the marsh plants and a carbon source as well as a substrate for sulfate reducing bacteria. The mature marsh in the anaerobic zone would generate sulfide ions which in turn are effective in liberating cyanide from stable base metal complexes, precipitating metals as sulfide precipitates and enhancing the rate of natural degradation of cyanide from the effluent. A marsh would also treat the raw sewage along the lines described by Black (undated) and Wile et al. (undated). Because metal cyanide effluents seem to inhibit plant growth in ponds it may be necessary to limit the use of wetlands for polishing of such effluents after natural degradation or pretreatment by aeration, etc. has done the bulk of the work. The applicability of an engineered nutrient-enriched marshland ecology as a means of polishing cyanide leach effluents is hypothetical and remains to be demonstrated.

5.8 Example 8: Removal of Selenium and Uranium by Algae

Brierley and Brierley (1981) showed how various aquatic species including filamentous algae *Spirogyra* and *Oscillatoria* and the benthic alga *Chara* remove metals such as selenium and uranium from a New Mexico uranium mine effluent. Concentrations of soluble uranium, selenium, and molybdenum were not diminished in mine water by passage through a series of impoundments.

Particulate concentrations of the mine water were reduced at least tenfold however. Since uranium, selenium, and molybdenum were associated with the suspended particles, reduction in the concentration of total suspended solids also reduced the concentration of these contaminants in the final effluent.

Large populations of microorganisms were reported by these authors in the water pumped from the uranium mines and in the water passing through the pond system. Of particular interest was the presence of sulfate-reducing bacteria believed to be *Desulfovibrio* and/or *Desulfotomaculum*. Laboratory experiments indicated that sulfate-reducing bacteria may play a role in the removal of uranium, selenium, and possibly molybdenum from solution.

6 Conclusions

1. There are numerous cases where biological systems (either plants, algae, fungi, bacteria, biomass and enzymes, etc. or combinations thereof) have been shown to remove metals and other contaminants, such as sulfate and cyanide, from effluent solutions.

2. Application of some biological processes in chemical reactor process type reactors may require residence times that would necessitate very large reactors, i.e., a high capital cost. There is a feeling among workers in this field that kinetics may be improvable through further research and development.
3. Applying integrated biological processes which include plants, algae, fungi, and bacteria, simultaneously under both aerobic and anaerobic conditions in artificial marshes, i.e., in engineered ecologies, may be one way to enhance the scope of applicability of biological processes for effluent treatment.
4. Initially, such systems may best be applied to polish effluents rather than for primary treatment.
5. Biological systems applicable for a particular job will be site-specific. For example, wetlands lose significant volumes of water through evapotranspiration. If water conservation is a primary goal, then a marshland polishing system probably would be counterproductive. However, in parts of the world, where there is a net precipitation, a marsh may also help to dispose of excess water.
6. Climatic constraints undoubtedly present a limiting factor in applying passive biological systems. What happens, for example, during the winter freeze or during a spring runoff? There are, however, many parts of the world where the winters never freeze and where there is no spring runoff.
7. If water is to be recycled to the metallurgical operation, then the effect of the biological treatment system on water quality must be taken into account. For example, will decaying biomass cause oxygen deficiencies, BOD, or add humic acids to the water which present problems in the metallurgical process when the water is recycled?
8. Biological process knowledge in the field of mine effluent treatment is in its infancy. Further work is bound to prove rewarding. Industry and academic organizations are be encouraged to devote additional resources to work in this direction. As is usual in emerging technology, practical results may not be immediate and industry should not underestimate the time and effort needed to develop useful and reliable systems.

References

Black SA (undated) The use of marshlands in waste water treatment – novel concept. Ontario Minist Environ. 135 St. Clair Av West, Toronto, Ontario M4V 1P5

Brierley CL, Brierly JA (1981) Contamination of ground and surface waters due to uranium mining and milling,vol 1. Biological processes for concentrating trace elements from uranium mine waters. Bur Mines Miner Res Contract Rep J0195033, Oct 1981

Brierley JA, Brierley CL, Goyak GM (1985) AMT-BIOCLAIM: A new waste water and metal recovery technology. Int Symp Biohydrometallurgy, Vancouver, 21–24 Aug 1985

ENDS (ed) (1985) ENDS Report 125, June 1985, pp 13–15

Ernst WJ (1976) Biological treatment of waste Waters from a lead-zinc mine-mill operation. MSc Thesis, Univ Missouri, Rolla

Gordon JA (1983) Iron, manganese, and sulfide mechanics in streams and lakes: a literature review. Rep Water Qual Branch, Div Water Resour, TVA, Chattanooga, TN Contract TU 30538A

Holm JD (1984) Pres Randol Int Ltd. Worksh Water management and treatment for minings and metallurgical operations. Beaver Creek, Col, Oct 1984

Holm JD, Guertin dF (1984) Theoretical assessment and design considerations for passive mine drainage systems. Forest Guertin, Col School of Mines, Golden

Holm JD, Sullivan T, Stenulson BK (1979) The restoration of Peru Creek. Col Dep Health, Water Qual Control Div, Oct 1979. 1313 Sherman Street, Denver CO 80203

Holm JD, Bishop MB, Tempo K (1984) Passive mine drainage treatment – Promoting natural removal mechanisms for acidity and trace metals from mine drainage. Rep Col Dep Nat Res, Mined land reclamation Div, 1313 Sherman Street, Denver CO 80203-2273

Imai K (1985) Utilization of sulfate-reducing and photolithothropic bacteria in biohydrometallurgy. Int Symp Biohydrometallurgy, Vancouver, 21–24 Aug

Jennett JC, Wixson RG (1977) Water quality studies, chap 6: The Missouri lead study. An interdisciplinary investigation of environmental pollution by lead and other heavy metals from industrial development in the new lead belt of Southeastern Missouri. Univ Miss Rep Period May 1972 to May 1977

Lapako K, Eger P (1981) Trace metal removal from mining stockpile runoff using peat, wood chips, tailings, till, and zeolite. Symp Surface mining hydrology, sedimentology and reclamation, Lexington, Kentucky, Dec 1981

Manahan SE (1984) Environmental chemistry, 4th edn. Willard Grant, Boston MA, 612 pp

Mudder TI Whitlock JL (1984), Biological treatment of cyanidation wastewaters, SME-AIME Annual Conf Feb 26, 1984, Los Angeles, California

Randol Int Ltd. (1985) Water management and treatment for mining and metallurgical operations. Eight vol Multi-Client Ref Randol Int Ltd., 21578 Mountsfield Drive, Golden CO 80401. (303) 526–1626

Randol Int Ltd. (1987) Innovations in gold and silver recovery – Phase III. Multi-client study by Randol Int Ltd., 21578 Mountsfield Drive, Golden CO 80401. (303) 526–1626

Simovic L, Snodgrass WJ, Murphy KL, Schmidt JW (1984) Development of a model to describe the natural degradation of cyanide in gold mill effluents. Univ Ar Conf Cyanide and the environment, Tuscon, Ar, 11–14 Dec 1984

Skerritt RM, Lester (1979) The microbiological control of mine waste pollution. Miner Environ 1 2

Spisak JF (1985) Biotechnology and minerals – Legitimate challenge or costly myth. In: Spisak JF, Jergenson GV II (eds) Frontier technology in mineral processing. SME-AIME Publ, Feb 1985

Townsley CC, Ross IC, Atkins AS (1985) Biorecovery of metallic residues from various effluents using filamentous fungi. Int Symp Biohydrometallurgy, Vancouver, 21–24 Aug 1985

Whitlock JL, Mudder TI (1985) The homestake biological treatment process: Biological removal of toxic parameters from cyanidation waste waters and bioassay effluent evaluation. In: Lawrence, RW et al. (eds), Fundamental and applied biohydrometallurgy, Proc 6th Intern Symp Biohydrometallurgy, Vancouver, BC, Canada, August 21–24, 1985

Wile I, Palmateer G, Miller G (undated) Use of artificial wetlands for waste water treatment. MOE Rep can be requested from Mrs Ivy Wile, Assistant Director, Div Hazardous Contaminants, Ministry of the Environment, 40 St. Clair Av West, Toronto MVP 1P5

Wixson BG, Davies BE (1985) Frontier technology for environmental cleanup, Chap 4. In: Spisak JF, Jergenson GV II (eds) Frontier technology in mineral processing. AIME-SME Publ, Feb 1985

Ecological Engineering: Biological and Geochemical Aspects Phase I Experiments

M. KALIN and R.O. VAN EVERDINGEN[1]

Abstract

Perpetual operation of acid mine drainage treatment facilities after mine closure imposes economic burdens and environmental liabilities which are difficult to overcome. The main goal of ecological engineering research is the development of methods which facilitate the establishment of vegetation cover types to serve as self-maintaining acid drainage treatment systems. Summarized below are the objectives and results of experiments on tailings and in the laboratory for five aspects relevant to the development of close-out scenarios for acid-generating tailings sites in temperate climates.

1. On dry surface areas with sparse vegetation covers, terrestrial moss covers curtail infiltration of precipitation. Vascular plant covers investigated had no amelioration effect on the underlaying tailings. Introduction and promotion of terrestrial mosses require further research.
2. Wetlands established in waterlogged, submerged sections and seepages reduce water infiltration and acid generation, as well as removing metals. Hand-transplanted cattails showed a 200% increase in number after 1 year. Hydroponic methods tested to date indicated little promise. Seed germination of some wetland plants was reported on neutralized pyrrhotite. Iron removal in seepages was found to be catalyzed by the surface provided by organic matter. Two species of aquatic mosses can serve as biological polishing agents where flow rates are low.
3. The precipitates in settling and polishing ponds from treatment plant operation have to be maintained in an alkaline state. Some species of *Charophytes*, attached macroscopic algae, have been tested extensively for these conditions and appear to be suitable.
4. Relative abundance values for ^{18}O in sulfate produced at one tailings site indicate that a major proportion of the sulfate oxygen is derived from water molecules rather than from molecular oxygen, suggesting that the oxidation process proceeds partly under anaerobic, water-saturated conditions. The water is slowly moving through the tailings. It is likely that the oxidation of sulfides in the tailings can be stopped only if the tailings are permanently submerged in completely stagnant water.
5. Data from freezing gauges and a thermistor cable installed in the tailings indicate that frost penetration in the tailings is minimal, suggesting that the rate of sulfide oxidation may not be reduced significantly during the winter.

[1]Boojum Research Limited, 139 Amelia Street, Toronto, Ontario M4X 1E6, Canada

1 Introduction

One of the most challenging aspects of reclamation is the maintenance of stable and enduring environmental protection against water quality degradation from acid generation by sulfide-bearing mining wastes in temperate climates.

In Canada, it has been estimated that at least 9000 ha are covered by sulfide tailings, located mainly in the provinces of New Brunswick, Quebec, Ontario, Manitoba, British Columbia and the Territories (RATS, 1984). Given that acidic effluents are continuously being released from these waste sites throughout the country, long-term environmental protection is required. Currently, operation of acid mine drainage treatment facilities, long after mine closure, imposes economic burdens which arise from the continuous demand for maintenance. These burdens are difficult ones for mining companies and governments to carry. Self-maintaining ecological systems, capable of treating acidic water and progressively curtailing the acid generation from the wastes, could provide an economically sound method for long-term environmental protection.

Many ecological studies of plant colonization patterns have indicated that, on waste sites, it is likely that species richness and the extent of the colonization will not change for decades or millennia (Kimmerer 1981; Bauer 1973; Morrison and Yarranton 1973). Ecological projections indicate that improvements due to natural recovery are unrealistic, despite some initial evidence of primary natural colonization of waste sites. The establishment of species, tolerant to the chemical and physical characteristics of the waste sites, appears to be a random process of chance. Self-sustaining covers on waste areas are often limited to localized areas of the site where conditions happen to be favorable for germination and establishment (Kalin and Caza, 1983). Therefore, methods have to be developed to provide suitable conditions, unilaterally favorable for self-sustaining vegetation covers comprised of those highly tolerant species. These covers would be the mainstay of self-maintaining mine drainage treatment systems.

The general concepts for the development of ecological engineering and biological polishing methodologies and procedures for treatment systems have been formulated (Kalin 1986a; Kalin and Smith 1986). The target of ecological engineering research is the development of methods facilitating the establishment of self-maintaining vegetation covers for the different conditions encountered at acid-generating waste management areas. These methodologies can then be applied to promote vegetation covers and wetland communities which would function as biofiltration systems.

Biofiltration and its applications to the treatment of mining effluents is not novel, and has been reviewed by Barth (1986). He suggests that the use of processes associated with biofiltration "may offer a low cost, highly effective, maintenance-free approach to treating some types of tailings seepage." Although several examples of working biofiltration systems are presented, he cautions that until the knowledge on biofiltration by wetlands increases, wide-scale application of biofiltration is premature.

Summarized in this chapter is on-site experimental research and concurrent laboratory studies initiated on two tailings sites. The objectives and results of the ongoing investigations are given. Detailed descriptions of the methods employed are not presented.

Technical and scientific information further describing the processes at work, together with on-site applications of the methods developed will, in the near future, demonstrate ecological engineering measures which can provide an acceptable alternative to the presently available treatment options for acid mine drainage.

2 Tailings Site Description

One of the experimental sites is a pyrrhotite-covered tailings area in Sudbury (central Ontario), where milling operations have ceased. Neutralized mine water covers part of the site. This pond receives acid water pulses during precipitation events from the pyrrhotite-covered area. A seepage, emerging at the foot of one of the tailings dams, is used as a study area to determine those factors controlling biological polishing processes, i.e. removal of metals.

A second experimental site is a waste management area located in the Red Lake area (northern Ontario). Operation of the zinc and copper concentrator ceased 4 years ago, after a 10-year life span of the mine. Reclamation of tailings, containing 4% pyrrhotite and 34 to 41% pyrite, was implemented recently. During operation of the mine, a small lake adjacent to the tailings became acidified. The original decant pond, containing sludges from water treatment and fines, receives drainage and run-off from the elevated tailings area. A complete close-out of this site is being implemented, using ecological engineering measures.

3 Requirements for Ecological Treatment Systems

Various surface and seepage characteristics, as well as specific behavioral aspects of the tailings mass, are primary considerations during the development of a close-out scenario for a waste management area. The tailings mass characteristics are related to the pattern of tailings deposition within the geomorphological and hydrological setting of the site.

A tailings area can be classified generally into three surface aspects: (1) vegetated, reclaimed surfaces with vegetation-free acidic areas; (2) waterlogged, submerged sections and seepages from dams; and (3) settling and polishing ponds remaining from the treatment plant operation.

The behavioral aspects of the tailings mass relevant in the long term are; (4) oxidation and acid-generating zones within the tailings mass and the hydrogeochemical characteristics of the tailings; (5) the seasonal variations in the oxidation process.

Fulfillment of the overall objectives of a self-maintaining close-out requires development of specific ecological systems for the three surface types (1 through 3 above), while decisions concerning the locations of these systems within the waste management area have to be based upon the behavior of the acid-generating tailings mass (4 and 5 above). The specific objectives addressing the three surface types of the waste management area are briefly outlined.

1. *Dry Areas with Sparse or no Vegetation Cover.* Vegetation covers are required which would reduce infiltration of precipitation. Field observations by M. Kalin on several acid-generating tailings sites consistently indicate that covers of terrestrial moss might serve this role. This was also suggested by Barth (1986). A moss cover has no roots, compared to vascular plants which penetrate the tailings and thereby increase porosity and permeability. Terrestrial mosses have shallow, rhizoid mats which serve to anchor the moss and provide moisture storage, particularly on tailings surfaces which are prone to severe seasonal droughts. Thus, a fraction of the precipitation is retained in the moss and rhizome layer, not infiltrating the tailings.

2. *Waterlogged, Submerged Sections and Seepage Areas.* Wetland vegetation covers increase transpiration and thereby counteract infiltration. That wetlands are an effective means of removing water is suggested by water economy evaluations of plant communities by Larcher (1980). He reports that in a wet meadow, 135% of total precipitation per year is transpired, compared to that of a dry grassland, where only 30% is transpired. A wetland would result in a net export of water from the site.

Furthermore, wetlands will promote the accumulation of organic matter. This layer will ultimately provide reducing conditions, which would cause some decrease of sulfate in the water. The amount of acidity increase would therefore be less than in the absence of a wetland cover. Reducing conditions curtail the activity of *Thiobacilli*. A schematic representation of the wetland development and its ultimate self-maintaining state is outlined in Fig. 1. Colonization of the wetland by appropriate algal and aquatic moss species should remove metals from the water, as some species have demonstrated a capacity for biological polishing.

3. *Settling and Polishing Ponds from Previous Treatment.* These areas are often very alkaline and filled with precipitates and sludges generated by the treatment system to be decommissioned. These areas need to be maintained so that potential acidification with remobilization of metals does not occur. Biological polishing agents are required which are tolerant to the alkaline conditions and capable of actively maintaining water at neutrality.

The requirements outlined above have to be connected to the behavioral aspects of the tailings mass and the rationale for their respective importance is outlined below.

4. *Acid Generation Within the Tailings Mass.* The generation of sulfuric acid during oxidation of pyrrhotite- and pyrite-bearing mine tailings is generally assumed to be the result of microbial action under aerobic conditions. When acidification proceeds to a pH lower than 3, the ferric iron produced will remain in solution and act as an additional oxidizing agent. However, oxidation may take place not only under the aerated, alternating wet and dry conditions in the relatively narrow zone of water-table fluctuation, but also under continuously saturated conditions, where water moving through the tailings will supply oxygen and remove oxidation products. It is reasonable to assume that the conditions of

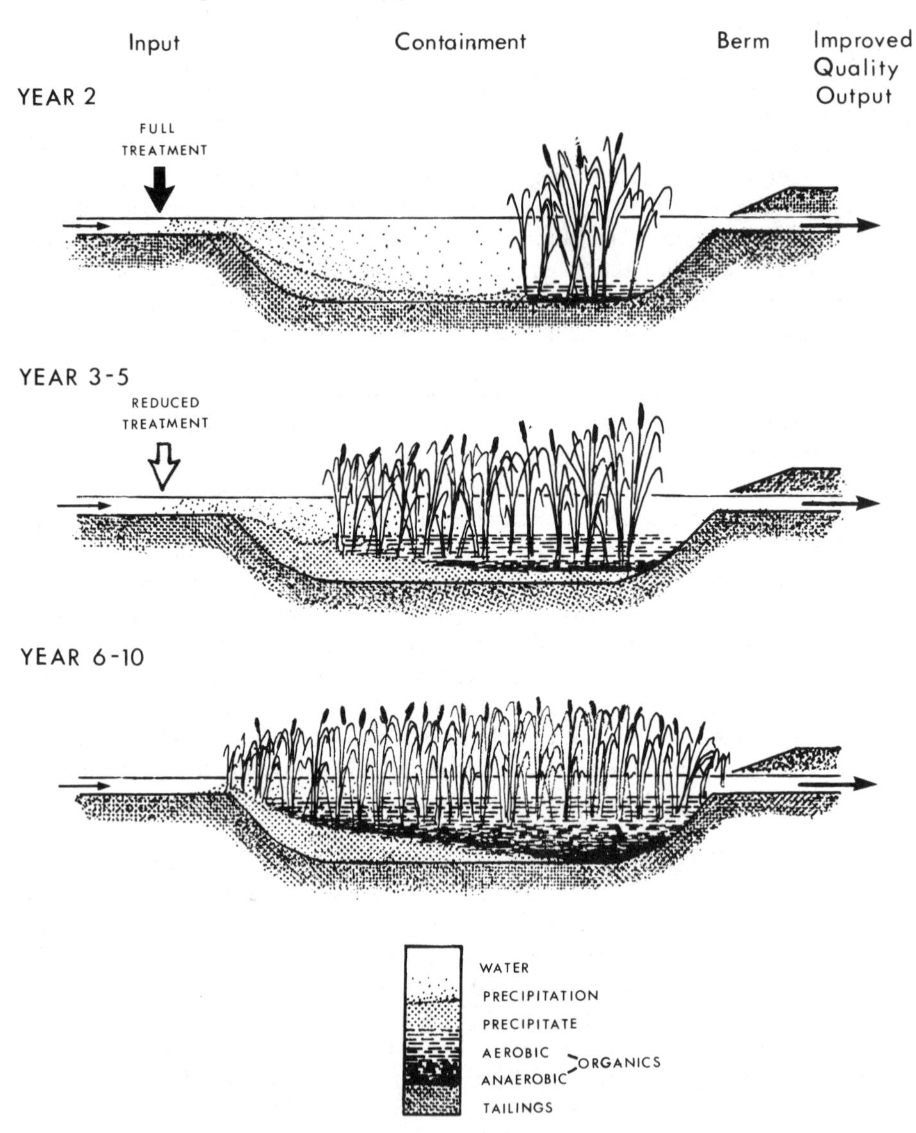

Fig. 1. Schematic development of wetlands on waste sites. The first 2 years might require some neutralization treatment to assist in the establishment of suitable indigenous species in the containment structure. As establishment of the wetland vegetation increases throughout years 3 to 5, the treatment can be reduced and is gradually withdrawn within years 6 to 10 when the system is self-sustaining. The anticipated sediment development is depicted in the enlarged *insert* providing an anaerobic, aerobic and finally precipitate layer underneath the treated water

water saturation and oxygen availability vary widely throughout the mass of a tailings deposit. Due to the varied conditions prevailing in different portions of the tailings mass, reaction and acid-generation rates will vary widely.

The conditions under which the oxidation proceeds (aerobic or anaerobic, sterile or bacterially mediated) determine the relative proportions of the sulfate oxygen that are derived from (1) atmospheric (or dissolved) oxygen and (2) water molecules (Taylor et al., 1984a, 1984b; van Everdingen and Krouse, 1985). In the first of two possible oxidation pathways, described by the reaction

$$2FeS_2 + 7O_2 + 2H_2O \rightarrow 2Fe^{2+} + 4SO_4^{2-} + 4H^+ \quad (1)$$

only 12.5% of the sulfate oxygen would be derived from water molecules, the remaining 87.5% from atmospheric or dissolved oxygen.

In the second pathway, described by the reaction

$$FeS_2 + 14Fe^{3+} + 8H_2O \rightarrow 15Fe^{2+} + 2SO_4^{2-} + 16H^+ \quad (2)$$

essentially all the sulfate oxygen could be derived from water molecules. A similar end product might result from oxygen-isotope exchange between the water and an intermediate partly-oxidized sulfur species.

The relative abundances of oxygen-18 (^{18}O) in atmospheric oxygen and in meteoric water (rain and snowmelt) are very different, particularly at higher latitudes. Therefore, ^{18}O abundance values for the product sulfate can be used to estimate the proportions of the sulfate oxygen that were derived from the two sources. This, in turn, will give an indication of the conditions under which the oxidation process operates, and also of the types of remedial measures that may be required to either reduce oxidation rates or treat the oxidation products, or both.

Relative abundance values for the following stable isotopes can provide useful information in the investigation of oxidation processes in tailings deposits:

1. ^{34}S and ^{18}O in dissolved sulfate in water samples collected from piezometers completed in, below and around the tailings;
2. ^{18}O and ^{2}H in the water molecules of these samples;
3. ^{34}S in the metal sulfides contained in the tailings.

The ^{18}O values for the sulfates should reflect the proportion of water oxygen incorporated in the sulfates. Comparison of the ^{34}S values for the sulfates with those for the sulfides should indicate whether any significant fractionation of sulfur isotopes occurred during the oxidation process. The relationship between ^{2}H and ^{18}O values for the waters should give an indication of the source of the water (rain or snowmelt).

The dominant reaction in the oxidation of the tailings can then be determined from a comparison of the relationship between the ^{18}O values for the sulfates and those for the water in the samples, with similar data for:

a) Dissolved marine sulfates (van Everdingen et al., 1982);
b) Acidic spring waters (van Everdingen et al., 1985);
c) Acid mine waters (Taylor et al, 1984a); and
d) Pyrite oxidation experiments (Taylor et al., 1984b).

5. *Seasonal Changes in Acid Generation.* Temperature fluctuations in, and freezing of the tailings have not been documented earlier for the tailings deposit under investigation. Seasonal variations in temperature are to be expected; they will cause seasonal variations in the rates of oxidation and acid generation. The tailings might even freeze, which would tend to stop the oxidation process. It will be useful to compare those variations with the seasonal growth cycles in wetlands and other biological treatment systems because, during spring freshet, biological polishing systems may have to accommodate the release of stored acid (and dissolved metals) generated during the winter.

A reasonable assumption might be that reduced biological polishing capacity during the winter will be compensated for by strongly reduced oxidation and acid-generation rates. In general, however, these processes are accompanied by the generation of heat, which will delay cooling and might prevent freezing of the tailings. In addition, freezing-point depression due to the generally high dissolved-solids content of the pore water may further delay freezing of the tailings.

In northern Canada, several acid-generating tailings deposits are located in areas with perennially frozen ground (permafrost). In principle, the possibility exists that the oxidation process in those deposits will be slowed down or even stopped by the development of permafrost in the tailings. If this were the case, then it would be of considerable interest to investigate various possible ways of inducing the development of perennially frozen conditions in those deposits.

4 Experimental Results and Ongoing Investigations

Work relevant to each of the areas for which the requirement for ecological treatment systems have been outlined is presented.

1. *Dry Areas with Sparse or no Vegetation Cover.* A vegetation cover, after 4 years of reclamation, was evaluated in detail with respect to interactions of the cover with the underlying tailings. Three types of grass covers could be differentiated (sparse, background and dense) which grow on a 30- to 40-cm-thick fill layer. The thickness of the fill layer, depth to oxidation layer, depth of root penetration and thickness of organic layer were measured. In addition, measurements of pH on the surface, in the fill and in the tailings were recorded. Temperature was determined on the surface and in the tailings. All measurements were performed in pits dug to the depth of tailings in the various surface-cover types.

An analysis of the results indicated that there was no relationship between the vegetation type, the thickness of the fill and the depth at which the tailings were acidic (pH < 3.5). The tailings were consistently acidic 2 to 3 cm below the fill, and only in areas with a gravel cover alone were the tailings acidic at a shallow depth (gravel cover depth 10 cm as compared to 30 to 40 cm of fill). The temperature on the surface of the vegetation was lowest in the dense grass cover (19°C) as compared the other cover types (23° to 28°C). However, the temperature in the tailings below did not vary between cover types (11° to 15°C).

The vascular plant species composition was determined based on collection of species most frequently occurring on the tailings after 4 years of reclamation. The composition was typical for waste lands, consisting of sedges (*Carex aquatilis* c.f.) and grasses such as wild rye, fuzzy wild rye, orchard grass and quack grass (*Elymus* sp., *Agropyron repens* c.f., *Dactylus glomerata*), foxtail barley (*Hordeum jubatum*), smooth brome grass (*Bromus inermis*), woolgrass or bog cotton (*Scirpus cyperinus*), birdsfoot trefoil and Alsike clover (*Lotus coniculatus* and *Trifolium hybridum*).

Annual above-ground biomass production for the sparse grass cover was 15 g m^{-2}, and 53 to 57 g m^{-2} for the background and dense cover respectively. A one-time application of fertilizer (21–7–7) of 6.25 g m^{-2} in the beginning of the growth season resulted in a net increase of biomass of 84 to 110 g m^{-2}. However, fertilization of the terrestrial moss cover, applied at the same rates, did not result in any noticeable changes in the cover extent or its growth.

Field and greenhouse experiments addressed methods of introduction of various terrestrial moss species to the pyrrhotite surface. In the field, two types of transplanting techniques were tested. Clumps of mine slimes bearing protonemata of *Leptobryum pyriforme* were dug into the surface of loose pyrrhotite. The inoculated mine slimes were slurried by circular motion during rain into the broken pyrrhotite. The pyrrhotite was amended with either sand or sandy topsoil.

The slurry technique indicated some promise in conjunction with sand-amended pyrrhotite. The greenhouse experiments produced similar results with respect to both introduction and amendment techniques (Kalin 1985).

2. *Water-Covered, Waterlogged and Seepage Areas.* Experiments were implemented to ascertain some of the factor(s) which control cattail (*Typha latifolia* and *T. glauca*) growth and stand expansion on alkaline mine slimes covering pyrrhotite tailings. It was found that the ratio of the number of juvenile individuals to the number of mature individuals around the cattail stand perimeter was lower in stands where the compaction of the waste was high and the rhizomes were buried deeper. A rhizome depth of greater than 10 cm had lower numbers of juvenile plants (Kalin and Buggeln 1986a).

Transplanting methods for introduction of cattails in areas free of any vegetation were tested. Mature cattails from various sources (stands in gravel pits, in dense cattail meadows, stands growing in tailings, stands growing in mine slimes) were transplanted by hand (individuals or clumps) and mechanically by a front-end loader. Hydroponic means of transplanting were tested, where root stocks were suspended in fishing nets and mature cattails fastened in rafts.

The results from the hand-transplanted cattails in the mine slimes after overwintering and growth for one season are summarized in Table 1. A total of eight stands were established, however the markings on the plants did not always survive the overwintering and data could not be obtained (Kalin 1986a). The data in Table 1 pertain to the marked plants.

Hand and mechanical transplanting were tested in acid water with pH values ranging over one growing season from 2.9 to 5.7, with a seasonal mean zinc concentration of 221 ± 79 mg l^{-1} and mean copper concentrations of 21 ± 36 mg l^{-1}.

Table 1. Hand-transplanted cattail survival and growth 1985/1986

Stand No.	Number transplanted	Number overwintered	Total plants June/Sept.	Increase over one season (%)
6	10	8	18/27	238
7	10	9	21/27	200
8	10	8	14/21	162
10	10	7	16/21	200

Such high concentrations of metals, though not unusual for waste management areas, present extremely unfavorable growth conditions.

Two months following transplanting, there was little difference between the two methods. The number of new shoots produced by mechanical and hand-transplanted populations had increased by 32 to 35%. The control (natural environment) populations transplanted by the same methods increased by 36 and 42% respectively. Populations transplanted to mine slimes at a shallow depth increased on average 100% within 2 months.

Hydroponic methods yielded less encouraging results under both acid (pH 2.5) and alkaline (pH 8–10) conditions, although shoots did develop initially. None of the hydroponics survived the overwintering from the 1985 experiments, since both alkaline and acidic test sites had dried up by the end of the growing season. In experiments carried out in 1986, the control populations in fishing nets initiated shoots after 2 months, and rafts showed some survival in alkaline water.

Seeding would clearly be one of the most desirable methods of introducing wetland vegetation. Seeds collected from a variety of semi-aquatic plants growing in a polishing pond were tested for germination. The seeds were subjected to various pregermination treatments (freezing, desiccation and storage at room temperature). Seeds from an aquatic grass, a sedge and cattails stored at room temperature showed best germination, whereas seeds from sedges and a rush failed to germinate, regardless of the pretreatment.

Seeds which showed positive germination after pretreatment were further tested on pyrrhotite, amended with either dewatered sewage sludge, with $Ca(OH)_2$ (pH 3.0, 7.0 or 10.0), or left unamended (pH 2.3 to 2.5). No seeds germinated on sewage-amended, pH 3.0 amended or unamended pyrrhotite. Excellent germination (50 to 100%) occurred with the grass and a sedge on pH 7.0 to 10.0 amended pyrrhotite. Cattail seeds, collected from different habitats (acid, alkaline or natural environment), however, showed varying degrees of germination rates (Kalin 1986c).

A seepage creek was analyzed in detail for its chemical, physical and biological characteristics. Nine stations at 50-m intervals along the creek were monitored monthly from June to August, and at irregular intervals for a complete year. Based on the data collected, the following processes can be described.

Organic matter (living and dead) in the seepage creek provides a catalytic surface for the precipitation of iron hydroxide from the water. Steady iron removal occurs over the upper 250 m of the creek. The hydraulic retention time of the creek increases as the water enters a broad expanse (=pond) mainly colonized by bent grass (*Calamagrostis* ssp.). A 6.5-fold reduction in total soluble iron occurs across this area along with a concomitant drop in pH of two full units (Kalin 1986b).

In the grass-covered, broad expanse, estimates of organic matter and iron precipitate were made. The results indicated that in dense or medium dense areas (live grass cover), an average of 8 g (air dry) iron precipitate was associated with 1 g (air dry) dead organic matter. In areas with sparse growth, 1 g dead organic matter was associated with 22 g precipitate. Dead organic matter estimates for the densely and medium growing areas were 3.8 to 4.8 kg m^{-2} and for the sparse 1.9 kg m^{-2}. These values suggest that there may be an upper limit of precipitate at which continuous growth of grass might be inhibited.

Laboratory studies of the cation exchange and binding capacities for metals of acidophilic aquatic moss species *Drepanocladus fluitans* and *Sphagnum riparium* were carried out. These species are tolerant to high acidity, evident from their colonizing inactive tailings areas with pH 2.5 to 3.0.

Growth rates of shoot for both species were 3 cm $month^{-1}$. The biomass increase differed between the species, due to differences in morphology and growth habit. *D. fluitans* produced 2.1 g m^{-2} $month^{-1}$ dry w. as compared to *S. riparium* which produced 17.2 g m^{-2} $month^{-1}$. Under field conditions, it can be expected that the respective dry matter production will range from 20 to 170 g m^{-2} $month^{-1}$ in the growth season.

The cation exchange experiments indicated that 1 g (dry w.) of moss can remove 3 μg nickel from 100 ml of seepage water (pH 3 to 4) with concentrations of 0.1 mg l^{-1} of nickel. Similar amounts of removal were observed with Cu and Mg. Increasing the iron concentrations in the seepage water (up to 36 mg l^{-1}) inhibited the metal uptake (Kalin and Buggeln, 1986b).

3. *Settling and Polishing Ponds from Previous Treatment.* Some species of Charophytes, a group of macrophytic attached algae, have been tested in alkaline waters of settling and polishing ponds of eight uranium sites and in ponds from two nickel operations (Kalin 1985). Growth tests in the field and the laboratory indicate that the algae concentrate uranium, radium 226, nickel, copper and zinc under alkaline conditions (Kalin and Smith 1986). The algae appear to provide pH stabilization due to accumulation of calcium carbonate on its surface. Physiological studies (Lucas 1985) indicate that this algal group can utilize bicarbonate at pH 8.5 to 10.0 during photosynthesis, after depleting carbon dioxide. This results in maintenance of neutral water, if populations can be developed with high densities.

Ponded areas of tailings sites of several abandoned gold mines have been colonized by Charophytes to varying degrees. The growth upon transfer to settling ponds from some of these populations was considerably better than from populations previously tested. Factors controlling population densities and growth in the field are being investigated at present. Experiments on inactive tailings ponds are being developed to address stimulation of the growth of the algae.

4. *Acid Generation Within the Tailings Mass.* Results of stable-isotope analyses on a number of water samples from the experimental site in the Red Lake area are shown in Fig. 2 and 3. The results of the analyses are expressed as per mil deviations from the two commonly used standards: Standard Mean Ocean Water, or SMOW, for ^{18}O and ^{2}H; Cañon Diablo Troilite, or CDT, for ^{34}S.

The initial results indicate that:

1. The water found in, below and around the tailings is derived primarily from infiltration of rainfall, not snowmelt (see Fig. 2);
2. From 43 to 83% of the oxygen incorporated in the dissolved sulfate in the samples is derived from H_2O molecules, suggesting that;
3. Between 35 and 80% of the dissolved sulfate is produced by a reaction that involves only H_2O oxygen (2) (see Fig. 3) not free or dissolved oxygen, or a reaction path that allows exchange of oxygen isotopes between the water and one or more partially-oxidized sulfur species;
4. Oxidation of the tailings is taking place under both aerobic and anaerobic conditions, with the probable involvement of *T. ferrooxidans* or similar bacteria; and
5. Sulfide oxidation and acid generation can likely only be *prevented or stopped if the tailings are permanently submerged in completely stagnant water.*

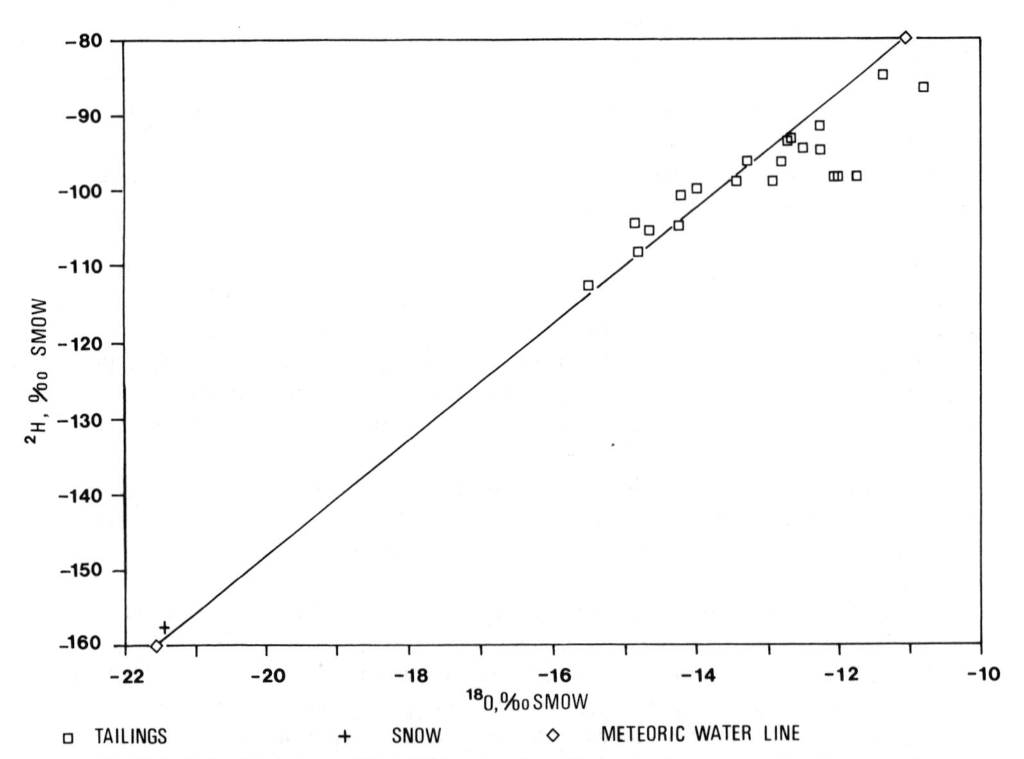

Fig. 2. Relationship between ^{2}H and ^{18}O values for tailings samples, compared to the meteoric water line for waters from the Canadian Shield

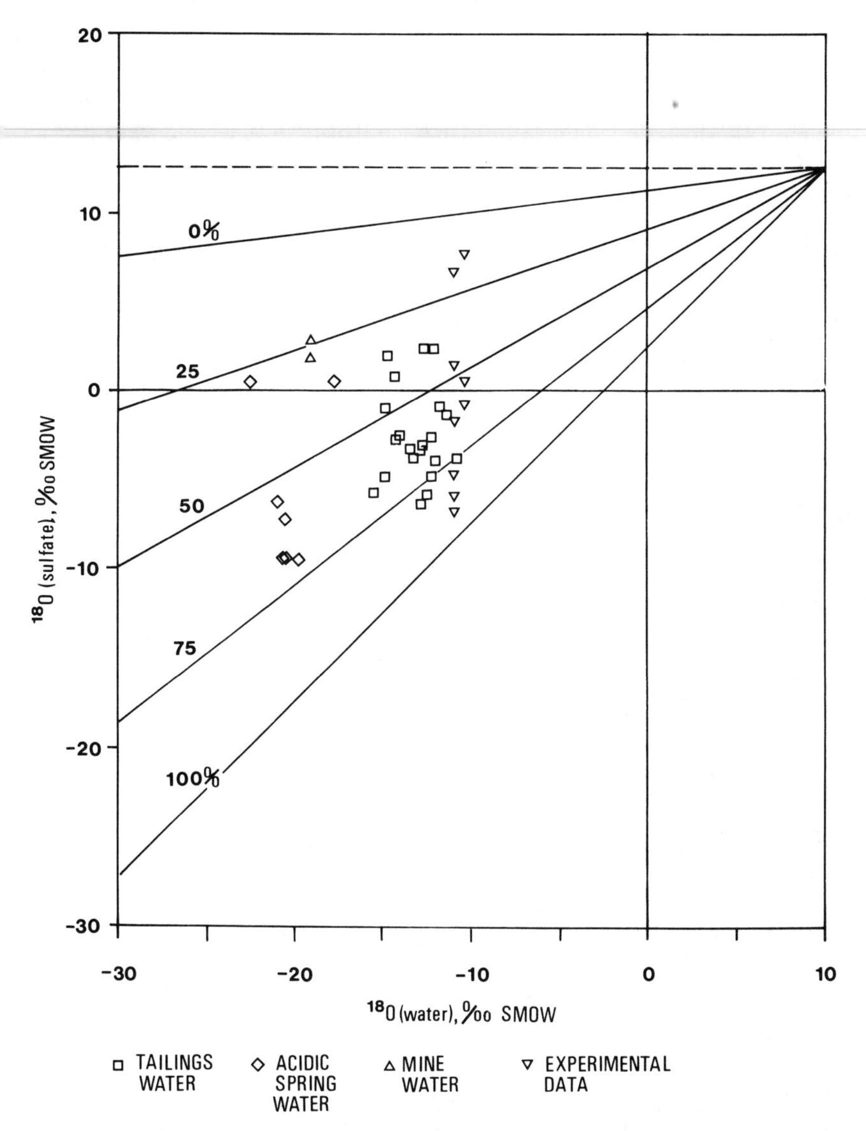

Fig. 3. Relationship between ^{18}O(sulfate) and ^{18}O(water) values for various acidic waters and pyrite-oxidation experiments. ◊ Acidic spring waters (van Everdingen et al., 1985); Δ mine waters (unpublished data); ∇ experimental data (Taylor et al. 1984b). Lines labelled *O–100%* indicate expected ^{18}O (sulfate) vs ^{18}O (water) relations for increasing contributions by reaction (2). *SMOW* = Standard Mean Ocean Water; *broken line* = atmospheric O_2

5. *Seasonal Changes in Acid Generation Rate.* Seasonal temperature variations and the possible freezing of the tailings in the experimental site in the Red Lake area are being investigated using a thermistor cable and a freezing gauge (Banner and van Everdingen, 1979) installed in the tailings in the fall of 1986. A second freezing gauge was installed outside the tailings area to enable a comparison between the behavior of the tailings and that of natural soil in the surrounding area.

Preliminary data from the freezing gauges and the thermistor cable suggest that freezing does not penetrate into the saturated portion of the tailings. Initial cooling to −1.3°C at a depth of 23 cm was reversed when snow cover was established; in early January, 1987, temperatures ranged from near 0°C at the ground surface to +8.6°C at 2.43-m depth. The data likely reflect the combined effects of heat generation, and insulation provided by the 43- to 50-cm-thick snow cover.

5 Conclusions

The conclusions drawn from the results presented are discussed in the context of the requirements to be achieved for each of the five components relevant for the development of ecological engineering treatment systems.

1. For the dry areas, with sparse or no vegetation cover, terrestrial moss establishment and growth would be desirable. Although some methods to introduce acid-tolerant moss indicate promise, further extensive work is required. Different densities of vascular vegetation covers on tailings appear insignificant with respect to curtailment of the acid generation underneath the covers. Four years after reclamation, the species composition of the cover has changed. The area has been colonized by indigenous species, representative of waste sites. Fertilization measures did increase the biomass of vascular plants while the growth of terrestrial moss was not affected. Greenhouse and field tests of introduction methods for terrestrial mosses indicate that modification of physical conditions (moisture, structure of the surface etc.) will likely promote moss development.

2. Wetland development methodology requires a better understanding of the conditions under which isolated pockets of the waste sites are colonized, particularly by cattails. One factor inhibiting the expansion of stands is burial of rhizomes under compacted material deeper then 10 cm.

Direct germination and establishment of seeds by some wetland species does occur under alkaline conditions (on mine slimes) or on neutralized pyrrhotite. The results suggest that seedling establishment under neutralized or alkaline conditions is the origin of the isolated populations on waste sites.

Mature cattails, transplanted by hand or mechanically, survived and continued to grow under both alkaline and acidic conditions. Mechanical transplant methods are restricted to areas where equipment access is possible, however, growth was comparable to hand transplanting. The hydroponic transplanting methods tested require further development.

Investigations of seepages indicate that the initial iron removal occurring as the water emerges under neutral conditions is due to a catalytic process, related mainly to the surface area of biomass. Active uptake by algae, although occurring, is secondary; the precipitation processes dominate. Therefore, the continued growth of wetland species is of importance in providing a surface area for iron removal. Biological polishing of seepages could be achieved in two steps. At the head of the seepage a neutral ecosystem is required to remove iron, followed by an acid-tolerant system for metal binding.

The growth rates and cation exchange capacity of two species of acid-tolerant aquatic mosses indicate that high metal removal could not be achieved by these two characteristics alone. However, in ecosystems with low flow rates, these species' contribution to metal removal might be significant through their provision of nucleation sites for adsorption processes.

3. Field and laboratory results indicate that the Chara Process has the potential to provide a buffering capacity for the water and to fix alkaline precipitates in settling and polishing ponds.

4. A working hypothesis for the determination of the oxidation process and its oxygen source derived, based on isotope ratios, will be tested in the future. If the trends indicated so far persist, i.e. water in the tailings originates mainly from rainfall and not from snowmelt, and that oxidation of iron sulfides occurs both aerobically and anaerobically, then realistic, site-specific predictions of active areas of acid generation within the tailings mass will be possible.

The geochemical data from an extensive sampling program of piezometer water in various locations of the same tailings mass and outside the tailings impoundment, support the hypothesis that different oxidation processes are occurring in the tailings.

5. The temperature behavior of the tailings mass suggests that, at best, a slight cooling can be expected in the winter. Since the frost penetration is considerably smaller than expected, only an insignificant reduction in the acid generation rate can be expected.

Acknowledgements. The financial support of BP Resources, Selco Division; Denison Mines Ltd.; Falconbridge Ltd., Sudbury operations; Inco Ltd., Sudbury operations; and Rio Algom Ltd., is gratefully acknowledged. The financial support of the Department of Energy Mines and Resources Canada, CANMET – Biotechnology, and the National Research Council, who furnished several contracts and grants, was essential to this research and is acknowledged and appreciated. Isotope analyses for this project were performed in the Stable Isotope Laboratory, Physics Department, The University of Calgary.

The contribution of Martin P. Smith to this research was integral to its success, and Jennifer Reuben's assistance in the preparation of the many reports and the final manuscript is gratefully acknowledged.

References

Banner JA and van Everdingen RO (1979): Frost gauges and freezing gauges. Inland Waters Directorate Tech Bull No 110, pp 18

Barth RC (1986): Reclamation Technology for Tailings Impoundments. Part I: Containment. Mineral & Energy Resources, *29* no 1. p 1-25

Bauer HJ (1973): Ten years' studies of biocenological succession in excavated mines of the Colonge Lignite District. In: Ecology and Reclamation of Devastated Lands. Vol. *1* Ed RJ Hutnik & G Davis Gordon & Breach, Science Publications, p 271

Kalin M (1985): The Chara Process – Biological economization of mill effluent treatment. pp 57; CANMET BIOTECHNOLOGY CONTRACT REPORT # DSS 07SQ.23440-4-9185

Kalin M (1986a): Ecological Engineering: Evidence of a viable concept. In: Fundamental and Applied Biohydrometallurgy. Process Metallurgy 4, Elsevier Ed RW Lawrence, RMR Branion and HG Ebner, p 489

Kalin M (1986b): Ecological Engineering: Tests of concepts and assumptions. Year 1: Ecology pp 88 Report under Contract DSS # 23 SQ 23440-5-9119 to Inco Ltd within the CANMET Reactive Acid Tailings Program

Kalin M (1986c): Developing a technique for reclamation of high iron sulphur tailings by establishing a marshland: Phase 1. Selection of suitable plant materials and methods of establishment. pp 43. Report under Contract DSS # 23SQ23440-5-9163 to Falconbridge Limited within the CANMET Reactive Acid Tailings Program

Kalin M and Buggeln R (1986a): Cattail stand development on base metal tailings areas. Proc of Canadian Land Reclamation Association, Vancouver. In press

Kalin M and Buggeln R (1986b): Acidophilic aquatic mosses as biological polishing agents. pp 17; CANMET BIOTECHNOLOGY CONTRACT REPORT # DSS 07SQ.23440-5-0236

Kalin M and Caza C (1982): An ecological approach to the assessment for vegetation cover on inactive uranium mill tailings sites. In: Management of wastes from Uranium Mining and Milling. IAEA, Vienna, STI/PUP/622. ISBN 92-0-020282-9, p 385-402

Kalin M and Smith MP (1986): Biological Polishing agents for mill waste water. An example: *Chara*. In: Fundamental and Applied Biohydrometallurgy. Process Metallurgy 4 Elsevier Ed. RW Lawrence, RMR Branion and HG Ebner p 491

Kimmerer RW (1981): Natural revegetation of abandoned lead and zinc mines (Wisconsin). Restoration and Management Notes, *1*, p 20

Larcher W (1980): Physiological Plant Ecology. Springer Verlag, Berlin, Heidelberg, New York. pp 272

Lucas WJ (1985): Bicarbonate utilization by Chara: A Reanalysis. In: Inorganic Carbon Uptake by Aquatic Photosynthetic Organisms. Proceedings of an Intl Workshop on Bicarbonate Use in Photosynthesis. Aug. 18-22, 1984. University of California, Davis. p 229 to 245

Morrison RA & Yarranton GA (1973): Diversity, richness and evenness during a primary sand dune succession at Grand Bend, Ontario Can J Bot *5*, p 2401

RATS, Department of Energy, Mines and Resources, CANMET (1984): Sulphide tailings management study, Monenco Limited

Taylor BE, Wheeler MC and Nordstrom DK (1984a): Isotope composition of sulphate in acid mine drainage as measure of bacterial oxidation. Nature, *308*, 538-541

Taylor BE, Wheeler MC and Nordstrom DK (1984b): Stable Isotope geochemistry of acid mine drainage. Experimental oxidation of pyrite. Geochim. Cosmochim. Acta, *48*, 2669-2678

van Everdingen RO and Krouse HR (1985): Isotope composition of sulphates generated by bacterial and abiological oxidation. Nature, *315*, 395-396

van Everdingen RO, Shakur MA and Krouse HR (1982): Isotope geochemistry of dissolved, precipitated, airborne and fallout sulfur species associated with springs near Paige Mountain, Norman Range, NWT, Canada. Can J Earth Sci *19*, 1395-1407

van Everdingen RO, Shakur MA and Michel FA (1985): Oxygen- and sulphur-isotope geochemistry of acidic groundwater discharge in British Columbia, Yukon and District of Mackenzie, Canada. Can J Earth Sci, *22*, 1689-1695

Part II Restoration and Revegetation

Rehabilitation Measures at the Rum Jungle Mine Site

J.R. HARRIES and A.I.M. RITCHIE[1]

1 Location of the Rum Jungle Mine and History of the Project

The Rum Jungle mine site is about 80 km south of Darwin in the tropical Northern Territory of Australia (see Fig. 1) where there is a well-defined wet season when much of the rain falls as localised thunderstorms. The average rainfall in the area is 1.5 m, most of which is confined to the period October to March. The mean temperature in the region is about 28° C with the mean monthly temperature varying between 29.6° and 25.1° C. The East Branch of the Finniss River which passes through the mine site usually begins to flow about December and stops flowing by about the end of June.

Uranium was discovered in 1949 by a Mr. White at the site of what was later to become White's orebody. As this orebody lay beneath the East Branch a diversion channel was dug to the south of the orebody to carry the wet-season flow of the river. Open cut mining of White's deposit was started in September 1954 and completed in 1958 with the open cut at a depth of 100 m, the overburden being deposited in a dump on the south bank of the diversion channel. Dyson's orebody, a uranium deposit discovered in 1951 about 1 km to the east of White's, was extracted by open cut mining between 1957 and 1958.

Between 1960 and 1963 uranium ore was extracted by open cut mining from the orebody at Rum Jungle Creek South which was about 6.5 km south of the Rum Jungle site. In 1963 the Intermediate orebody, which was only some 500 m west of White's orebody, was also mined by open cut operation. This was a copper orebody, not uranium or uranium/copper as were the other orebodies. The high-grade copper ore was processed in the Rum Jungle treatment plant but much of the low-grade sulphide and oxide ores from the orebody were put in a pile where an attempt was made to extract the copper by heap-leaching techniques.

Until 1961, tailings material was pumped as a slurry to a disposal area to the north of the treatment plant; solids settled out and the supernatant followed natural drainage pathways to the East Branch. Some bunds were built from time to time to reduce the extent to which tails material was washed into the East Branch. After 1961, tails were pumped into Dyson's open cut until 1965, and from then until 1971 into White's open cut.

Initially highly acid (pH 1.5) waste liquor from the treatment plant was also discharged to the tailings area after first passing through copper launders if copper

[1]Environmental Science Division, Australian Nuclear Science and Technology Organisation, Private Mail Bag 1, Menai, NSW 2234, Australia

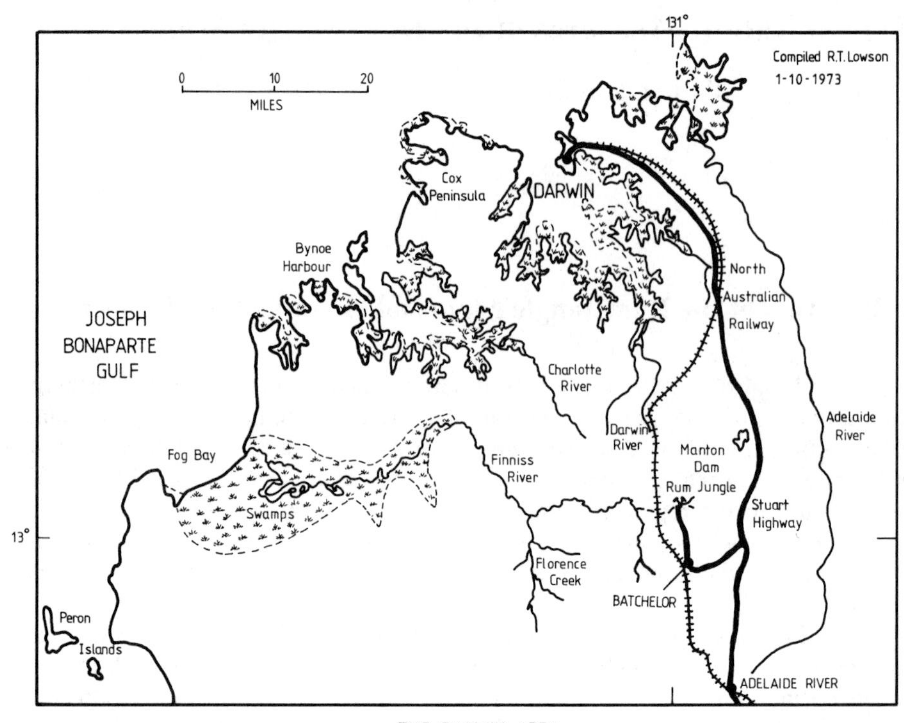

Fig. 1. Map of the Darwin area, showing Rum Jungle and Batchelor

concentrations warranted this. From 1961, when tailings went into Dyson's open cut, waste liquors went to a holding dam in the bed of the East Branch and were released down the river in the wet season. From 1967, part of the waste liquor was used in leaching operations on the copper heap leach pile while the rest was pumped to White's open cut which was used as a reservoir.

When mining of White's was completed in 1958 the open cut was allowed to fill with water. Measurements on the water body in 1959 showed that it was acidic and contained copper at a concentration of about 4 ppm. The decrease in water quality that occurred subsequently could have been caused either by chemical action in the pyritic shales of the open cut walls or by seepage from pyritic waste piles bordering the open cut.

In 1971, when treatment of the ore was completed, the mine was closed down. On closure some 2 million t ore had been treated and about 3530 t yellowcake and 20000 t copper extracted. The main features at the mine site when it was abandoned are shown in Fig. 2. In addition to the features at the mine site there were an overburden dump and water-filled open cut at Rum Jungle Creek South, a similar dump and open cut at Mt. Burton 7 km west of the mine site where a small uranium orebody was extracted in 1958 and at Mt. Fitch, a similar distance north-east of the mine site, an orebody exposed by removal of the overburden in 1969 but not mined. However, the impact of these outlying areas on the environment was small

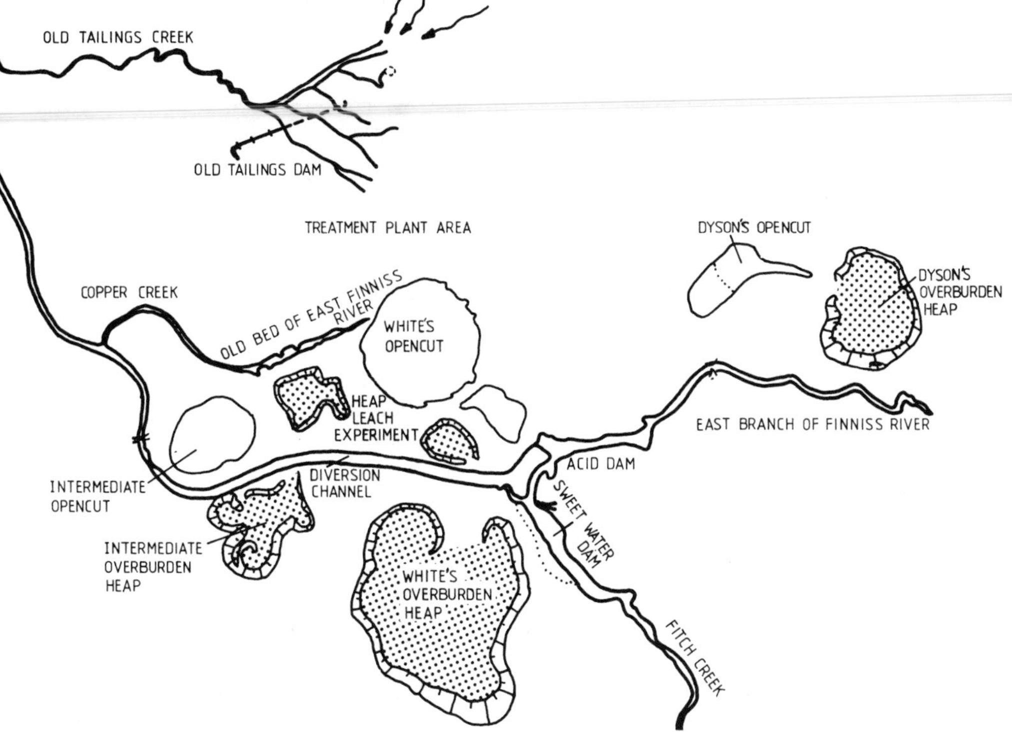

Fig. 2. Map of the Rum Jungle mine site

compared to that of the Rum Jungle mine site which will be the focus for discussion here. More details of the Rum Jungle area and the Rum Jungle mining operation can be found in Lowson (1975), Barrie (1982) and Ritchie (1985).

2 Environmental Impact and Sources of Pollution

2.1 Early Indications

After the mine closed in 1971 it was apparent that aquatic flora and fauna in the East Branch below the mine site were adversely affected because of the high trace metal content of the water. The dominant metals were copper, manganese and zinc, most of which were being solubilised in the mine waste residues at the site by bacterially catalysed oxidation of the pyritic material in these residues (Dugan, 1972, Brierley 1978, Harries and Ritchie 1982). During a period in 1973–74, which covered the 1973–74 wet season an extensive series of measurements were carried out in the Rum Jungle area to estimate the trace metal and radiobiological pollutant levels and to identify the major sources of these pollutants (Davy 1975). Subsequently some more detailed measurements were carried out as described below.

2.2 Waste Rock Dumps and Copper Heap Leach Pile

2.2.1 Description

White's dump, which was the most uniformly shaped of the dumps, was some 13 to 18 m above the original soil surface and sloped gently towards the centre where a drainage channel collected runoff water and discharged it northward into the East Branch. Most of the top surface of the dump was smooth and well graded but some 20% consisted of loosely piled heaps. The sides were steep (~30°) with, towards the base, a high proportion of large boulders which probably resulted from end tipping of the waste material from the top of the dump. It is therefore quite likely that there are regions of relatively high permeability along the base of parts of the dump. The overburden consists of carbonaceous shales and graphitic schists interspersed with dolomite; the chemical composition is given in Table 1 (Lowson 1975). The top of the dump was largely devoid of vegetation although there was some grass and trees growing in the north-west corner prior to rehabilitation.

Table 1. Some physical and chemical properties of the dumps and piles at the Rum Jungle mine site prior to rehabilitation

Entity	Physical features			Water catchment areas (x 10^{-4} m^2)		Chemical Composition ($mg\ g^{-1}$)			
	Area (x 10^{-4} m^2)	Volume (x 10^{-5} m^3)	Mass (x 10^{-6} t)	Sides	Principal runoff channel	S	Cu	Mn	Zn
White's	26	40	8	6	16	32.7	0.86	0.99	0.11
Intermediate	7	8	1.6	2	2.9	30.6	2.0	0.27	0.25
Dyson's	8.5	12	2.3	2.6	4.8				
Copper heap leach pile	3	2	0.37				9		
White's North[a]	4	1.7	0.34						

[a]Material from White's open cut dumped on the north side of the diversion channel opposite White's dump.

Intermediate dump was smaller and had a rougher top surface than White's. The area at the south-west corner of the dump, at 20 to 25 m above the original soil surface, was significantly higher than the rest of the dump and consisted of two lifts. The rest of the top surface of the dump was generally about half that height above ground surface. The sides of the dump were as steep as those on White's with the same preponderance of large material towards the base. As with White's, much of the runoff water from the top surface drained into a principal runoff channel which discharged northwards into the East Branch. The overburden consists of generally the same shales, schists and dolomitic material as White's overburden; its chemical composition is also given in Table 1. The dump was essentially devoid of vegetation of any sort.

As Dyson's overburden had been dumped on the side of a hill it was of much more variable depth than White's or Intermediate which were dumped on relatively flat ground, but it was about 20 m above the original ground surface along the southern sides which faced towards the East Branch. As with the other two dumps the top surface was shaped such that most of the runoff collected in a principal runoff channel that discharged into the East Branch. The dump material is a mixture of shales, schists and dolomite similar to the other two dumps and, although there are no detailed data on the chemical composition of the dump, it is probably similar to White's and Intermediate. There was more vegetation on Dyson's prior to rehabilitation than on the other two dumps.

Table 1 summarises the physical features of the copper heap leach pile and the three waste rock dumps. This table shows that, although the copper heap leach pile was much smaller (about 10 m high and 2.8×10^4 m^2 in area) than the waste rock dumps, it had a higher copper concentration and was a significant source of copper pollution when the mine site was abandoned. It was constructed in 1965. In early 1970 copper concentration in the pregnant liquor had dropped to 0.5 g l^{-1} (Andersen et al. 1966; Andersen and Allman 1968) which was about the copper concentration found in springs and ponds near the base of the pile in the survey carried out during the 1973–74 wet season.

2.2.2 Oxidation of Pyritic Material

The mechanism that solubilises metals in the waste rock dumps and in the copper heap leach pile is essentially that used to extract metals such as copper, nickel and uranium from low-grade ores (Murr 1980; Torma and Bosecker 1982; Ralph 1985) and can be summarised by the following equations:

$$FeS_2 + {}^7/_2 O_2 + H_2O \rightarrow FeSO_4 + H_2SO_4; \quad (1)$$
$$FeSO_4 + \tfrac{1}{2} H_2SO_4 + \tfrac{1}{4} O_2 \rightarrow \tfrac{1}{2} Fe_2(SO_4)_3 + \tfrac{1}{2} H_2O; \quad (2)$$
$$FeS_2 + Fe_2(SO_4)_3 + 3O_2 + 2H_2O \rightarrow 3FeSO_4 + 2H_2SO_4; \quad (3)$$
$$MS + Fe_2(SO_4)_3 + {}^3/_2 O_2 + H_2O \rightarrow 2FeSO_4 + H_2SO_4 + MSO_4; \quad (4)$$

where M stands for some metal. Many of these steps are catalysed by bacteria which increase the rate at which the reactions proceed (Brierley 1978). This is particularly so of the second step where the chemical reaction the absence of bacteria proceeds only slowly at the pH~2–3 typically found in leach piles and piles of pyritic mine wastes. Bacterially catalysed oxidation of some metal sulphides proceeds directly in reactions analogous to reaction (1) rather than by reactions of the type described by Eq. (4) (Soljanto and Tuovinen 1980; DiSpirito and Tuovinen 1981).

The most closely studied bacteria associated with the oxidation of pyritic material is the genus *Thiobacillus* but some attention has been given to moderately thermophilic *Thiobacillus*-like bacteria and to the extremely thermophilic *Sulfolobus* genus (Norris and Barr 1985; Marsh et al. 1983; Brierley et al. 1978) which oxidises sulphidic material at temperatures frequently found in heap leach and mine waste piles, but too high for the *Thiobacillus* spp. to remain viable. Similarly, some work has been reported on *Leptospirillum ferrooxidans* (Norris 1983) which catalyses the solubilisation of iron pyrites and appears to do so under conditions of low pH that inhibit *T. ferrooxidans*. Acidophilic heterotrophs are also frequently found in leach piles and waste rock piles (Goodman et al. 1981a, Harrison 1984), but little is understood of the role, if any, they play in the oxidation of pyritic material (Brierley 1978). Goodman et al. (1983) reported that solubilisation of a metal sulphide occurred under micro-aerophilic conditions in the presence of bacteria but not in a sterile sample. Further work is therefore needed to elucidate the mechanism of bacterially catalysed pyritic oxidation and so enable proper assessment of the effectiveness of the means being employed to reduce the environmental impact of pyritic oxidation in mine wastes.

A sampling program carried out by the University of N.S.W. School of Biotechnology between 1975 and 1979 showed that *Thiobacillus* spp. was common in material from the waste rock dumps and, as is to be expected, from the heap leach pile (Goodman et al. 1981a). Acidophilic heterotrophs were also common. Sampling was limited to the first 4 to 5 m of the dumps where little change in bacterial type or numbers was found with depth. Similarly, there were no marked changes in the bacterial population between wet and dry seasons.

2.2.3 Release of Pollutants from Waste Rock Dumps

Average copper concentrations in runoff during the 1973–74 wet season were estimated to be about 1 mg l^{-1} for White's, 18 mg l^{-1} for Intermediate and 0.2 mg l^{-1} for Dysons with total loads in runoff for that year of 2.0 m rainfall estimated to be 0.2 to 0.3, 0.6 to 1.2 and 0.01 to 0.02 t respectively, depending on the assumptions made on the runoff fraction.

Flat-vee weirs were subsequently constructed on the principal runoff channels on both White's and Intermediate dumps to measure the fraction of the rainfall that appeared as surface runoff and the level of pollutants in runoff water. The runoff fraction for White's dump was 13% during the 1975–76 wet season and 18.6% during the 1976–77 wet season, giving an average for the two seasons of 15.6% (Daniel et al. 1982, Harries and Ritchie 1983a). The approximate average concentrations of pollutants in the runoff were copper 8 g m^{-3}, zinc 1.3 g m^{-3}, manganese 0.8 g m^{-3} and sulphate 0.8 kg m^{-3}. The annual loads in runoff from White's dump were estimated to be about 40, 0.4, 0.06 and 0.04 t of sulphate, copper, zinc and manganese respectively, for the average rainfall of 1.5 m.

Springs appeared at various parts of the dumps and heap leach pile as the wet season progressed. These were most marked on Dyson's and White's dump where they appeared at the base of the dumps on the southern and north-eastern sides respectively. Flow started generally about January in the middle of the wet season and continued through to May or June and sometimes July. Average copper concentrations estimated for spring waters and groundwaters associated with the overburden dumps in the 1973–74 wet season were 104, 215 and 1.7 mg l^{-1} for White's, Intermediate and Dyson's respectively (Conway et al. 1975).

The total flow of seepage water from the springs at the base of White's dump near its north-east corner was less than 5% of the rainfall on the whole dump during the wet season. The average sulphate concentration in this seepage water was 17 kg m^{-3} and the average metal concentrations were copper 80 g m^{-3}, zinc 45 g m^{-3} and manganese 35 g m^{-3}. Figure 3 shows pollutant concentrations in spring water for a number of different wet seasons.

The annual evaporation loss from White's dump was estimated to be in the range 25 to 35% of the incident rain, depending on the details of the rainfall record (Daniel et al. 1982). Evaporation and runoff account for between 40 to 50% of the water falling on White's dump as rain. This leaves 50 to 60% of the incident rain which must percolate through the dump. Of this, less than 10% appears as seepage from the base of the dump and the rest must pass through the original ground surface and become part of the groundwater. If the concentrations of pollutants in spring water are assumed representative of pollutant concentrations in the water passing through the base of the dump, then the total pollutant loads released to groundwater by this dump in a typical wet season of 1.5 m rainfall were 3800, 18, 10 and 8 t of sulphate, copper, zinc and manganese respectively. These estimates

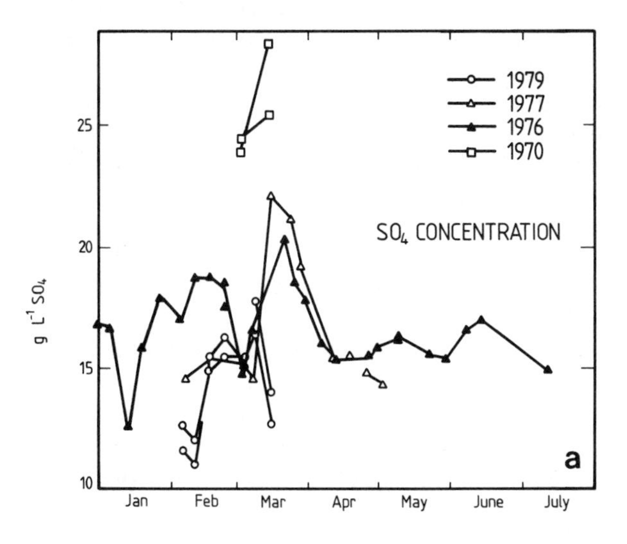

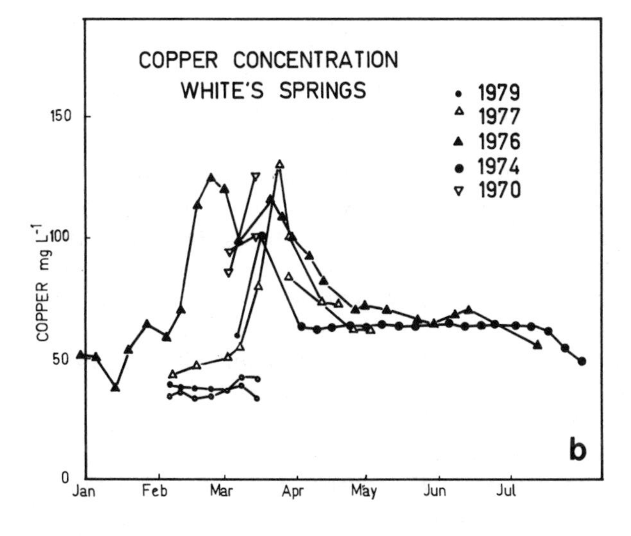

Fig. 3.a,b. Variation of pollutant concentration in spring water from White's dump with wet season and time in the wet season. **a** Sulphate concentrations; **b** copper concentrations

assume that the pollution load arose from rainfall on the whole surface of the dump. If it is assumed that there was little infiltration of rain falling on the sides of them dump, the the pollution loads will be proportionately smaller (see Table 1).

Measurements on Intermediate dump showed that it had evaporation rates and runoff fractions very similar to those of White's (Daniel et al. 1982). Average concentrations in runoff from Intermediate were 1480, 33, 2.6 and 2.1 g m^{-3} for sulphate, copper, zinc and manganese compared to 22000, 210, 52 and 52 g m^{-3} in seepage water leading to annual load estimates in runoff of 22, 0.5, 0.04 and 0.03 t and loads to groundwater of 1200, 12, 3 and 3 t respectively. It is apparent that, as for White's dump, the pollution loads in runoff were only a few percent of the load presented to groundwater.

The relative insensitivity of pollution levels in runoff to the rainfall pattern on the dumps during the wet season was consistent with transport to the surface of soluble salts produced by pyritic oxidation within the dump but inconsistent with dissolution of salts generated in situ at the top surface of the dump. It became apparent (Harries and Ritchie 1983a) that the oxidation process was not a surface phenomenon but most likely involved a large fraction of the volume of the dumps. This picture was borne out in part by the finding (Goodman et al. 1981a) that soluble salt levels, bacterial types and bacterial populations did not change significantly in the top 5 m or so of the dumps.

2.2.4 Microenvironment in Waste Rock Dumps Before Rehabilitation

The oxidation of the pyritic material in the waste rock dumps is an exothermic reaction and the resulting temperature rise was used to determine the location and rate of oxidation. A series of probe holes was installed in White's and Intermediate dumps at the locations shown in Fig. 4. At some locations there were two different types of probe holes, identified as "n" and "g". Gas concentration measurements were carried out in g holes, neutron moisture metre measurements in n holes and temperature measurements in all holes. The temperatures as a function of depth in the various probe holes are shown in Fig. 5 (Harries and Ritchie 1983b). Temperatures below about 5 m showed little variation with the wet/dry season cycle. The distributions of heat sources (oxidation rates) which would produce the observed temperature distributions have been calculated (Harries and Ritchie 1981).

The oxidation of pyritic material to produce sulphates requires a supply of oxygen to the oxidation sites. Oxygen and carbon dioxide levels within the dumps were measured in the field (Harries and Ritchie 1985b). In some locations (B, X and W) the oxygen concentrations decreased monotonically from the surface indicating that transport of oxygen into the dump was predominantly by diffusion from the top surface. The oxygen concentration at these locations was less than 1% by volume at all depths below 5 m. At other locations (A, Y and Z) the oxygen concentration, after first decreasing from the surface, passed through a minimum and increased again towards the bottom of the dump. The increased level of oxygen near the bottom of the dump indicates that oxygen was being transported by convection. The high temperatures at depth in the dump at these locations created a chimneylike effect, drawing air in from the side of the dump and up through the

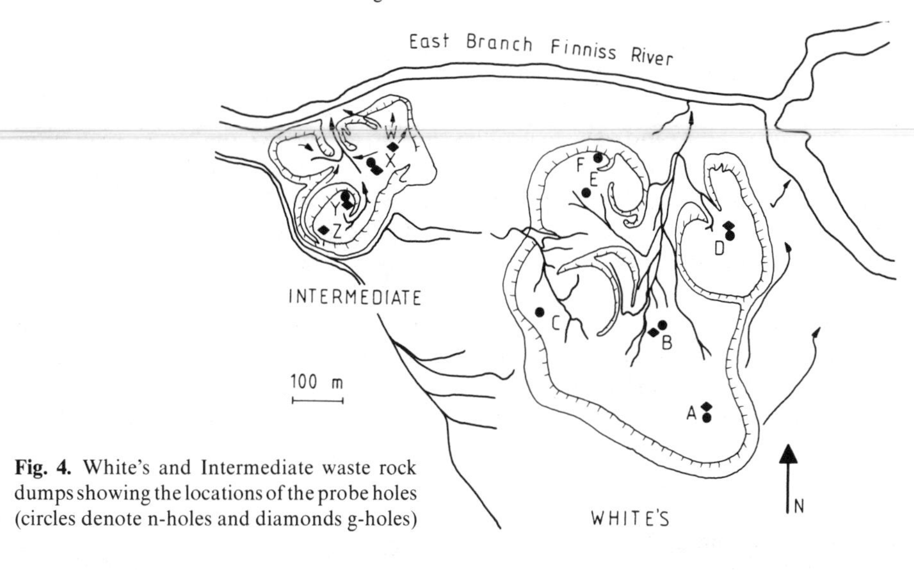

Fig. 4. White's and Intermediate waste rock dumps showing the locations of the probe holes (circles denote n-holes and diamonds g-holes)

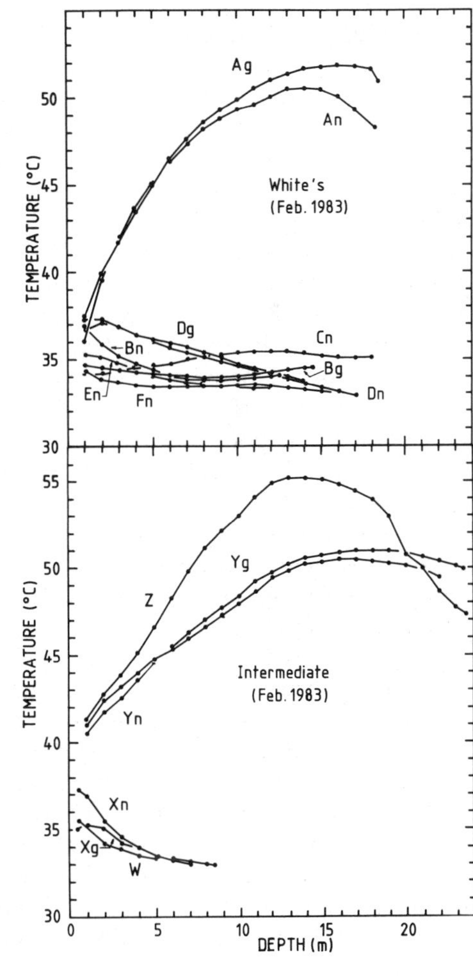

Fig. 5. Temperature profiles in White's and Intermediate waste rock dumps

hot part of the dump. At a few locations (D and W) the oxygen concentrations at a given depth varied dramatically over a period of a day. This variation correlated with the diurnal changes in atmospheric pressure (Harries and Ritchie 1985b).

The temperature and oxygen profiles showed clearly that the pyritic oxidation rate in these dumps was limited by the rate of supply of oxygen and that oxygen was supplied to oxidation sites deep within the waste dumps by a variety of gas-transport mechanisms. The principal gas transport mechanisms in the dumps before rehabilitation can be summarised as follows:

1. Convection was the most important way of transporting oxygen into those regions deep within the dumps where there was a high rate of pyritic oxidation;
2. Diffusion was the dominant process away from the edges of the dumps; and
3. Advection, driven by variations in the atmospheric pressure, occurred along some high permeability paths.

2.3 Tailings Disposal Area

Between 1954 and 1961 some 640 000 t of tailings material was deposited in an area of about 30 ha, leading to a covering varying in depth between about 0.3 to 2 m. A significant quantity of this material was washed into the East Branch where it has been deposited on its bed and banks. Possible pollutants from the tails material are trace metals such as copper and manganese and the radionuclide ^{226}Ra and its daughters such as ^{210}Pb. Radium in the tails also gave rise to radon and its daughters in the air in the environs of the tailings disposal area.

Experiments carried out in the 1973–74 survey indicated that tailings material in the tailings area leached annually at the rate of about 0.0074, 0.0054 and 0.5 g kg^{-1} of copper, manganese and SO_4 compared to 8.5×10^{-6}, 1.9×10^{-3} and 0.16 g kg^{-1} respectively, from tailings material taken from the bed of the East Branch (Conway et al. 1975). These experiments were predicated on the assumption that bacterially catalysed oxidation of the pyritic material occurred during the dry season, an assumption not supported by later work which showed that there was no bacterial activity in the tailings material (Babij et al. 1981). However, the results of Conway et al. still allowed reasonable estimates to be made of the quantity of pollutant leached out of the tailings disposal area.

Water samples taken from the East Branch both above and at the mine site during the 1973–74 wet season were analysed for ^{226}Ra and ^{210}Pb, together with samples taken from creeks at the mine site, including Old Tailings Creek which drains the tailings disposal area (see Fig. 2). Only one sample contained levels of ^{226}Ra exceeding the 10 pCi l^{-1} limit recommended by the ICRP for drinking water and this sample was from Old Tailings Creek where levels ranged from 0.5 to 12 pCi l^{-1} (Davy and O'Brien 1975a). An estimate of the radium intake of a hypothetical group of people living in the Finniss River area downstream of the East Branch confluence indicates an intake of 8.2 nCi yr^{-1} which exceeds the ICRP recommended value of 8 nCi yr^{-1}. However, a large portion (3 nCi yr^{-1}) of this intake arises from estimated radium levels in buffalo meat. These levels require substantiation by direct measurement if land use restrictions are contemplated. At that time it was estimated that a substantial quantity of radium (~ 90 Ci) had been released into the

Finniss River as a consequence of process water disposal practices in the period 1954–67. However, this estimate was based on experimental data involving the then standard method for radionuclides of radium in water (ASTM, 1976) which was subsequently found to give erroneous results for process water (Levins et al. 1978). A better estimate of the amount of radium contamination carried by process water is now 10 Ci (D. Davy, pers. commun.). In any case, it is likely that the practices in the period 1954–67 dominated the quantity of Ra released in the period from 1967 on (Davy and O'Brien 1975b). If this is so, rehabilitation of the mine site will not have a large effect on the dose commitment of groups living in the Finniss River plains.

Although measurements of the emanation rate of radon from the tailings disposal area have been carried out and yield an average figure of 2 Bq m^{-2} s^{-1} (Davy et al. 1978) no extensive measurements have been carried out of radon and radon daughter concentrations in air. Such measurements have been carried out in the relatively nearby South Alligator tailings area where the average emanation rate is 6 Bq m^{-2} s^{-1}, and radon and radon daughter concentrations averaged over weather conditions (which approximate conservatively the average meteorological conditions) yield a figure of 80 mWL over the tails area (Whittlestone and O'Brien 1983). It is therefore reasonable to estimate a figure of about 25 mWL for average levels over the Rum Jungle tailings disposal area. This compares with 10 mWL recommended as the upper level for continuous exposure for a member of the public.

2.4 Open Cuts

The physical dimensions and data on the water quality of the open cuts are presented in Table 2. Although the dissolved oxygen in White's open cut had a marked clinograde at about 8 m from the surface (Conway et al. 1975, Babij et al. 1981), the pH and concentrations of dissolved solids such as Fe, Cu, Mn, Zn and SO_4 did not vary significantly between near-surface and the rest of the water column. There was no discernible clinograde in Intermediate open cut or the much shallower Dyson's open cut.

Abundant populations of *T. ferrooxidans* and autotrophic sulphur-oxidising bacteria in the sediments of both Intermediate and White's open cut indicated the microbial degradation of sulphidic material (Babij et al. 1981). The physico-

Table 2. Some physical and chemical properties of the water-filled open cuts at the Rum Jungle mine site prior to rehabilitation

Entity	Physical features			Chemical composition ($mg\ l^{-1}$)				
	Area ($\times 10^{-4}$ m^2)	Volume ($\times 10^{-5}$ m^3) After mining	Prior to rehabilitation	pH	Cu	Mn	Zn	SO_4
White's	10.5	35	27	2.6	60	230	6	8200
Intermediate	4.4	9.7	11	3.5	70	60	7	3100
Dyson's	7	9.2	4.5	3.1	7	20	-	1500

chemical conditions of the water column and the sediments are different in the two open cuts and vary with time in different ways. This indicates that the two open cuts are behaving differently. Although the water quality of the open cuts shortly after the mine site was abandoned probably reflected the use of Dyson's and White's for tailings disposal, the use of White's as a reservoir for process water, the drainage of polluted surface water into the open cut from adjoining sulphidic areas as well as the degradation of sulphidic material on the walls and bottoms of the open cut, water quality in the long term depends on a number of processes that are not well understood. One of these is the interaction between local groundwater and water in the open cut. In assessing the pollutant load presented by the open cut to the East Branch Conway et al. (1975) estimated seepage rates from the open cuts to be about 5, 5 and 35% for White's, Intermediate and Dyson's respectively. The figure for Dyson's represents overflow through a drainage pipe into the East Branch during the wet season rather than seepage into groundwaters represented by the other two figures.

2.5 Estimates of Pollution Load

Table 3 contains estimates of the pollution loads presented to the East Branch by the principal sources at the mine site in a typical season of 1.5-m rainfall. In the 1973–74 survey, the quantities of copper, manganese and zinc released from the mine site area were estimated using three independent methods; measurement of flow rate and concentrations in the East Branch below the mine site; measurement of flow rate and concentration at each of the three tributaries of the East Branch at the mine site; and estimates of loads from individual sources at the mine site. The three methods produced results which were in general agreement (Conway et al. 1975). This gave confidence in the estimates for different pollutant sources. Table 4 shows the annual flow and annual load of Cu, Mn, Zn and SO_4 measured at a point in the East Branch just downstream of the mine site during the period 1964

Table 3. Estimated annual pollution load from principal sources at the Rum Jungle mine site for a year of average rainfall, 1.5 m

Source	Load (t)			
	Cu	Mn	Zn	SO_4
Dumps				
White's	18.5	8.0	10.0	3890
Intermediate	12.5	3.0	3.0	1230
Dyson's	0.1	2.0	0.1	1040
Copper heap leach	24	–	0.4	520
Opens cuts				
White's	6.0	23	0.6	810
Intermediate	2.5	2.0	0.3	110
Dyson's	0.8	2.3	–	170
Others				
Tailings area	3.5	2.6	–	240
Bed of old acid dam	–	9.0	–	–

Table 4. Annual pollution loads carried by the East Branch immediately prior to and after abandonment of mine site

Season	Rainfall (mm)	Period of flow	Flow (x 10^{-6} m^3)	Pollution load (t)			
				Cu	Mn	Zn	SO_4
1969–70	896	Dec-May	7.0	44	46	–	3300
1970–71	1611	Nov-Aug	33.2	77	110	24	12000
1971–72	1542	Nov-July	30.9	51	64	19	6600
1972–73	1545	Dec-July	26	45	49	16	5500
1973–74	2000	Nov-Sept	97	130	100	40	13000

to 1974. This spans the time immediately before and just after abandonment of the mine site in 1971.

Table 3 shows clearly that the waste rock dumps and the copper heap leach pile were the dominant sources of pollution, with the tailings disposal area responsible for only about 5% of the total pollutant load. Consideration of the simple fact that the annual sulphate load from White's waste rock dump used up only about 0.5% of the sulphur content of the dump implied that the waste dumps would continue to be a source of acid and trace metal pollution for some hundreds of years. More detailed mathematical modelling of the gas-transport process that supports pyritic oxidation within the dumps (Ritchie 1977; Davis 1983; Davis and Ritchie 1986, Davis et al. 1986) predicted that the dumps would act as pollution sources for about 500 years, and that the pollutant production rate would decrease by about a factor of two between 25 and 100 years after creation of the dumps and by about a factor of 20 between 25 and 500 years.

Acid and trace metal production from the tailings disposal area could also be expected to last for 100 years or so. On the other hand, this area would have continued to be a source of radon, essentially unchanged over a period of a 1000 years or so, if it had remained in the state left when the mine site was abandoned.

3 Rehabilitation Strategy and Rehabilitation Measures

The Rum Jungle Rehabilitation Project was established in 1982. An agreement between the Commonwealth of Australia and the Northern Territory Governments specified the objectives and strategies of the project.

The objectives were:

1. A major reduction in pollution in the watercourses feeding the river and in particular the reduction of the annual average releases of copper, zinc and manganese into the river by 70, 70 and 56% respectively, as measured at the junction of the river with the Finniss River.
2. Reduction in public health hazards and in particular reduction of radiation levels at the site at least to the standards set out in the Code of Practice on Radiation in the Mining and Milling of Radioactive Ores published by the Australian Government Publishing Service in 1980.

3. Reduction of pollution in the water contained in the open pits known as White's and Intermediate.
4. Aesthetic improvements including revegetation.

The cost of $(Aust)16.2 million (1982 dollars) was funded by the Commonwealth Government and the work was managed by the Northern Territory Government. A Liaison Committee with joint Commonwealth and Northern Territory Government representation was set up to oversee the rehabilitation program. Of the $16.2 million, $5.3 million was for earthworks and $5.4 million was for the construction and operation of a water treatment plant. A more detailed breakdown of the estimated cost for individual parts of the rehabilitation program is given in Table 5.

The rehabilitation strategy was (Allen 1985):

1. To cover the overburden dumps with a seal to reduce the ingress of water.
2. To collect the tailings and cover them with enough material to reduce radon emanation.
3. To remove the copper heap leach pile and dispose of the material above the water table in a location with minimal water ingress.
4. To treat the water in the open cuts to raise the pH and reduce the trace metal concentrations to a recreational standard where limited eye, nose and ear contact with the pit water would be permissible.
5. To clean up and revegetate other areas.

Table 5. Estimated costs (June 1982) of rehabilitation work at the Rum Jungle mine site

Category	Item	Cost (Aust. $,000)
Earthworks	Copper heap leach pile	578
	Tailings dam	1497
	Dyson's open cut	310
	White's open cut	77
	Intermediate open cut	–
	Dyson's overburden dump	300
	White's overburden dump	696
	White's north overburden dump	328
	Intermediate overburden dump	217
	Acid/sweet water dams	191
	Other areas	333
	Site establishment, including protective fence	731
Construction and operation of treatment plant	Water treatment plant	2020
	Discharge pipeline to White's open cut	80
	Construction camp	280
	Chemicals for water treatment	2300
	Operation of treatment plant	740
Project management	Camp services, accommodation, etc.	922
	Monitoring	171
	Engineering and management	3416
	Site services	415
Contingency		600

The project commenced in mid-1982 and engineering works are expected to be completed on schedule in June 1986. First treated was White's dump where earthworks to reshape and cover the dump were completed in 1984. In 1984 the tailings and the heap leach material were removed and a start made on the treatment of the water in the open cuts. Dyson's and Intermediate dumps were reshaped and covered in 1985. A program to monitor the effect of various parts of the rehabilitation has been developed and is funded to June 1988. Some aspects of this program could well continue for a number of years.

The bulk of White's North dump was lifted and deposited as a fillet round the base of White's dump to be treated as part of that dump. White's, Intermediate and Dyson's were reshaped to create a stable landform. This involved reducing the slope on the top surfaces to less than 5° and installing engineered drainage channels. The sides of White's and Intermediate dumps were reduced to slopes of less than 1 to 3 (horizontal) and drop structures were installed on White's to carry the water to toe drains. A significant amount of earth moving was carried out on Intermediate to make it of a more uniform height. This involved reducing the height of much of the south-west portion from its original of more than 20 m to about 14 m. The top and side surfaces of the dumps were covered by three layers, a low-permeability sealing layer (1A material), a moisture-retention layer (1B material) and an erosion-protection gravel layer (2A material). On the top surface of the dumps the cover consisted of a minimum 225 mm of 1A material, 250 mm of 1B material and 150 mm of 2A material respectively. On the side of the dumps the erosion-protection layer was a rock mulch (3A material). Erosion control banks were constructed on the top surface. Finally, the dumps were revegetated to provide additional erosion protection.

The material from the tailings area and the heap leach pile was collected and deposited in Dyson's open cut. The tailings material and affected subsoil from underneath the tailings were placed at the bottom in the open cut and the top surface of the tailings material was sloped to discharge water. The fine-grained and wet nature of the tailings necessitated the use of swamp dozers to place the material. The tailings material was then covered with a geotextile fabric and a 1-m-thick rock blanket to carry away any ground- or seepage water. The material from the heap leach pile and the worst affected subsoils were placed on top of the rock blanket where it was above the water table. The top fill in Dyson's open cut projected above the original ground surface and was sealed by a covering of three layers similar to the covering on the overburden dumps. Erosion control banks and drains were installed on the surface similar to those on the tops of the rehabilitated overburden dumps.

The area from which the tailings were removed was shaped to allow drainage to a central drain, topsoiled and revegetated. Topsoiling and revegetation were also used to rehabilitate other areas but in some cases, such as the base of the heap leach pile where there was considerable trace metal and acid contamination of the remaining soil, a pore breaking-drainage layer was laid down and covered with 1A, 1B and 2A material prior to vegetation.

The water in White's and Intermediate open cuts was treated by hydroxide precipitation to raise the pH and so remove the trace metals. White's water was pumped through the treatment plant, treated and returned to the open cut where natural stratification separated the treated from the untreated water. On the other hand, treatment of the water in Intermediate open cut was carried out in situ by the addition of some 670 t of lime and aerated to ensure mixing. After settlement, metal hydroxide precipitates were pumped from the bottom of the open cut to the treatment plant for dewatering. Filter cake from the treatment plant was buried. It is intended to divert the East Branch through the two open cuts and use the annual wet-season flow to prevent a future buildup of acidity and trace metal pollution.

Although the actual costs of implementing some of the rehabilitation works have differed from the estimated costs, the overall cost will be very close to the budgeted $(Aust)16.2 million (1982 dollars). Similarly, some details of the rehabilitation works have been changed but the overall strategy and scope of the work has remained the same. For example, the sides of Dyson's waste rock dump have not been sealed and rock-mulched as were the sides of the other two waste rock dumps, and portions of White's North dump have been left in situ to be sealed in the same way as the tops of the waste rock dumps rather than be transported in their entirety to White's and treated as part of that dump. The program of monitoring has also been expanded.

4 Post-Rehabilitation Monitoring

Monitoring of the effectiveness of the rehabilitation is an important part of the project. The overall success of the project is being determined by the improvement in the water quality of the water in the East Branch downstream of the mine site. In addition, individual aspects of the project are being monitored. Rehabilitation is still in progress and no definitive statement of the effectiveness of the project can yet be made. Nevertheless, it is informative to discuss the monitoring program and the results to date.

4.1 Water Quality

Water quality in the East Branch and other surface waters is being measured by the Water Division of the Northern Territory Department of Transport and Works. Water samples are collected at regular intervals from a gauging station on the East Branch downstream of the mine site and daily composite samples prepared. These are analysed for pH, specific conductance, calcium, magnesium, copper, manganese, zinc, nickel, cobalt and sulphate. Daily river discharges are obtained and the daily loads of the various pollutants calculated.

The copper, manganese, zinc and sulphate loads in 1982–83 and 1983–84 appear to be less than those expected on the basis of the earlier 1969–74 data. For example, the copper and sulphate loads in 1983–84 were 28 and 3600 t for a flow of 45×10^6 m^3 (Johnston and Alcock 1984) compared with 45 and 5500 t for a flow of 26×10^6 m^3 in 1972–73 (Table 4). This could be due in part to a reduction in pollutant generation rates in the waste rock dumps in the decade from the early 1970s to the early 1980s or to the effective removal of White's dump as a pollution source in the 1983–84 wet season.

It is apparent that the concentrations of the various pollutants in the river change in different ways throughout the wet season and the total loads vary from wet season to wet season in a way that is not simply related to the total flow. These effects are likely to be even more marked in the coming years as the rehabilitation measures improve the quality of local groundwater and reduce pollutant levels in the bed of the East Branch.

Hence, it could be some years before the effectiveness of the rehabilitation can be quantified unequivocally. To assist in the assessment of the effectiveness of the rehabilitation measures, the Water Division has installed some 27 shallow (1–2 m) and 23 deep (15–65 m) wells around the mine site between 1983 and 1985. It is too early yet to discuss results obtained from the groundwater studies in detail but it is clear that there are two main aquifers in the area: a near-surface, unconfined aquifer about 2 m deep and an aquifer in weathered rock about 20 to 30 m from the surface. It would appear that most of the pollutants are carried by the near-surface aquifer which discharges into surface water in the area (private communication Appleyard S 1983; Salama R 1985).

4.2 Overburden Dumps

As the overburden dumps are the major source of pollution at the mine site, assessment of the effects of reshaping and sealing the dumps is an important part of the monitoring program. Although the seal was designed primarily to prevent the ingress of water, it should also reduce the transport of oxygen into the dumps and furthermore, since the oxidation process in these dumps is oxygen rate-limited, reduce the pyritic oxidation rate and hence the pollutant production rate.

Lysimeters have been placed in White's and Intermediate dumps beneath the clay seal to collect any water which percolates through the seal. The lysimeters consist of 200-litre drums with suitable plumbing to allow water collected in them to be pumped out and measured. Preliminary results show that the amount of water percolating through the clay layer was only about 2.9% of the incident rain in the 1983/84 wet season and about 2.5% for the 1984/85 wet season. This is much less than the 50 to 60% estimated as the fraction of rain percolating through the dump prior to rehabilitation (Daniel et al. 1982) and close to the design figure of 5%.

The three probe holes in White's fitted with gas ports remained largely undisturbed throughout the reshaping and emplacement of the seal on that dump, whereas the large amount of earth moving carried out on Intermediate meant that a new set of probe holes had to be installed. This was done in November-December 1985 and it is too early to present results on the effect of rehabilitation on temperatures and gas composition in Intermediate. Therefore, the following discussion is limited to White's where data have been collected for 2 years following installation of the clay seal.

There was a rapid decrease in the oxygen concentration in the interstitial gas near the surface at each hole as soon as the clay seal was put in place (Harries and Ritchie 1984, 1985a). At location B the seal was put in place in late October 1983 and the oxygen concentration at all depths at that location had decreased to less than 1% by January 1984. Although the clay cover was put in place near hole A at the end of September 1983, the area within about 1 m of hole A was left uncovered until December 1983 when the clay seal was placed around the hole. By the end of January 1984 the oxygen concentrations in the top 5 m of the dump at this location had decreased to less than 1% (Fig. 6). There was a slower decrease at depth and even at the end of March 1984 the oxygen concentration was still about 4% at a depth of 12 m. The rate of decrease below about 7 m in hole A was consistent with the stopping of convective transport and in situ usage of the oxygen in the interstitial gas.

The low oxygen concentrations showed that the seal had effectively stopped the transport of oxygen into the dump. However, measurements in August 1984 showed that the near-surface levels of oxygen had increased and the oxygen concentrations were changing over a period of a day. A detailed study in October 1984 showed that this diurnal variation was closely correlated with changes in atmospheric pressure. The increase in the oxygen levels beneath the seal indicates that some cracking of the clay seal had occurred, presumably because of drying out during the dry season. The oxygen levels remained high during the following wet season indicating that wetting did not close these cracks. It should, however, be

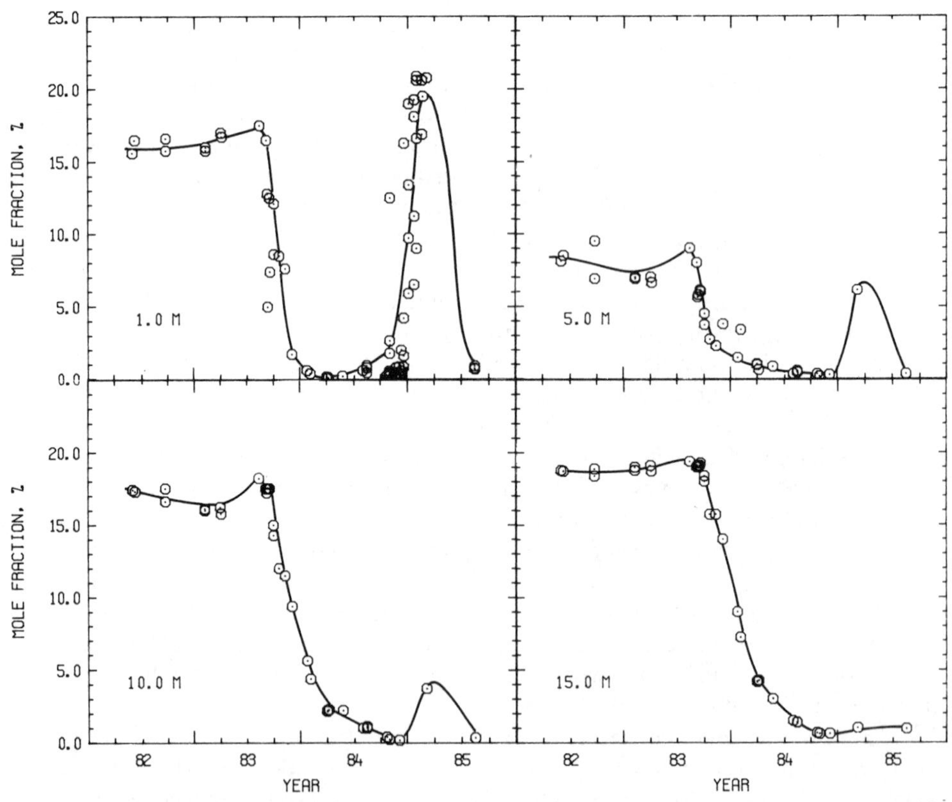

Fig. 6. Change in oxygen concentrations in hole A on White's dump. The dump was reshaped in September 1983 and the cover around the hole completed by December 1983. The variability post-August 1984 is due to diurnal atmospheric pressure changes; the *line* is an "eye-ball" fit through the points

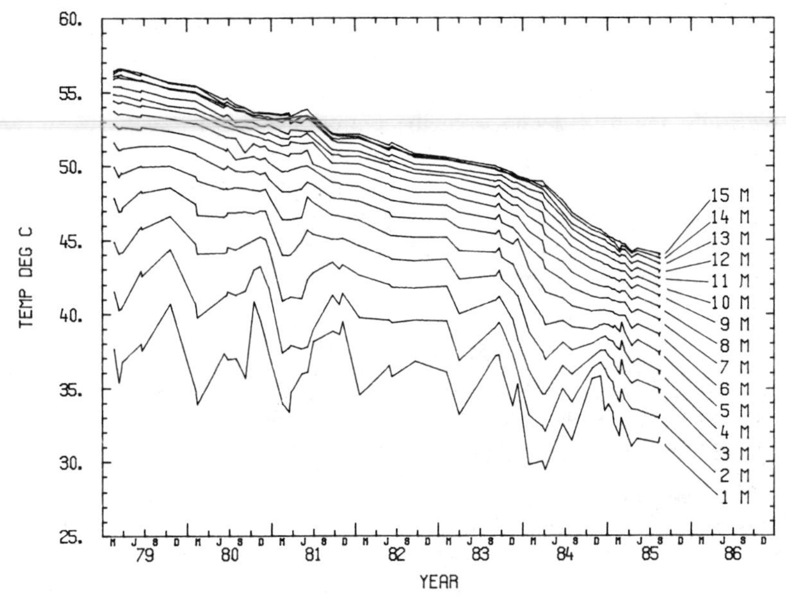

Fig. 7. Temperature variation in hole An at various depths pre- and post-rehabilitation

stressed that, although the transmission rate of gas through the seal appears to have increased, the transmission rate of water still remains unchanged.

The oxygen levels near the base of hole Ag have begun to increase indicating the possible re-emergence of convective transport in this region of the dump. However, the oxygen levels are still much less than they were before rehabilitation.

Holes Bg and Dg showed a somewhat similar behaviour; the oxygen levels decreased following reshaping of the dump and emplacement of the cover. By August 1984, oxygen concentrations in all three holes were responding to atmospheric pressure changes.

Figure 7 shows the change in the temperature in hole An over the past 4 years. The temperatures in hole An were decreasing before rehabilitation commenced but the rate of decrease has increased since the cover was put in place.

5 Conclusions

The measures adopted to rehabilitate the Rum Jungle mine site represent one way of solving the problem; the monitoring program which is an important part of the rehabilitation program will quantify the effectiveness of such measures. There are, therefore, useful lessons to be learned that could be applied in the rehabilitation of other mine sites as many of the pollutant generation and pollutant transport mechanisms are not well understood. This is particularly so of pollutant generation in pyritic mine wastes which are the major source of pollution at Rum Jungle.

As the rehabilitation program will not be completed until mid-1986 it is too early to draw conclusions on improvements in water quality, but some useful tentative conclusions can be reached on the effect of the cover on the waste rock dumps. The cover has proved effective in reducing the infiltration rate of water from about 50–60% of incident rain to about 3%. Hence, it would be reasonable to expect a ten-fold decrease in the pollution loads from the waste rock dumps. However, although the cover was initially very effective in reducing the ingress of air, it has proved less effective after exposure to its first dry season. The effect of the increased air transmission rate on oxidation rates within the pyritic waste material still remains to be quantified but it is clear that, after 2 years, these remain lower than pre-rehabilitation levels.

Acknowledgements. A study that has covered so many years has involved a number of people who have contributed to the project. Of special note is Mr. J. Daniel, now retired, whose hard work and dedication we would like to acknowledge. We would also like to thank Mr. N. Conway, Mr. J. Casteleyn, Mr. M. Blackford, Mr. M. Hyde for their assistance over a lengthy period and Dr. J. Bennett and Mr. Bowdler for their contributions more recently. The useful discussions with and assistance from colleagues in the Rum Jungle project team, the N.T. Department of Transport and Works, the N.T. Department of Mines and Energy and the Commonwealth Department of Resource and Energy are much appreciated. Finally, we would like to thank our Division Chief, Mr. D.R. Davy, for his continued support.

References

Allen CG (1985) The rehabilitation of Rum Jungle. Northern Terr Dep Mines Energ, Darwin, April 1985

American Society for Testing and Materials (1976) Standard method for radionuclides of radium in water. Annu Book ASTM Stand pt 31, pp 677–681

Andersen JE, Allman MB (1968) Some operational aspects of heap leaching at Rum Jungle. Australas Inst Min Met Proc 225:27–31

Andersen JE, Herwig GL, Moffitt RB (1966) Heap leaching at Rum Jungle. Aust Min 58 (4):35–39

Babij T, Goodman AE, Ralph BJ (1981) Ecology of Rum Jungle, pt 3. Leaching behaviour of sulphidic waste material under controlled conditions. Aust Atom Energ Commiss Rep AAEC/E520

Barrie DR (1982) The heart of Rum Jungle. Barrie, Batchelor, Australia

Brierley CL (1978) Bacterial leaching. CRC Crit Rev 6:207–262

Brierley JA, Norris PR, Kelly DP, LeRoux NW (1978) Characteristics of a moderately thermophilic and acidophilic iron-oxidising *Thiobacillus*. Eur J Appl Microbiol Biotechnol 5:291–299

Conway N, Davy DR, Lowson RT, Ritchie AIM (1975) Sources of pollution. In: Davy DR (ed) Rum Jungle environmental studies, chap 6. Aust Atom Energ Commiss Rep AAEC/E365

Daniel JA, Harries JR, Ritchie AIM (1982) Runoff and seepage from waste rock dumps containing pyritic material. IE Aust Conf Hydrology and water resources, 11–13 May 1982, Melbourne, pp 28–32

Daniel JA, Harries JR, Ritchie AIM (1983) The quality and quantity of runoff and ground water in two overburden dumps undergoing pyritic oxidation. Australian water and waste water association Conf, 4–9 Sept 1983, pp 24–1

Davis GB (1983) Mathematical modelling of rate-limiting mechanisms of pyritic oxidation in overburden dumps. Thesis, Univ Wollongong, Aust

Davis GB, Ritchie AIM (1986) A model of oxidation in pyritic mine wastes – I Equations and approximate solution. Appl Math Modelling 10:314–322

Davis GB, Doherty G, Ritchie AIM (1986) A model of oxidation in pyritic mine wastes – II Comparison of numerical and approximate solutions. Appl Math Modelling 10:323–330

Davy DR (ed) (1975) Rum Jungle environmental studies. Aust Atom Energ Commiss Rep AAEC/E365

Davy DR, O'Brien BG (1975a) Radiological aspects. In: Davy DR (ed) Rum Jungle environmental studies, chap 9. Aust Atom Energ Commiss Rep AAEC/E365

Davy DR, O'Brien BG (1975b) The fate of the discharged metals. In: Davy DR (ed) Rum Jungle environmental studies, chap 8. Aust Atom Energ Commiss Rep AAEC/E365

Davy DR, Dudaitis A, O'Brien BG (1978) Radon survey at the Koongarra uranium deposit. Aust Atom Energ Commiss Rep AAEC/E459

DiSpirito AA, Tuovinen OH (1981) Oxygen uptake coupled with uranous sulfate oxidation by *Thiobacillus ferrooxidans* and *T. acidophilus*. Geomicrobiol J 2:275–291
Dugan PR (1972) Biochemical ecology of water pollution. Plenum, New York
Goodman AE, Khalid AM, Ralph BJ (1981a) Microbial ecology of Rum Jungle, pt 1. Environmental study of sulphidic overburden dumps, experimental heap leach piles and tailings dam area. Aust Atom Energ Commiss Rep AAEC/E531
Goodman AE, Khalid AM, Ralph BJ (1981b) Microbial ecology of Rum Jungle, pt. II. Environmental study of two flooded open cuts and smaller, associated water bodies. Aust Atom Energ Commiss Rep AAEC/E527
Goodman AE, Babij T, Ritchie AIM (1983) Leaching of a sulphide ore by *Thiobacillus ferrooxidans* under anaerobic conditions. In: Rossi G, Torma AE (eds) Recent progress in biohydrometallurgy. Assoc Miner Sarda, Iglesias, p 361
Harries JR, Ritchie AIM (1981) The use of temperature profiles to estimate the pyritic oxidation rate in a waste rock dump from an open cut mine. Water Air Soil Pollut 15:405–423
Harries JR, Ritchie AIM (1982) Pyritic oxidation in mine wastes: its incidence, its impact on water quality and its control. In: O'Loughlin EM, Cullen P (eds) Prediction in water quality. Aust Acad Sci, Canberra, p 347
Harries JR, Ritchie AIM (1983a) Runoff fraction and pollution levels in runoff from a waste rock dump undergoing pyritic oxidation. Water Air Soil Pollut 19:133–170
Harries JR, Ritchie AIM (1983b) The microenvironment within waste rock dumps undergoing pyritic oxidation. Recent progress in biohydrometallurgy. Assoc Miner Sarda, Iglesias, p 377
Harries JR, Ritchie AIM (1984) The effect of rehabilitation on the oxygen concentrations in waste rock dumps containing pyritic material. 1984 Symp Surface mining, hydrology, sedimentology, and reclamation, 2–7 December, Univ Lexington, Kentucky, p 463
Harries JR, Ritchie AIM (1985a) The impact of rehabilitation measures on the physicochemical conditions within mine wastes undergoing pyritic oxidation. Int Symp Biohydrometallurgy, Vancouver, 22–24 August 1985
Harries JR, Ritchie AIM (1985b) Pore gas composition in waste rock dumps undergoing pyritic oxidation. Soil Sci 140:143–152
Harrison JP, Jr (1984) The acidophilic thiobacilli and other acidophilic bacteria that share their habitat. Annu Rev Microbiol 38:265–292
Johnston GM, Alcock JF (1984) East Finniss River and Finniss River pollution study. Water Resources Division, Dept Mines and Energy, Darwin, Australia
Levins DM, Ryan RK, Strong KP (1978) Leaching of radium from uranium tailings. Proc OECD/NEA Sem Management, stabilisation and environmental impact of uranium mill tailings, Albuquerque, New Mexico, July pp, 271–285
Lowson RT (1975) The geography, geology and mining history of Rum Jungle. In: Davy DR (ed) Rum Jungle environmental studies, chap 2. Aust Atom Energ Commiss Rep AAEC/E365
Marsh RM, Norris PR, LeRoux NW (1983) Growth and mineral oxidation studies with *Solfolobus*. In: Rossi G, Torma AE (eds) Recent progress in biohydrometallurgy. Assoc Miner Sarda, Iglesias, p 71
Murr LE (1980) Theory and practice of copper sulphide leaching in dumps and in situ. Miner Sci Eng 12:121–189
Norris PR (1983) Iron and mineral oxidation with *Leptospirillum*-like bacteria. In: Rossi G, Torma AE (eds) Recent progress in biohydrometallurgy. Assoc Miner Sarda, Iglesias, p 83
Norris PR, Barr DW (1985) Growth and iron oxidation by acidophilic moderate thermophiles. FEMS Microbiol Lett 28:221–224
Ralph BJ (1985) Geomicrobiology and the new biotechnology. Dev Ind Microbiol 26:23–59
Ritchie AIM (1977) Heap leaching: a gas diffusion rate-limited model. Aust Atom Energ Commiss Rep AAEC/E429
Ritchie AIM (1985) The Rum Jungle experience; a retrospective view. Aust Sci Mag 4:22–30
Soljanto P, Tuovinen OH (1980) A microcalorimetric study of U(IV)-oxidation by *Thiobacillus ferrooxidans* and ferric-ion. In: Trudinger PA, Walter MR, Ralph BJ (eds) Biogeochemistry of ancient and modern environments. Springer, Berlin Heidelberg New York, p 529
Torma AE, Bosecker K (1982) Bacterial Leaching. In: Bull MJ (ed) Progress in industrial microbiology, vol 16. Elsevier, Amsterdam, p 77
Whittlestone S, O'Brien BG (1983) A radon survey of tailings dumps at Moline and South Alligator, Northern Territory. Aust Atom Energ Rep AAEC/C35

Mine Tailings Reclamation. Inco Limited's Experience with the Reclaiming of Sulphide Tailings in the Sudbury Area, Ontario, Canada

T.H. PETERS[1]

1 Introduction

In 1975, an inventory of mine wastes throughout Canada was undertaken by the Canada Department of Mines, Energy and Resources. In all, 718 mine sites were assessed as to the area, type and vegetation cover associated with the waste material. It estimated that there were approximately 45,300 ha of land affected by waste rock, overburden or tailings (Murray, 1985). It is of interest to note that 7,300 ha or about 16% of this land had been revegetated to some extent.

Studies indicate that approximately 9,000 ha (20%) of this total are sulphide tailings (Phinney 1984). The Copper Cliff Tailings Area of Inco Limited at Sudbury, Ontario, presently covers approximately 1,100 ha and is currently being expanded to cover double this area.

In the Sudbury area, Inco Limited is involved in a progressively expanding program, initiated over 70 years ago, to modify the physical effects of mining, concentrating, smelting and refining on the local environment. Historically, the impact of the various facets of these operations, along with that of lumbering, has made Sudbury one of the most ecologically disturbed sites in Canada. The reclamation of these lands which have been subjected to various industrial stresses continues to progress. Inco Limited's experience in tailings reclamation and revegetation and the problems encountered is but one aspect of the total current program in the "Greening of Sudbury" which is underway.

2 Historical Ecology of the Tailings Area

In his report of a geological survey of the area in 1886, Hooker stated.

"There appears to be little land suitable for cultivation. The thin layer of earth covering the hills has in places, suffered from fires, permitting it to be readily washed away and exposing the bare rock underneath".

The original topography of the area was undulating with small valleys in between rounded hill tops. The vertical difference between the top of the hills and the valley bottoms seldom exceeded 60 m. Glaciation was responsible for many of the geomorphic formations and soil types currently found in the area.

[1]P.O. Box 607, Copper Cliff, Ontario, POM 1NO, Canada

The natural soils in the area are humo-ferric podzols with glacial surface deposits of water-modified tills, lacustrine silts and sands located in the valleys (Peters in Sheehan et al. 1981). These soils are shallow and frequently are less than 1-m thick over bed rock.

Many of the valleys contain lakes, swamps and, in some cases, areas sufficiently well drained to allow the development of a forest cover of native trees and shrubs. Drainage is generally poor in these valleys resulting in a soil which is slow to warm up in the spring. This lack of drainage has also hindered the development of the chemical and physical structure, the porosity, aeration and nutrient capacity of the soil. Soil pH values are 6.0 or less (Heale 1980).

The climax vegetation of the area was significantly influenced by these factors. Although there were local variations, the climax vegetation for the area was classified as being in the Laurentian Upland section of the Great Lakes-St. Lawrence Division of the Hemlock-White Pine-Northern Hardwood Forest (Braun, 1950).

The natural ecosystem in the area was a pine forest. Three species, eastern white pine (Pinus strobus L.) red pine (Pinus resinosa Ait.) and jack pine (Pinus banksiana Lamb.) grew on the soil types which were suitable to the particular species. Eastern hemlock (Tsuga canadensis Carr.) and white spruce (Picea glauca Voss.) were the other main coniferous species found in the climax forest. Logging operations commenced on a large scale around 1872 (de Lestard 1967) with white pine being the most valuable economic species. Although there is some natural regeneration by the pine species, the principal current forest cover is made up of American white birch (Betula papyrifera Marsh.), trembling aspen (Populus tremuloides Michx.) and large tooth aspen (Populus grandidentata Michx.) (Rowe 1972).

The mean annual temperature for the Sudbury area is 4.4°C (Chapman and Thomas 1968). The mean annual precipitation is 0.84 m. There is an average frost-free period of 112 days and an average annual growing period of 183 days.

3 The Mine Tailings Problem

Nickel and copper sulphide ores were discovered in the Sudbury District during the building of the first transcontinental railway in 1883. The International Nickel Company, now known as Inco Limited, was formed by the merger of several smaller companies in 1902. The Company has grown over the years. At the present time, Inco Limited is the largest mining, milling smelting and refining complex in the non-Communist world. In the Sudbury area, the Company operates nine mines, two mills, one smelter, one sulphuric acid plant, one copper refinery and one nickel refinery. A total work force of approximately 8,500 is currently employed in these operations. Altogether 14 elements are extracted from the ores mined. The list includes nickel, copper, sulphur, gold, silver, cobalt, platinum, osmium, iridium, selenium, tellurium, palladium, rhodium and ruthenium.

The rated capacity of the two mills, Clarabelle and Frood-Stobie, is 53,524 t day^{-1}. At full production, the milling and concentrating is a continuous process operating 24 h day^{-1}, 7 days a week. Production operations vary with market requirements.

In the late 1920s, the decision to utilize the current technology of the day and to eliminate the discharge of sulphur dioxide at ground level was made. This change from heap roasting in roast yards to roasting in Herreschoff Roasters and reverbratory furnaces to eliminate the sulphur was incorporated into the new smelter. This smelter commenced operations in 1930 and the resultant sulphur dioxide was discharged through high chimneys. The feed for the Herreschoff Roaster had to be a finely ground concentrate. The concentrating of the ore meant that large quantities of finely ground waste rock, known as mill tailings, had to be disposed of by storage. The finding of disposal sites for tailings storage became a necessity. Nearby valleys located between rock hills provided these areas and the rock hills acted as dams or as buttresses for dams.

The tailings are transported as a slurry (35% solids: 65% water) from the mill in large diameter pipes under pressure to the tailings area. The discharge pipes are located on and around the outside perimeter of the dams. After discharge, as the tailings flow by gravity towards the centre of the disposal site, the solids settle out. The water then collects in the low portion of the area from where it is decanted by siphoning. This water is either used again in the tailings circuit or passed through a water treatment plant before it is returned to a natural stream.

Currently, dams are built by first constructing a starter dam using waste mine rock of a suitable size. The height of the dam is then increased by raises of 3 m to 5 m using the on-site tailings and by stepping in towards the centre of the area with each raise. Dam heights vary with local requirements, but are currently built to a height of 50 m in the Copper Cliff Tailings Area.

This tailings disposal site has been developed in sections over the years. Each section was given an alphabetical designation as it was constructed. At present, the A, CD, and M sections have reached their final elevation and are at various stages of the revegetation process. The P and O sections are almost completely filled with tailings and a start has been made on their revegetation. The new R section is becoming the receiving site for the major portion of the tailings now being produced.

The approximate size of the sections are as follows:

A, CD and M sections	485 ha
P and O sections	600 ha
R section	1,120 ha
	2,205 ha

There are two smaller areas at Frood-Stobie and Levack Mine which would total approximately 80 ha.

Since the composition of the tailings at any particular location will vary with the source and nature of the ore being mined at any given time, it is almost impossible to give an average composition. However, the following general figures may be used (Montreal Engineering Company, 1975; Wilson, unpublished 1976):

Mineralogical composition of copper cliff tailings. (sample size not reported)

Feldspar	+50 %
Amphiboles (chlorite)	20 %
Quartz	10 %
Pyroxenes	7 %
Biotite	7 %
Pyrrhotite	5.6%
Magnetite	0.6%
Pentlandite	0.5%
Chalcopyrite	0.3%

Trace element analysis of tailings (available concentrations).

Range found in 12 samples:	Single sample at each site:
pH	3.7–6.2
Cu	1–81
Ni	1–87
Fe	59–441

Expressed in ppm on a dry weight basis.
Technique: 2.5% acetic acid leach for 30 min.

The Copper Cliff tailings disposal area is located adjacent to and on the west side of the former town of Copper Cliff which is now Ward 8 of the City of Sudbury. Additional residential subdivisions have developed around this tailings area in recent years. The prevailing winds in the area are northwest in the winter and southwest in the summer. Dustfall sampling stations are located at different locations to measure dust accumulations on a monthly basis (Browne 1984).

In the early developmental stages, when the tailings levels are low, the surrounding hills and forest cover minimize the effect of the winds and therefore limit the amount of dust generated from the tailings beaches. As the elevation of the disposal area increases, the protection provided by the natural topography and forest cover progressively decreases until a level is reached where it is non-existent.

The necessity of controlling the dust from this source became apparent to the Company in the late 1930s. In addition to being a nuisance to the residents of Copper Cliff, the dust contaminated certain electro-metallurgical refining processes and reduced the effectiveness of lubricants used to minimize normal wear in operating machinery.

Various research projects were undertaken, both in the laboratory and in the field, during this period to stabilize the surface of the tailings. At first, various stabilizers such as bituminous sprays, limestone chips, water glass and other chemical sprays, a timed-water spray through an irrigation system to keep the surface moist, were tried. Different patterns of snow fences were put in place in an attempt to minimize surface winds in the most exposed areas but had little or no effect. All of the above proved to be ineffectual or uneconomical. In the mid-1940s, some experimental seedings of grass were attempted but were unsuccessful.

In 1953, a small test plot (5 × 5 m) was planted on a slope in the D area. One grass, Canada Blue (Poa compressa L.) in the mixture survived and persisted for several years, even though partially buried by drifting tailings. This led to a revegetation research program.

4 The Early Revegetation Research Program

A preliminary review of the success of the small 1953 plot indicated that two factors were mainly responsible. These were: pH adjustment of the substrate and the location of the plot relative to the prevailing wind which minimized the impact of drifting tailings. Depending on the velocity of the wind, the airborne tailings either bury or by abrasion cut the young plants off just above the ground level.

In the late summer of 1957, a plot about 2 ha in size was prepared for seeding. Agricultural limestone was spread at the rate of 11.2 ha^{-1} and was incorporated into the top 15 cm of the tailing by discing. Just prior to seeding, 4–12–10 fertilizer was applied at the rate of 560 kg ha^{-1} and harrowed in.

The solution of the problem of the selection of species to be used in the grass-seed mixture was based on the following three criteria:

1. Only species known to be adapted to the climatic conditions of the Sudbury area, based on the actual observation of their presence (e.g. in undisturbed roadsides, in old, unused bush haulage trails, in long abandoned pastures) were to be included.
2. All of the species to be used had shown that they were capable of growing in locations where the soil was low in nutrients, had a less than neutral pH and was not always well drained.
3. The seed was commercially available.

In addition to the Canada Blue grass (Poa compressa L.), the following were selected for inclusion in the mixture: Timothy (Phleum pratense L.), Crested Wheat (Agropyron crestatum Gaert.), Red Top (Agrostis gigantea Roth.), Park Kentucky Blue (Poa pratensis L. var. Park). The mixture was seeded at the rate of 56 kg ha^{-1} using Fall Rye (Secale cereale L.) seeded at the rate of 60 kg ha^{-1} as a companion crop. Brome grass (Bromus inermis Leyss.) was then hand seeded at the rate of 11 kg ha^{-1} and brushed into the surface using a light drag of tree tops.

This crop survived the winter in good shape. Encouraged by this result, an additional 3 ha. were seeded in the spring of 1958, using oats (Avena sativa L.) as the companion crop. The same grass-seed mixture was used with the addition of two legumes. Approximately one-half the area received the seed mixture containing Yellow Blossom Sweet Clover (Melilotus officinalis Desr.) and the other half was seeded with a mixture containing alfalfa (Medicago sativa L.). The sweet clover grew well in 1958 and 1959 but did not regenerate itself after that. On the other hand, the alfalfa growth in the first 2 years was not too impressive, but after a slow start, it became well established and persists to this day. As a matter of interest, the surviving plants are used as a gene bank by a plant breeder who is attempting to increase the hardiness in the alfalfa varieties which he is developing.

The success of these seedings led to progressively larger seedings in 1959 and 1960, with the result that the whole CD area, containing 170 ha was seeded by the end of the latter year. Fortunately, the time required for the preparation of the area for seeding, delayed the seeding until late July. The time delay was caused by the size of the area which was to be seeded and the time required to prepare it with the size and power of the few tillage machines which were available. Later, when more equipment had been acquired, spring seedings were attempted. These failed more often than not due to the long hot days in early July when the companion crop had not reached sufficient height to shade the grass seedlings. This condition led to high rates of evaporation of moisture from the soil and transpiration from the plants at a time when there usually was insufficient readily available soil moisture near the surface of the tailings to meet the demand.

Irrigation, both planned and on an emergency basis, besides proving too costly, did not ensure uniform growth when seedings were carried out in the spring.

5 Research Investigations Over the Years

Since the program's inception, research into methods to improve the procedures used to establish a vegetative cover on the tailings by the most economical method have continued and are continuing. When and where possible, a preliminary assessment of a new procedure is tested in the greenhouse. As in other fields of endeavour, successful greenhouse or laboratory tests do not always achieve or produce similar results under field conditions.

In greenhouse tests, the addition of manganese in the fertilizer increased the vigour of young grass seedlings, yet when included at the same rate as a fertilizer additive in the field, no significant difference, either visual or measurable, was observed.

Different methods, rates of application and timing of limestone applications have been tried over the years. In the greenhouse tests, an application of limestone at the rate of 11 t ha^{-1} on the fresher unoxidized tailings immediately after seeding appeared to be beneficial in the early growth phase. This was probably due to the increased neutralization of the acid salts in the immediate vicinity of the seedling. In the field, again no significant difference was discerned.

Greenhouse growth tests are regularly made on rates of limestone application for new areas as they reach their final elevation and are to be seeded during the current growing season.

The use of waste by-products of the cement, paper and chemical industries, as well as sewage sludge from municipal disposal plants, to assist in the establishment of vegetation has been investigated at different times over the years. They were generally beneficial, as far as growth was concerned, but economically impractical due to the added transportation and labour costs. Physically, these additional costs were attributable to their bulk, the distance from the source and the material-specific problems encountered in handling and spreading them in the field. Their low nutrient value per unit, even though the material was supplied at no cost at the source, was another factor in their rejection.

The experience gained from the successes and failures since the start of the program some 30 years ago has resulted in the formulation of the following guidelines:

1. The seeding should be established on that portion of the area which is closest to the prevailing wind during the growing season to minimize the covering or injuring of the young plants by drifting tailings. Subsequent seeding should progressively advance across the remaining tailings from the original seeding.
2. Agricultural limestone, as required, should be applied prior to seeding. The major portion of the total amount should be applied at least 6 weeks prior to seeding. This permits sufficient time for the reaction to raise the tailings pH to approximately 4.5–5.0. The balance should be applied immediately prior to the pre-seeding cultivation.
3. In the Sudbury area, late summer is the best time to seed grasses. After July 21 the rate of success of seed germination and seedling establishment is enhanced due to more suitable temperatures and the increased availability of moisture from precipitation.
4. Although the late summer is the optimum time for seeding grasses, the short period which is left of the growing season is insufficient for legume seedlings to establish a sufficiently deep and strong root system to withstand the heaving effects of the repeated freezing and thawing of the tailings surface the following spring.
5. The use of a companion crop, to reduce the surface winds and to provide some shade, is essential.
6. Nitrogenous fertilizer should be applied, at low but sufficient rates, several times as required during the establishment period to ensure maximum uptake.
7. The surfacing of slopes with south and southwesterly exposure with 15 cm of clay to ensure an adequate supply of moisture for growth is worthwhile.
8. The use of a mulch on the slopes, preferably one containing straw, to provide shade for the seedlings and to reduce evaporation of soil moisture during the critical period of seedling establishment has proven to be essential.
9. When clay is used to surface a tailings dam outside slope, the whole slope should be covered from top to bottom. The addition of a clay topdressing on a tailings slope will physically reduce the porosity of the slope's surface. In all cases, it is essential to maintain the structural integrity of the dam and no treatment which will change a dam's structural strength should be employed in revegetation. If required, adequate diversion ditches for safe drainage of the slopes should be included as part of the surfacing program.

6 Inco Limited's Current Tailings Revegetation Program

The current program used by Inco Limited to establish grass on tailings as the first step in the revegetation program is based on the experience gained in the past. These have changed and will change in the future as the basis of knowledge gained from experimental tests is incorporated into the procedures.

The pH of tailings varies with the amount of oxidation which has taken place since they were deposited. Therefore, early in May, pH tests are made at 60-m intervals of the area to be seeded. Based on the information thus obtained, agricultural limestone is spread at rates from 10 to 35 t ha^{-1} and disced into the tailings surface.

Recent experience indicates that the best time to start seeding is immediately following the first heavy rain after July 20. Just prior to seeding time, an additional 4 to 5 t ha^{-1} of agricultural limestone is spread and disced into the tailings surface. Then 450 kg ha^{-1} of 5–20–20 fertilizer are broadcast and then disked or harrowed, as conditions dictate, into the top 5 to 8 cm of the surface. A conventional farm seed drill follows seeding of 94 kg ha^{-1} of grass seed mixture (see below) and an additional 390 kg ha^{-1} of 5–20–20 fertilizer along with 60 kg ha^{-1} of Fall Rye (Secale cereale L.). This is followed by a Brillion seeder which compacts the seed bed as it places an additional 22 kg ha^{-1} of grass seed in the ground.

After the grass germinates, light topdressings of nitrogen are applied, as required, during the balance of the season and the following spring. Limestone and fertilizer are applied in subsequent seasons based on visual assessment of the growth or on soil tests.

Legumes are introduced into the resultant sward during the first or second spring after the original seeding, using a powertill seeder. Birdsfoot trefoil (Lotus corniculatus L.) is the current legume being used, although some experimental work is being done with alsike (Trifolium hybridum L.). Legume seed is always innoculated before seeding.

Inco Limited Tailings Grass Seed Mixture.
25% Canada Blue Grass (Poa compressa L.)
25% Red Top (Agrostis gigantea Roth.)
15% Timothy (Phleum pratense L.)
15% Kentucky Blue Grass var. Park (Poa pratensis L.)
10% Tall Fescue (Festuca arundinacea Schreb.)
10% Creeping Red Fescue (Festuca rubra L.)

The percentage indicates the proportion by weight of each species in the seed mixture.

7 The Development of Tree Cover

The need to establish a vegetative cover on the tailings surface, as the forerunner to the development of a total ecosystem was proven by the voluntary invasion of the locally indigenous deciduous trees into the grassed areas in the early to mid-1960s. The birch trees (Betula papyrifera Marsh.) germinated in the small vacant spaces between the clumps of grass, where and when microclimatic conditions along with suitable soil moisture and plant nutrients combined to provide the required environment for seedling development. Trembling aspen (Populus tremuloides Michx.) and willow (Salix spp.) constituted less than 10% of the total voluntary tree seedlings in the early stages of this transition forest cover phase of a developing climax ecosystem for the site. It is of interest to note that the trembling aspen, due to its vegetative method of propagation, spreads over an annually increasing area and has become the numerically dominant species in localized sites. A few oaks (Quercus rubra L.) have appeared and were probably introduced by birds or rodents transporting the acorns from adjacent hills. Test plots of hybrid poplar have not produced as much growth as anticipated. However, plots of black locust (Robinia pseudo-acacia L.) have shown outstanding growth. Their ability, as legumes, to fix atmospheric nitrogen is reflected in their vigorous growth in a substrate which is notably nitrogen deficient, but also benefits other tree species growing adjacent to them.

The successful invasion of the deciduous trees into the tailings area, prompted an investigation of the adaptability of coniferous species to this environment. The climax vegetation of the area had been principally red and white pine and they had been lumbered off in the 1880s. Replicated plots to assess the adaptability of five coniferous species: jack pine (Pinus banksiana Lamb.), red pine (Pinus resinosa Ait.), scots pine (Pinus sylvestris L.), white spruce (Picea glauca Voss.) and black spruce (Picea mariana Mill.) were established in 1972. The jack pine (Pinus banksiana Lamb.), as expected due to the nature of the substrate, generally proved to be the most adaptable. Seven years after planting, individual trees had reached heights of 3–3.6 m and had produced seed cones. Seed from these cones have been sown and the seedlings planted on the grassed tailings and other areas being reclaimed with reforestation.

The success experienced to date with the coniferous species, may indicate that the planting of these more valuable economic species may lead to the reclaimed area becoming a managed forest as part of its end use.

8 Development of the Wildlife Management Area Program

In addition to the invasion of various species of flora, a more rapid invasion of many different species of fauna occurred. With the establishment of vegetation, herbivorous insects, such as grasshoppers, spittle bugs, etc., colonized the grassed areas. These rapidly attracted the meadow birds and gulls which prey on these insects. The grass and grain seed attracted field mice and voles which in turn attracted owls and the many species of raptors (hawks) which have been observed.

Waterfowl and shorebirds, in increasing numbers and species, began to use the area as a stopover during their spring and fall migrations. Some stayed over during the summer to raise their broods of young on the ponds.

During the period from 1973 to 1981, 92 different avian species were identified during their visits. These included 19 species of waterfowl, 26 species of shorebirds, 3 species of gulls, with the remainder being various meadow- and wood-habiting species.

The small mammals which had taken up residence in the tailings attracted fox and coyotes to the area.

In 1974, after consulting with local Rod and Gun Clubs, a decision was made to develop the tailings into a Wildlife Management Area. Negotiations commenced with the Ontario Ministry of Natural Resources to have the area so designated. A wildlife biologist was retained by the Company to act as a consultant to prepare a plan of development. The plan was completed in 1976. Since that time, different research programs have been undertaken to determine the methods to be employed and the path to be followed in this development. Research to ascertain whether or not the trace elements common to the tailings accumulate in the tissues of the plants, insects or small mammals which inhabit the area, has been carried out. The results of these tests carried out over a period of several years indicated that these elements would not pose a problem in the food chain (Rose 1981; Cloutier 1984).

In June 1985, in cooperation with the Ontario Ministry of Natural Resources and the Copper Cliff Rod and Gun Club, 60 fledgling Canada Geese (Branta canadensis L.) were released in the CD area of the tailings. They stayed in the general vicinity of the whole tailings area all summer and left during the fall migration with other geese. All of these geese had been banded prior to their release in order that accurate records of those who return can be maintained.

9 Other Forms of Plant Life Present

Although they are not as readily noticeable and are often overlooked, the part that moss plays as a tailings cover is nonetheless important in places where the microenvironment inhibits the growth of the higher plants. These sites are usually on the beaches adjacent to the ponds or in small localized spots within the tailings (McLaughlin, 1982). Their role as an oxygen barrier and a precipitation interceptor is significant.

The beaches, which vary in size and exposure with the pond water levels, also support, under certain conditions, bullrushes (Scirpus lacustris) and cattails (Typha latifolia). Introduced into the CD pond in 1974 and 1975 (Wilson, unpublished 1976), they have spread extensively and have provided a haven for waterfowl plus improving the aesthetic appearance of the pond.

In 1983, a research program to investigate the introduction of aquatic plants to improve the quality of the water in the ponds was initiated by Inco Limited. The preliminary findings are promising, although more time will be required before a final assessment can be made (Kalin, unpublished 1985).

10 Research Related to Soil and Plant Development

The CD area, which was the first to be revegetated, has been the subject of many studies related to soil and plant, as well as ecosystem, development. Watson (unpublished 1970) commented on the high surface and reflected temperature of artificial waste areas. This reflected heat from the tailings' surface necessitates the use of fast-growing companion crops and proves the necessity of establishing some vegetative covering on the surface to permit natural forest regeneration. Labine

(1971), in his study of monoliths of the CD area "soils", found the development of an organic horizon (A°) of 2–3 cm and the beginning of a podzolic profile. It was found that drainage at different slope levels quantitatively affected soil profile development and the formation of iron pan at different depths (Swanson, unpublished 1977). This iron pan layer generally limited root penetration to cracks in the formation. However, single hair roots penetrated the iron pan from the thick root masses which developed in some locations adjacent to the upper surface of the iron pan layer.

The physical ability of soil to retain moisture and nutrients is essentially a function of particle size. Tailings, which are generally deficient in clay-sized particles, behave like a sandy loam and may be prone to moisture deficiency (Dimma 1981). Lacking colloidal moisture absorption, it would appear that the water which is retained in tailings has resulted from capillary action, which is most pronounced in the silt-sized particles (Pitty 1979).

Due to the fixation of phosphates by the high level of iron oxides found in the tailings, the amount of phosphorus which is actually available for plant growth may be significantly lower than analytical results indicate (Dimma 1981).

The information gained from this research and other investigations, which are mentioned in this chapter, was used in developing new or modifying existing procedures.

11 Tailings Revegetation and Effluent Water Quality

At the start of the program some 30 years ago, the main objective was to stabilize the tailings surface to prevent the dust storms which originated from there under certain weather conditions. This program was extended with the use of chemical sealants on the temporarily unused portions of the active tailings area in recent years.

At this time, as close-out scenarios and end-use plans are developed for the tailings area, another factor, the quality of drainage and seepage water from these areas, must be considered. The establishment of the vegetation on the tailings impacted on this factor in three ways:

1. By interception of precipitation;
2. By transpiration;
3. By the formation of an oxygen-consuming barrier.

The percentage of the rainfall intercepted will vary with the density of the vegetative cover and the period, intensity and total precipitation of the individual rainstorm. Generally, dews are almost 100% intercepted, while the percentage of water intercepted in a heavy driving rainstorm is minimal. The precipitation intercepted evaporates directly to the atmosphere and does not reach the ground surface to become either seepage or runoff. Vegetation stems act as air passages in snow accumulations which increase the area exposed to the evaporation process, particularly on the warm sunny days of late winter, and thus helps reduce seepage and runoff water.

Transpiration of water by vegetation has the greatest impact on reducing the amount of water available for seepage. Two of the many factors which affect the rate of transpiration are plant density and species variation. These have a bearing on the species selection and the density and combination in which they are planted for achieving the climax vegetative cover. On the Copper cliff tailings this will be a mixture of coniferous and deciduous trees.

The development of a thick root structure, along with the accumulation of decaying vegetation on the surface, has set up an oxygen-intercept layer, which reduces the amount of oxygen penetrating into the sulphide tailings and thus reduces the rate of acid formation. This, in turn, reduces the amount and rate at which residual trace elements become available for inclusion in the seepage water.

12 Cost of Tailings Revegetation

The cost per hectare to revegetate tailings varies with the area to be revegetated and the treatment required. This makes the detailing of precise costs in a concise form impractical, if not impossible.

The following cost evaluation of tailings disposal and revegetation as part of the integrated operation at Inco Limited is shown in Table 1 (Browne, 1984).

Table 1. Cost evaluation of tailings disposal in comparison with plant operations (percent of total cost)

Mining, milling, smelting and refining	100	–	–	–
Milling, smelting and refining	50	100	–	–
Milling (inc. tailings)	10	20	100	
Tailings disposal	1	2	10	100
Reclamation and dust control	0.15	0.3	1.5	15

13 Present and Future Concerns

The need for additional research to solve the problems of phosphate availability and the recycling of plant nutrients tied up in plant debris by encouraging microflora and microfauna activity are but two of the present problems needing attention.

Process changes of the future are always of concern as the need to reduce sulphur dioxide emissions by the smelter means that increased amounts of the sulphur-bearing portion of the ore will be rejected in the concentration process and transported to the tailings area. This will necessitate the development of new procedures to revegetate higher sulphide tailings.

Conclusions

The revegetation of tailings and the development of the Wildlife Management Area will continue at the Copper Cliff Tailings Area of Inco Limited.

Although achieved by different methods, as the situation required, the principles to be followed, based on the experience to date, are similar to those found in other areas (Bradshaw, 1978). They are as follows:

1. Establishment of initial plant communities using available species that are tolerant of drought, low soil pH, poor soil texture, the lack of organic materials and nutrients, and other factors characteristic of metal-extracted tailings.
2. Modification of the local plant microclimate to benefit plant establishment.
3. Establishment of soil invertebrate and microbial communities to decompose naturally accumulating organic matter and to assist in the building of a soil.
4. Establishment of essential nutrient cycles.
5. Establishment of a vegetative habitat suitable for wildlife colonization.
6. By manipulating species competition, the establishment of climax plant communities for the area.

In all cases, the need to follow the old adage "adapt not adopt" the methods and practices employed by others in this field, will, due to the heterogeneous nature of tailings from site to site, assist in the solving of a particular problem in revegetating sulphide tailings as has been demonstrated in the success achieved in the revegetating of the Copper Cliff Tailings Area.

References

Bradshaw AD, Humphries RN, Johnson MS, Roberts RD (1978) The restoration of vegetation on derelict land produced by industrial activity. In: Holdgate MW, Woodman MJ (eds) The Breakdown and restoration of ecosystems. Plenum, New York London, pp 249–278

Braun EL (1950) Deciduous forests of eastern North America. Hafner, New York

Browne C (1984) Stabilization of mine tailings at INCO's Sudbury operations with the use of vegetation. Tech Rep, Dep Metall Eng, Queen's Univ, Kingston, Can 25 pp

Chapman LJ, Thomas MK (1968) The climate of Northern Ontario. Climatol Stud, Dep Transport Meteorol Branch, Toronto, Can

Cloutier NR (1984) Metals (Cu, Ni, Co, Fe, Zn, Pb,) and Ra-226 levels in meadow voles and vegetation from nickel and uranium mine tailings in Northern Ontario. M Sci Thesis, Biol Dep, Laurentian Univ, Sudbury

de Lestard JPG (1967) A history of the Sudbury Forest District. History series. Sudbury District Rep 21

Dimma DE (1981) The pedological nature of mine tailings near Sudbury, Ontario. M Sci thesis, Queen's Univ, Kingston, Ontario, 184 pp

Heale EL (1980) Effects of nickel and copper on several woody species. M Sci thesis, Univ Guelph, Ontario, 220 pp

Labine CL (1971) The influence of certain seeded grasses on the evolution of mine tailings. B Sci thesis, Biol Dep, Laurentian Univ, Sudbury, Ontario

McLaughlin BE, Crowder AA, Rutherford GK, van Loon GW (1982) Site factors affecting semi-natural herbaceous vegetation on tailings at copper cliff. Reclamation and revegetation research, vol 2

Montreal Engineering Company (1975) Report on sulphide mine tailings in Noranda, Timmins and Sudbury Areas

Murray DR (1985) The challenge in reclamation. Proc 10th Annu Meet Canadian land reclamation association, Quebec, pp 1–8

Phinney KD (1984) Sulphide Tailings Management Study. Monenco Ltd.

Pitty AF (1979) Geography and soil properties. Methuen, London, 287 pp

Rose G (1981) Effects of smelter pollution on temporal changes in feather and body tissue metal content of ruffed grouse near Sudbury, Ontario. M Sci thesis, Biol Dep, Laurentian Univ, Sudbury, Ontario

Rowe JS (1972) Forest regions of Canada. Can For Serv Rep 1300, Dep Fish Environ Ottawa, Can, 172 pp

Sheehan PJ, Miller DR, Butler GC, Bourdeau Ph. (eds) (1984) Effect of pollutants at the ecosystem level, SCOPE. John Willey & Sons, Chichester New York Brisbane Toronto

Case 7.8, Rehabilitation of mine tailings – A case of complete ecosystem reconstruction and revegetation of industrially stressed lands in the Sudbury Area, Ontario, Canada. Peters, pp 403–421

Development of a Revegetation Programme for Copper and Sulphide-Bearing Mine Wastes in the Humid Tropics

I.M. ARCHER[1], N.A. MARSHMAN[1], and W. SALOMONS[2]

1 Introduction

Revegetation of mine wastes is an ecologically desirable component of post-mine rehabilitation. It is also often a legal requirement. Amongst other things it is useful in stabilizing the waste against both water and wind erosion, in increasing the biological productivity of the spoiled area and in reducing water pollution.

Worldwide mine waste revegetation techniques are very varied (Bradshaw and Chadwick 1980). Some of the reasons for this are differences in revegetation objectives, climate and disposal method, the volume, surface area, chemical and physical properties of the wastes and also in the level of understanding of the revegetation process by the revegetation practitioner.

The development of a successful and cost-effective revegetation programme has a number of stages. First, it is necessary to have an objective for the revegetation programme. Site-specific research is then advisable to define potential limitations to successful plant establishment and growth. This needs to be followed by trials to demonstrate practical methods to overcome these limitations and finally revegetated areas need to be monitored so that the techniques used are assessed for their suitability in achieving the original objective.

This chapter describes the development of a cost-effective method to revegetate areas covered with mine wastes in the humid tropics. The physical, chemical and biological nature of the waste material has been studied and the programme has progressed from the initial setting of objects through research, trial of methodology, implementation and appraisal stages. Particular emphasis is placed here on the chemical aspects of the study programme.

2 Background and Objectives

Little has been published about revegetation of mine wastes in the humid tropics. In this region moisture stress and temperature do not limit plant growth but nutrient deficiencies due to high leaching losses occur (Uehara and Gillman 1981). As a guide to mine waste revegetation useful information is available on acid

[1]Bougainville Copper Limited Panguna, NSP Papua, New Guinea
[2]Delft Hydraulics Laboratory, Haren, The Netherlands

sulphate soils and leached acid mineral tropical soils (Kittrick et al. 1982; Greenland 1981). Care is needed, however, in the use of this information as significant differences in soil properties occur between mine wastes and these soils.

2.1 The Mine and the Environment

Bougainville Copper Limited (BCL) operates a large open-cut porphyry copper mine at Panguna (Lat 6° 19′S, Long 155° 29′E), Bougainville Island, Papua New Guinea. The mine is situated in the headwaters of the Kawerong/Jaba river system at an altitude of 670 m. Panguna is surrounded by rugged, steep-sided ridges and valleys. The natural vegetation ranges from high montane rain forest at the mine site to lowland swamp in the lower Jaba river system.

The climate on Bougainville is typical of that of the humid tropics. Mean daily temperatures are high (23°C at Panguna and 27°C at the coast) with no seasonal variation. The mean annual rainfall at Panguna is 4,400 mm, with the wettest month being March (550 mm) and the driest being June (220 mm). There is high net annual precipitation as the mean annual evaporation rate (from a class A pan) is 1200–1600 mm, depending on altitude.

The mine commenced production in 1972 and up until December 1985 518 million tonnes (Mt) of ore had been treated in the concentrator. In 1985 50 Mt of ore were processed. The tailings (98.5% of the ore) are piped from the concentrator at Panguna 4 km away and discharged into the Kawerong River. About 40% of the material discharged into the Kawerong/Jaba river system has been deposited along the 30 km to the coast (Fig. 1). The remainder has formed a marine delta in Empress Augusta Bay. At present 18 km² of land is covered with tailings and the delta is a further 7 km² in area.

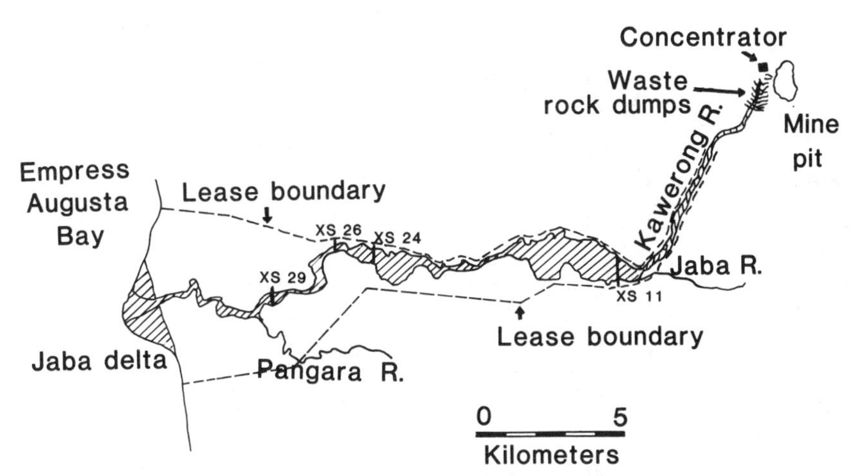

Fig. 1. General layout of the Panguna mine site and tailings disposal area. The *hatched area* represents the regions covered with tailings. The river system is divided into a number of reaches each separated by cross-sections (XS^s), four are shown

In addition to the ore, a further 350 Mt of waste rock (rock with sub-economic mineralization of recoverable Cu and Au) has been removed from the mine and dumped into the sides of the Kawerong Valley immediately adjacent to the pit. The surface area of the waste dumps is 2.5 km^2.

Over the remainder of the mine life there will be an increase in the surface area of all waste deposits requiring revegetation. To date most of the land used for waste disposal is still being used for that purpose or is being used for haul roads, building sites or other mine-related purposes. Consequently, the majority of the revegetation of spoiled areas remains to be carried out. It is Company practice, however, that revegetation be undertaken as soon as practicable.

2.2 The Mine Wastes

Four different types of material are recognized as requiring revegetation (Archer and Marshman 1984). They are the waste rock dumps, the tailings in the river system which will become freely draining and have a water table below the plant root zone, water-saturated tailings as exist on the Jaba Delta and finally disturbed natural sites that have resulted mainly from construction of final batters around the pit perimeter and general access roads. A description of the revegetation of this latter material is not given in this chapter.

The main rock types in the orebody are andesite, biotite granodiorite, biotite diorite and leucrocratic quartz diorite. Most of the material is unweathered and is composed of primary minerals. The main mineral types present are feldspars, quartz, biotite, chlorite and micas with some clay minerals, magnetite and iron pyrite. No significant amounts of carbonate minerals are present in the orebody. Chemically, the composition of the ore (Table 1) is similar to much other unweathered earth crust material (Lindsay 1979), but different to highly weathered mineral soils. Concentrations of potentially toxic heavy metals (Hg, As, Pb and Cd) are low and similar to other parts of the lithosphere. Also, there is a wide range of elements that are essential to plant growth.

Table 1. Particle size analysis of typical Bougainville mine wastes

Percent Passing				
Sieve size (μm)	Waste rock[a]	Tailings XS29	Tailings XS26	Tailings XS11
600	68	97	97	95
425	44	97	95	87
300	36	96	91	73
212	29	96	78	53
150	21	92	49	31
106	14	73	19	16
52	3	27	2	3

[a]Waste rock sizing for less than 2 mm fraction.

Waste rock is merely fractured rock. The average particle size of the material is large (< 5 cm), but the majority of the dumps contain up to 20% of minus 2-mm particles with a proportion of this material in the sub-sand size range (Table 2). Chemically, the waste is similar to that of ore except for lower concentrations of Cu, Ag and Au and higher concentrations of S.

Tailings are discharged as a slurry with 18% of the material coarser than 0.3 mm. In the river system there is sorting and the deposits contain a low proportion of silts and clays (Table 2). The median particle size of the deposits decreases along the river system from 0.4 mm to 0.1 mm. On the delta the deposits are finer with the median particle size ranging from 0.05 to 0.12 mm. Chemically, the tailings contain lower concentrations of Au, Cu, Ag and S than the ore but they contain reagents used for the recovery of Cu, Au and Ag. The reagents are lime ($Ca(OH)_2$), xanthates and frothers. The organic reagents do not affect revegetation, whereas the lime raises pH and available Ca.

Table 2. Total concentration of elements of ore processed in 1984

Element	(%)	Element	(ppm)
Al_2O_3	15.2	Ag	1.3
BaO	0.029	As	4
CaO	3.49	Au	0.48
Cu	0.42	Bi	20
F	0.052	Cd	< 0.5
Fe	6.1	Cl	610
K_2O	2.26	Co	20
MgO	2.7	Cr_2O_3	190
Mn	0.034	Hg	0.02
Mo	0.003	Pb	10
Na_2O	4.12	Sb	< 10
P_2O_5	0.21	Se	20
SiO_2	61.6	Te	10
S	0.67	V	240
SrO	0.042	Zn	250
TiO_2	0.58		

2.3 Revegetation Objectives

Clear objectives for revegetation were set at the commencement of the study. In setting the objective recognition was given to (1) the agreements between the Company and the Government; (2) traditional uses made of land by the village-based population on Bougainville; (3) cost effectiveness; and (4) the capability of the wastes to support plant growth.

The main objectives of the programme were to achieve rapid plant colonization with a self-sustaining plant community requiring no post-establishment maintenance. Where possible, species native to the island and used by the community were to be included. High importance was also placed on selection of plant species suited for growth on the wastes. Amelioration of toxicity or deficiency conditions was recognized as being potentially necessary but only after species selection trials had indicated the need.

The main advantages of these objectives are rapid development of stable waste deposits, potential for early return to the traditional landowners of land in a maintenance-free and ecologically stable state, improvement in soil fertility, potential for limited harvesting for village construction and firewood purposes and an improvement in the visual environment.

3 Site Trials

During the study, a range of pot and field trials were carried out. This was done to define limitations to plant growth and then to test practical methods to overcome these limitations. Successful development of a maintenance-free revegetation scheme depended upon having a detailed understanding of the physical, chemical and biological properties of the waste, of the way that these properties changed through time and of the way in which the wastes, plants and applied fertilizer and liming agents interact. This section describes these trials with each of the main factors that affect plant growth being discussed in separate sub-sections.

3.1 Sampling Techniques and Chemical Methods

All aqueous samples were collected, preserved and analyzed according to standard methods (APHA 1980).

There was rigorous replication of soils collected in the field to obtain representative samples that excluded surface organic matter, residual lime and fertilizer. Soils were sieved to less than 2 mm. Unless otherwise specified all methods were taken from or based upon American Society of Agronomy (1982).

Total N was determined by a Kjeldahl digestion method; total P was determined colorimetrically after digestion with $HNO_3/HCl/HClO_4/HF$; available P was extracted using the Olsen method; exchangeable K, Ca, Mg, Na, Mn and cation exchange capacity (CEC) were determined with an in-house method by leaching the soil with neutral ammonium acetate and then with NaCl, the NH_4+ was determined as a measure of CEC and the cations by atomic absorption spectrophotometry (AAS); exchangeable Al and H and total exchangeable acidity were determined by leaching with KCl, Al was determined by AAS, acidity by titration and H by the difference between the two; available Cu, Zn and Fe were extracted with a sodium citrate – EDTA solution at pH 8.5; total S was determined by the patented method of Leco Corporation using a Leco induction furnace and automatic S titrator; SO_4/S by extraction with boiling Na_2CO_3 and gravimetric determination of $BaSO_4$ (in-house modification), sulphide sulphur was calculated as the difference between total sulphur and SO_4/S; organic carbon by Walkey Black wet oxidation; available Mo and B were extracted with Tamm's solution and with hot water respectively, then they were determined by DCP; pH and conductivity were determined using a 1:5 soil:distilled water suspension.

Foliar samples were collected using standard techniques depending upon the species. They were washed six times in distilled water before drying for 48 h at 50°C in a forced air oven. Chemical analysis of plant material was based upon standard laboratory methods. Nitrogen was determined by Kjeldahl; Ca, K, Mg, Cu, Mn, Zn, Fe, Mo and P were determined after digestion with $HNO_3/HClO_4$; samples for S were digested in a saturated $MgNO_3$ solution and determined for SO_4/S. B was determined, after ashing at 500°C, colorimetrically.

In-house reference soils were used for quality control purposes. At least two reference samples were included in each batch of samples for analyses. International and in-house reference samples were used in every batch of foliar samples.

3.2 Physical Properties

Both waste rock and tailings are largely devoid of silt- and clay-sized particles and this affects a range of physical properties that are important in soils for plant establishment, root development and nutrient and water supply.

The bulk density of the mine waste is variable and depends upon the waste type, the degree of trafficking and location. Bulk densities of tailings range from 1.1–1.7 t m^{-3} and are generally suitable for plant growth and root development. In some cases, especially trafficked waste rock areas, ripping prior to revegetation is required.

Tailings are highly erosive, particularly under intensive rainfall. Soil conservation measures, i.e. grading, contouring and drainage works, are required before revegetation can occur. The water-holding capacity of the material is low. Water content vs pf (soil suction) curves have been prepared for a range of deposits. For tailing deposits from the river there is less than 20% water-filled pore space at 0.2 bar suction. Despite the low water-holding capacity moisture stress is not a problem for established plants on all except coarse waste rock. Moisture stress has been observed during seed germination and plant establishment. In cases of poor seedling establishment immediate re-sowing has been successful.

The hydraulic conductivity of the material is high. Vertical steady state infiltration rates of 17–83 mm h^{-1} have been measured in tailings immediately downstream to the Jaba/Kawerong confluence (Meynink pers. commun.). The infiltration rate of ripped waste rock deposits is also high.

The CEC of all waste material is low (Table 3). This is due to the lack of clay minerals and organic matter. This, coupled with the high infiltration rates, suggests high leaching of water-soluble compounds from the material.

High leaching rates were confirmed in a trial where tailings from the river system were placed in 200-litre drums with water collection ports at the bottom of the drum. The CEC of the tailings was 1.7 mEq.% and the pH 9.1. After a period of equilibration, a commercial N:P:K (14:14:14) fertilizer at 750 kg ha^{-1} and soluble, 20% B, as $Na_2B_4O_7{\cdot}4H_2O$ at 10 kg ha^{-1} were added to the surface and the leachate collected after each rainfall event over a 64-day period.

At the end of the trial 55% of the N, 65% of the K, 10% of the B but no P was recovered after allowance was made for the composition of both the leachate prior to fertilizer addition and the rainwater. The trial also demonstrated that nitrification was quick to occur as 62, 37 and 1% of the recovered N was present as NO_3^-, NO_2^- and NH_4^+ respectively.

Particular care is therefore required in the selection and application of fertilizer. Insoluble, slow-release compounds should perform better than water-soluble compounds. In the long term, however, an increase in soil organic matter is required to aid nutrient recycling, increase CEC and minimize leaching losses.

Table 3. Chemical properties of mine waste used for revegetation studies

Waste material			
	Waste rock[a]	Tailings XS29	Tailings XS26
N total %	ND[b]	< 0.003	ND
Organic carbon %	0.13	0.04	0.04
P Total ppm	1770	ND	670
P available ppm	10.3	5.6	0.6
K exch. mEq. %	0.14	0.07	0.07
Ca exch. mEq. %	1.4	0.45	1.5
Mg exch. mEq. %	0.33	0.11	0.16
Na exch. mEq. %	ND	< 0.02	0.01
H exch. mEq. %	1.7	ND	< 0.05
Al exch. mEq. ppm	261	119	< 1
Mn avail. ppm	6.6	4.9	1.2
Cu avail. ppm	263	98	71
Fe avail. ppm	265	ND	27
B avail. ppm	ND	< 0.2	ND
Mo avail. ppm	ND	0.7	1.2
S total %	0.85	0.23	0.20
SO_4/S total ppm	193	117	50
CEC mEq. %	6.5	3.0	1.5
Conductivity MS	ND	0.06	ND
pH	4.6	4.5	9.2

[a]The waste rock and tailings (XS26) were used for the rock phosphate trial (Sect. 4) and the tailings (XS29) for the nutrient addition pot trial (Sect. 3.3)
[b]ND; not determined.

3.3 Soil Acidity and Sulphide Oxidation

Mine wastes contain sulphide minerals which oxidize and produce acid in the presence of O_2 and water. Factors affecting sulphide oxidation on Bougainville as well as the effect of soil acidity on plant growth are discussed in this section.

The final pH of mine waste varies with a number of factors. The most important is the presence of O_2. In waterlogged material on the Jaba Delta, no acidification has been observed in 11-year-old tailings deposits. Associated with these waterlogged deposits is a healthy plant which has added organic matter to the sediment. The organic matter contributes to low sediment pore water dissolved oxygen (< 0.1 ppm) and redox potential (Eh).

Both Eh and pH have been measured in situ in freely drained acid tailings and in water-saturated neutral pH tailings. In the freely drained sediment Eh was high (550 to 800 mV), but was low in the water-saturated tailings (100 to -50 mV). Soil solution pH was 3.0–5.0 for the acid material and 6.3–7.0 for the two materials. This information has not been transferred to mineral stability diagrams because of the recognized difficulty of measuring Eh in the field (Stumm and Morgan 1980).

These data, together with the low measured dissolved oxygen concentrations in pore water, suggest that much of the tailings on the delta will not oxidize and hence revegetation can be carried out on neutral pH material. It is also noted that hydraulic conductivity, particle size distribution and sediment drainage pattern are all important physical factors that contribute to a high water table.

In freely drained aerated sediment sulphide content is important in affecting final sediment pH. This has been quantitatively observed in 2×1×1 m high lysimeters that were filled with tailings from XS11 and XS24 (Fig. 1) with sulphide sulphur content of 0.64 and 0.17% respectively. The tailings were maintained at field capacity except when sampled, every eighth week. At sampling, water was applied to the top of the lysimeters, allowed to infiltrate and then stand for 3 h before samples were taken. The water was then allowed to drain from the bottom.

For the tailings from XS11 the pH of the leachate remained neutral for the first 2 years and then decreased to 3.4 after 5 years (Fig. 2). The rate of fall decreased over time, the acid from oxidation being neutralized by gangue minerals with release of metal cations, some of which are important in affecting plant growth. The average concentrations (ppm) of the major cations leached in the fourth year were: Al(58); Cu(270); Fe(6); Mn(22); Ca(270); Mg,(73) and K(111).

Soil pH did not fall to the same level as leachate pH and in the fourth year was 4.0. This difference between soil pH and pore-water pH has been repeatedly found.

Sulphide sulphur concentrations in the sediment followed a similar pattern to leachate pH. Sulphide oxidation proceeded relatively slowly for the first 3 years but then increased (Fig. 2) so that after 5 years less than 40% of the original sulphide remained and total sulphide oxidation seems likely to finish within 7 years. These rates are similar to those observed in trials established by Hartley (1979), where sulphide oxidation was complete within 8 years (unpublished data).

In the tailings from XS24 leachate pH fell to 4.6 over 3 years but has subsequently risen to 4.9 (Fig. 2). In the fourth year sediment pH was 5.5. The sulphide content of the tailings fell more or less uniformly over the 5-year period and the composition of the leachate (ppm) was A1(2.5); Cu(17); Fe(0.2); Mn(2); Ca(30); Mg(7) and K(34). The lower sulphide content of these tailings has limited both the drop in pH and the release of cations.

The pH values observed in this trial may not be the same as would have occurred in the field. The actual water input in this trial was 12% of the annual

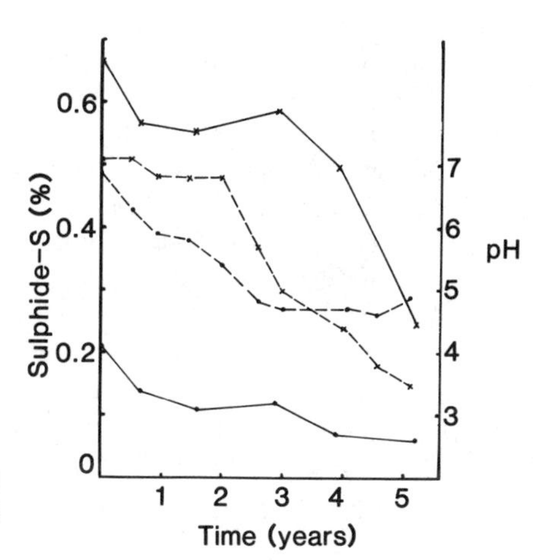

Fig. 2. Change in sulphide sulphur content (—) and leachate pH (- - -) in lysimeters containing tailings from XS11 (x) and XS24 (.)

average Panguna rainfall and acidic cations were not leached from the sediment as rapidly as in field. In the field the pH of mine wastes has seldom fallen below 4.5.

Sulphide oxidation is still continuing in this trial so it has not been possible to determine the final pH of the sediment. In the trial established by Hartley (1979) sediment pH rose after sulphide mineral depletion. Ten years after commencement of that trial the average pH of the sediment was 5.5 (unpublished data). Similar values have been observed in the field.

The differences in soil pH observed in mine waste of different sulphide content have important implications for revegetation. The sulphide content of waste rock is higher than that of tailings, also the sulphide content of the tailings deposits decreases down the river system from 0.5% S at the Kawerong/Jaba confluence to 0.2% S close to the coast. The waste with the higher sulphide content will become more acid and thus potentially more difficult to revegetate.

Plant growth is related to sediment pH. In waste that does not acidify plants rapidly establish themselves from water- and air-dispersed seed and require no maintenance. This currently occurs on the Jaba Delta. Initially, the material is colonized by the grasses *Phragmites karka* and *Paspalum vaginatum*. Through time a range of other species enter. These include *Nypa fruticans Sonneratia caseolaris, Terminalia brassii, Aeschynomene indica* and *Hibiscus tiliaceus.* No gross nutrient deficiencies nor any chemical toxicities have been observed in these water-saturated areas.

Poor plant growth can occur when waste material acidifies. *Saccharum robustum* naturally revegetates freshly deposited stable alkaline tailings deposits on the banks of the river, but dies out as the tailings become acid. Acid-tolerant species, however, once successfully established, are able to grow vigorously in waste at pH values of 4.5 provided that an adequate supply of nutrients exist.

Even with the use of acid-tolerant species, however, an acid substrate leads to poor plant establishment as has been observed both in the field and in a nutrient addition pot trial using the grass *Paspalum dilatatum* and the legume *Desmodium intortum.*

In the pot trial 2 kg of air-dried tailings from XS29 (Table 3) were weighed into 13-cm diameter pots. Lime [2 t ha^{-1}; $Ca(OH)_2$] was applied to half the pots and mixed into the top 8 cm and then the fertilizer treatments were applied. The pots were watered to field capacity gravimetrically and allowed to stand for 1 week prior to transplanting seedlings to the pots. Pots were re-randomized and brought to field capacity three times per week throughout the trial. Seven 2-week-old seedlings were transferred to each pot, the *D. intortum* seedlings having been inoculated with a commercial cowpea *Rhizobium* culture. After a further 2 weeks the plants were thinned to four per pot with the dead or smaller seedlings being removed. Plant tops were harvested 1 cm above the surface 12 weeks after transplanting.

A basal fertilizer of diammonium phosphate (250 kg ha^{-1}) was used for all pots. Each treatment also had one element added as analytical grade reagents. The nutrients and application rates (kg ha^{-1}, in elemental form) were NH_4NO_3(100), PO_4^{3-}(45), K^+(58), Ca^{2+}(143), Mg^{2+}(50), Zn^{2+}(2), MoO_4^{2-}(0.6), $B_4O_7^{2-}$(1), SO_4^{2-}(25) and Fe-EDTA (150 as Fe EDTA).

Desmodium intortum seedlings planted without lime did not grow and died by week 8. The plants suffered from intervenous chlorosis, followed by general chlorosis, necrosis and death. *Paspalum dilatatum* seedlings were generally more hardy but all except those receiving Fe EDTA died by the end of the trial. The seedlings became dull-coloured and the leaf sheaths tinged with purple anthocyanin before becoming necrotic. Apart from the FeEDTA addition, treatment did not appear to affect the rate of plant death or the symptoms exhibited by the plants.

Liming the acid tailings permitted plant establishment and growth. Nutrient addition significantly ($p < 0.01$) affected the tops dry weight of both *D. intortum* and *P. dilatatum*. The results are discussed in later sections.

The rapid death of all plants (excluding the *P. dilatatum* Fe-EDTA treatments) with the no-lime treatments suggests some pH-mediated toxicity rather than nutrient deficiencies. The original pH of the tailings was 4.5 but fell to 3.6. Corresponding to the very low pH were high levels of exchangeable Al (181 ppm) and available Cu (148 ppm). Mean CEC was 3.3 mEq. %, and was 61% saturated with Al. Thus, Al toxicity was a likely cause of plant failure (Kamprath, 1970; 1978). Observed foliar symptoms for the *P. dilatatum* and *D. intortum* were consistent with Al and Cu toxicity respectively (Robson and Reuther, 1981).

To overcome this pH-mediated toxicity, lime is applied prior to seeding of all acid Bougainville mine wastes. Soil pH is raised to 5.5–6.5. Four t ha^{-1} of limestone is added to waste rock and 2 t ha^{-1} to tailings 1 week prior to sowing. The particle size distribution of the lime is important. Lime coarser than agricultural grade limestone reacts slowly and at the above application rates does not raise the pH to the same level.

3.4 Nitrogen

Bougainville wastes are deficient in available N (Table 3) and a growth response to applied N occurred in the nutrient addition pot trial (Sect. 3.3). In the limed tailings there was a significant increase ($p < 0.05$) in tops dry weight for *P. dilatatum*. The dry weights of *P. dilatatum* was 1.6 g pot^{-1} compared to the control 1.0 g pot^{-1}, also foliar N was increased from 0.81% for the control to 1.08%. No other element significantly improved growth of the grass. This section describes the efficiency of utilization by plants of different forms of applied N and the way that plants have been provided with a long-term source of N on Bougainville.

A trial using limed acid tailings in freely draining pots was established to ascertain the utilization by plants of different forms of applied N. Nitrogen was applied (100 kg ha^{-1}N) as either NH_4Cl or $NaNO_3$ at 0 and 10 weeks. The pots were placed outside and two *Eucalyptus deglupta* seedlings sown in each pot. The trees were harvested at 22 weeks.

The trees receiving NH_4/N had a significantly higher dry weight ($p < 0.010$) at harvest (mean weight = 7.41 g) than those receiving NO_3/N (mean = 2.84 g). The trees receiving NH_4/N contained 54 mg of N compared with 3.8 mg for the plants receiving NO_3/N and total N utilization was low, 4.4 and 0.31% respectively.

The poorer growth and lower N utilization of the plants receiving NO_3/N was probably due to its greater leaching. It is therefore concluded that if soluble N fertilizer is to be used on Bougainville, NH_4/N is preferable to NO_3/N.

Both the positive response of *P. dilatatum* to applied N and the low utilization by *E. deglupta* of N in trials indicate that a continued input of plant available nitrogen is required for continued plant growth. *D. intortum* did not respond to applied N in the nutrient addition trial and this suggests that effectively nodulated leguminous plants can be successfully established in mine waste.

A wide range of tropical herbaceous legumes are used for agricultural purposes. Some of these species grow well in acid soils, and both seed and a suitable *Rhizobium* inoculum are commercially available. A selection has been tried and the most successful is used as the major component of the seed mix for revegetation. In the field the leguminous species grow rapidly and do not exhibit nitrogen deficiency symptoms. The main species used are *Desmodium intortum, D. uncinatum, Macroptilium atropurpureum, Stylosanthes guianensis, Calopognonium mucunioides* and *Lalab purpureus.*

Despite the success in the establishment of herbaceous legumes, leguminous shrubs and trees were required in the species mix to increase the ecological stability of the areas. Species of interest include *Acacia auriculiformis, A. angustissima, A. holosericea, Albizzia falcataria, A. chinensis, A. lebbeck, Cassia obtusifolia, Casuarina equisetifolia, C. gracillum, Crotalaria anagyroides, C. pallida* and *Sesbania grandiflora* and *Parasponia andersonii.* All of these species are capable of symbiotic nitrogen fixation. Many of the species grow well on Bougainville.

Initial attempts to establish these species on mine waste had variable success, possibly due to inconsistent natural nodulation. To overcome this variability the seed is inoculated prior to planting. In the case of legumes, *Rhizobium* cultures have been isolated from nodulated plants. The methods employed for collection, isolation, maintenance and inoculation of the cultures were taken from Vincent (1970) and Thompson (1980).

To ensure effective nodulation of the casuarinas, soil and nodules are collected in the field from the base of established trees and included in the potting mix used for tubing the seedlings.

Soil pH, available Ca and Mo are important parameters that contribute to effective nodulation in the field (Bramfield et al. 1983; Robson 1978). Effective nodulation occurs best in neutral to slightly acid soils that have high available Ca and Mo. These conditions are met on Bougainville due to (1) the liming of acid wastes, and (2) Mo is associated with chalcopyrite in the Panguna orebody.

3.5 Phosphorus

The phosphorus content of Bougainville mine wastes is high (Table 3), with waste rock generally having higher concentrations than tailings. The main consideration for revegetation therefore is of P availability, application rates and long-term supply. These points are discussed below and in Section 4.

In the nutrient addition pot trial (Sect. 3.3) addition of P to limed tailings significantly increased growth of *D. intortum* ($p < 0.01$). Tops dry weight was 16 times that of the control treatment. In all treatments, however, foliar P concentrations were low to deficient (0.09 to 0.12% P) with P addition increasing plant growth without increasing foliar P concentrations. Evidently, the wastes are deficient in available P and the rate of P application in this trial was insufficient to maximize plant growth.

Beckworth (1964) proposed the use of adsorption isotherms to measure the amount of P fertilizer required to bring soils to various levels of adequacy for crop production. He proposed that the measured quantity of P sorbed at a standard supernatant concentration of P is an estimate of the P requirement of a soil.

Adsorption isotherms were prepared using the method of Fox and Kamprath (1974). The wastes used were acid waste rock (−2 mm fraction), acid tailings (XS29) and neutral tailings (XS11 and XS26). The wastes had a varying potential to adsorb P (Fig. 3) with the acid wastes requiring higher applications of P. From the isotherms it was possible to calculate the fertilizer requirement for the top 10 cm of the soil. To do this it was assumed that a solution equilibrium P concentration of 0.3 ppm was required for optimal plant growth, and that 25% of the waste rock material was less than 2 mm. Measured soil bulk density values were used. The calculated P requirements were 200, 184, 43 and 20 kg ha^{-1} for waste rock and tailings from XS29, XS11 and XS26 respectively. The range is wide and the acid waste has a very high P requirement.

The actual response of plants to applied P was examined in a pot trial with *D. intortum* using acid tailings. The tailings were limed to 2 t ha^{-1}. Phosphorus was applied at a rate of 16, 32, 64, 128, 250, 500 and 1000 kg P ha^{-1} as equal quantities of $NaH_2PO_4 \cdot 2H_2O$ and $Na_2HPO_4 \cdot 2H_2O$. Plant tops were harvested at 8 weeks and then at 17 weeks with the roots.

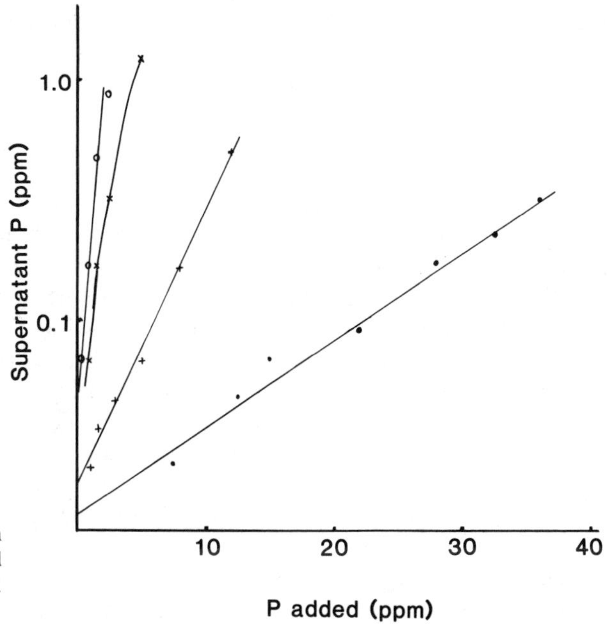

Fig. 3. Phosphorus adsorption isotherms for waste rock (•) and tailings from XS29 (+), XS11 (x) and XS26 (o)

Addition of P significantly affected top weights ($p < 0.01$) and root weight ($p < 0.01$), foliar nutrients and soil chemistry. For both harvests tops and root dry weight increased with P addition to the 250 kg ha^{-1} and 64 kg ha^{-1} treatment respectively, and decreased above these levels.

Statistically significant relationships were fitted to the data for both harvests. For the second harvest:

$$Wt = -11.65 + 15.96 \log P - 3.84 \log P^2 \qquad (r^2 = 0.92);$$
$$Pf = - 0.18 - 0.23 \log P \qquad (r^2 = 0.98);$$

where Wt is the tops dry weight (g), P is the P application rate (kg ha^{-1}) and Pf is the foliar P concentrations (%). From the relationships the P application rates necessary to give maximum top weights were 158 and 112 kg ha^{-1} at the first and second harvest respectively.

There was also a statistically significant ($p < 0.05$) second-degree polynomial relationship between top weight and foliar P concentrations. From the relationship for the second harvest, the critical foliar P concentration corresponding to 95% maximum growth was 0.25% and was achieved by a P addition of 64 kg ha^{-1} P. This foliar concentration is comparable to the critical foliar P concentration at the pre-flowering stage given by Andrew and Robins (1969) of 0.22%.

As expected, available soil P increased with log P addition ($p < 0.01$ for both harvests). For the second harvest available soil P corresponding to 64 kg ha^{-1} P addition was 12 ppm, a level in excess of that found for unfertilized mine waste.

This pot trial demonstrated that P fertilizer is necessary for high plant growth, but that large additions of soluble P fertilizer may decrease plant growth as well as being expensive and wasteful. The literature also indicates that difficulties arise in obtaining an adequate supply of P to plants in the presence of exchangeable Al, Fe, Mn and Cu (Smeck 1972; Lindsay 1979). Availability of P decreases as soils become more acid and the proportion of Al and Fe phosphates increases (Kamprath 1978; Bohn et al. 1979). This occurs as a two-step adsorption process (Barrow 1978) especially when P is added as soluble fertilizer.

3.6 Other Elements

A range of elements apart from N and P are essential for successful plant growth. These elements are of secondary importance, however, as far as affecting plant growth on Bougainville mine waste. Despite this, deficiency conditions have occasionally been observed or thought to be possible and controls implemented.

Potassium, Ca and Mg are essential plant nutrients that in mineral soils are supplied to plants via exchange sites from the weathering of minerals. The concentrations of these elements in Bougainville mine waste are high, consequently the rate of primary mineral breakdown is important. In the lysimeter experiments (Sect. 3.3) release of these cations was high and pH-related.

Of the three cations plants have the highest requirement for K. The soil chemistry of K is complex as after solubilization from primary minerals it can either be held onto exchange sites, fixed, especially onto 2:1 clay minerals (Tisdale and Nelson 1975), or leached. Tropical soils are usually low in exchangeable K (Graham and Fox 1971) and range from 0.1 to 4 mEq. %.

Bougainville waste has low exchangeable K, which is in part due to the low CEC of the material. Deficiencies, however, have not been observed when K is included in follow-up fertilizer treatments in the field.

Few soils contain insufficient available Ca per se for plant growth (Kamprath 1978). However, many soils in the humid tropics contain insufficient Ca to maintain a suitable degree of base saturation. This is important in affecting the availability of basic cations to plants (Tisdale and Nelson 1975). In such soils exchangeable Al dominates (Kamprath 1970), and the amount of Ca required for optimum plant growth increases with reduction in soil solution pH (Moser 1942).

Calcium is the main exchange-held cation in fresh tailings. This is lost due to leaching and acidification and both acid tailings and waste rock have low exchangeable Ca. All acid waste material is limed prior to seeding; this raises the level of exchangeable Ca. As a consequence, foliar Ca levels are acceptable (Andrew and Pieters 1972).

Magnesium can become deficient in tropical soils as a result of low reserves (Kamprath 1978), by high leaching (as seen in the lysimeter trials) or upon addition of large quantities of fertilizers and liming agents (Adams and Henderson 1962; Tisdale and Nelson 1975). In general, exchangeable Mg should exceed 10% of the CEC (Doll and Lucas 1983) and Mg deficiencies occur where it is only 4% of the CEC (Adams and Henderson 1962).

Magnesium may be limiting in fresh tailings. Through acidification and leaching it is likely to be deficient in both waste rock and acid tailings. In these wastes the exchangeable Mg is less than 4% of the CEC. Magnesium is applied to mine waste 3–6 months after seeding.

Boron deficiencies are most common in coarse, well-drained mineral soils. Boron uptake is also reduced by increasing soil pH (Tisdale and Nelson 1975) and occurs when soils are overlimed, possibly due to an unfavourable Ca:B ratio in plants (Tisdale and Nelson 1975).

On Bougainville mine waste boron deficiency is observed. Low foliar B values have been seen in plants not receiving B fertilizer. Boron is currently applied to all revegetation areas 3 to 6 months after planting.

4 Field Techniques

In the previous sections details of specific factors limiting plant growth on Bougainville have been described. There was, however, a need to combine all the measures proposed to control these limitations to test the overall effectiveness of the programme. This was done in field trials on both alkaline tailings and acid waste rock. In addition, it was decided to examine the effectiveness of rock phosphate (RP) as a long-term source of P. Previous trials had shown that soluble P fertilizers were not suited for providing continued plant growth without repeated applications. The RP used in this trial was a ground blend of Nauru and Christmas Island rock with 16% P and 1.6% citrate soluble P.

On waste rock (Table 3) RP was applied at 250 (RP250), 500 (RP1000) and 1000 kg ha^{-1} (RP1000) in factorial combination with lime at 2 t ha^{-1} (L2) and 4 t ha^{-1} (L4). The area was ripped with a tracked dozer, the treatments added and lightly raked 1 week prior to seeding. On tailings (Table 3) RP was added at 0, 250, 500 and 1000 kg ha^{-1}; no lime was added. The treatments were surface spread and incorporated into the sediment with a chisel plough immediately prior to seeding. The trial was replicated twice on waste rock and thrice on tailings.

A mixture of tropical grasses and *Rhizobium*-inoculated legume seed, 40 kg ha^{-1}, was surface spread and 250 kg ha^{-1} N:P (18:18) fertilizer applied. A further 250 kg ha^{-1} N:P fertilizer was applied 1 month later. After 3 months 500 kg ha^{-1} of (N:P:K:Mg:B, 6:18:12:2:1) fertilizer was applied. The vegetation was cut 5 cm above the ground at 29 (Hl) and 50 weeks (H2) after sowing for the trial on waste rock and at 19 (Hl) and 44 weeks (H2) for the trial on tailings. At both harvests foliar samples of *Macroptilium atropurpureum* and soil samples were collected.

In the trial on tailings there were no significant differences in plant wet weights between replicates or RP treatments. For all plots plant establishment and growth were excellent, with rapid ground cover achieved. Rock phosphate did not affect either foliar nutrient concentrations or soil analysis. At Hl, except for B, foliar nutrient concentrations were generally in the adequate to optimum range (Andrew and Pieters 1972). Foliar B levels (13 ppm) were sub-optimal. By H2, however, foliar N (2.0%), P (0.08%) and K (1.39%) had fallen and were deficient. There was no correlation between foliar P and soil P or pH.

On waste rock both lime and RP significantly affected biomass, soil chemistry (Table 4) and foliar nutrient concentrations (Table 5). Plant biomass was also well correlated with soil and foliar parameters.

The establishment and growth of the plants receiving L4 was excellent with complete ground cover by H 1. In contrast, the establishment and growth for L2 was

Table 4. Soil analysis from the rock phosphate field trial on waste rock, second harvest[a]

	2 t ha^{-1} Lime			4 t ha^{-1} Lime		
	Rock	Phosphate	Application	Rates	(kg ha^{-1})	
	250	500	1000	250	500	1000
P tot[b] %	1440	1615	1595	1570	1710	1750
P av[b] ppm	14.5	12.8	18.9	18.9	13.4	18.7
K ex[b] mEq. %	0.26	0.27	0.25	0.29	0.23	0.37
Ca ex mEq. %	3.45	3.83	2.95	6.35	12.05	8.20
Mg ex mEq. %	0.47	0.55	0.42	0.48	0.41	0.53
Na ex mEq. %	0.05	0.06	0.05	0.04	0.03	0.04
CEC mEq. %	8.9	9.1	8.4	9.5	8.9	9.3
pH 1:5	4.5	4.6	4.5	5.1	6.8	5.5
H ex mEq. %	1.67	1.24	1.32	1.11	<0.1	0.66
Al ex ppm	312	278	314	179	2	61
Mn ex ppm	19	16	12	10	6	8
OC[b] %	0.22	0.14	0.19	0.26	0.42	0.28
Cu av ppm	388	358	324	326	337	353
Zn av ppm	7	7	5	5	11	5
Fe av ppm	336	231	245	219	107	166
S tot %	0.80	0.83	0.51	0.93	0.60	0.88
SO_4/S tot ppm	481	545	465	1075	374	548

[a] Average of two replicates per treatment.
[b] tot = total; av = available; ex = exchangeable; OC = organic carbon.

Table 5. Foliar analysis of *Macroptilum Atropurpeum* from the rock phosphate field trial on waste rock, second harvest[a]

Treatments						
	2 t ha^{-1} Lime			4 t ha^{-1} Lime		
Analyses	Rock	Phosphate	Application	Rates	(kg ha^{-1})	
	250	500	1000	250	500	1000
N %	2.20	2.28	2.44	2.46	2.47	2.53
P %	0.09	0.11	0.12	0.13	0.11	0.15
K %	1.52	1.35	1.63	1.52	1.84	1.58
Ca %	0.81	0.78	0.74	0.73	0.76	0.82
Mg %	0.16	0.17	0.18	0.16	0.16	0.17
B ppm	12.3	17.5	19.3	23.5	27.0	25.0
Mo ppm	11.0	15.5	13.0	11.0	21.0	14.0
S %	0.18	0.18	0.18	0.17	0.19	0.21
Al ppm	408	607	408	473	414	426
Mn ppm	71	63	46	50	38	51
Cu ppm	89	135	84	105	101	104
Fe ppm	926	1395	937	986	1006	1045
Zn ppm	36	35	32	34	32	36

[a]Total nutrient, average of two replicates per treatment.

poor, ground cover patchy to sparse and the plants appeared "stressed". Foliar symptoms of P deficiency and Al toxicity were observed, particularly for the herbaceous legumes. Above ground biomass was significantly lower (p—0.01) than the L4 treatments, being 76% and 36% of the L4 treatments at H1 and H2 respectively. The higher lime rate increased foliar P but reduced foliar N, Al, Mn and Fe for both harvests and had no effect on foliar Cu (Table 5).

Irrespective of lime rate foliar N, K, Cu, Mg ,Mn, B and Zn were acceptable, whereas those for Fe and Cu were high (Andrew and Pieters 1972). Available soil P was higher for the L4 treatment (Table 4) at H2 but no different at H1. Soil pH was also higher.

Visually, it was difficult to detect the RP effect due to the large lime response. At H1 and H2 the plant yields, however, were significantly higher ($p < 0.01$) for RP1000. The RP500 treatment yielded more than RP500 at H2 but not at H1 (Archer and Marshman 1985). At H2 the yield from RP1000 was 36% higher than for RP500.

Foliar N and P concentrations followed a similar trend to plant weights (Table 5). Foliar P levels were highest for RP1000 but still below the critical levels ascertained by Andrew and Robins (1969) of 0.24%. Possibly P was limiting plant growth. Foliar N concentrations were satisfactory but increased with increasing RP application. Foliar metal levels were unresponsive to RP. Available soil P showed a similar trend to plant weight and foliar P.

A number of significant linear correlations were also observed between plant growth, foliar and soil parameters. Plant weights were related to both foliar N and P at both harvests, thus indicating the importance of these elements. There was, however, no significant correlation between foliar P and available soil P. Plant weights were also significantly correlated with pH and a number of soil parameters that are pH-dependent, i.e. exchangeable Al and Mn (negative correlations) and exchangeable Ca. No significant relationships were observed between plant growth and available soil Cu, Fe or Zn. This suggests that metal toxicity was not present. Both exchangeable Al and H were strongly correlated with pH, indicating the value of soil pH as an indicator of potential Al and H toxicity.

These field trials indicated that practical methods exist for the establishment of plants on both acid and alkaline mine waste, that both lime and RP have significant benefits to plant growth in acid waste and that no metal toxicities are evident other than those that can be controlled through the use of lime.

5 Revegetation Practice

The understanding that has been gained from the above study has enabled plants to be successfully established on over 60 ha of mine waste. On wastes that will acidify, the methods used (Archer and Marshman 1984) are the same as were used on the rock phosphate trial.

The composition of the species mix is central to successful revegetation on Bougainville. Most of the species used are sown as seed that is commercially available. The species have been chosen for their tolerance to acid soils that have low available P. Herbaceous legume species comprise the major component of the seed mix. Grass species form another large component. Shrub and tree species, including inoculated legumes, are also included, the seed of which is collected locally. In some areas tree species are also planted as seedlings.

Some amelioration of the waste material is recognized as essential with the main emphasis being placed on pH correction and application of P fertilizer. Amelioration includes soil conservation measures, ripping, liming and addition of rock phosphate. Fertilizer (N:P) is applied at sowing and at 1 month so that plants are able to vigorously establish. Additional fertilizer (N:P:K:Mg:B) is applied at 3–6 months to minimize nutrient deficiencies in the first year of growth. By then there is virtually complete ground cover.

As part of the overall revegetation programme all revegetation areas are monitored and assessed. The oldest revegetation areas are on waste rock and are 4 years old. Soil pH in these areas has fallen since the initial liming but the vegetation has remained vigorous and does not exhibit nutrient deficiency symptoms. Some change of species composition, however, has occurred. Initially, herbaceous legumes and grasses dominated. These species begin to be shaded out at about 6 months by the shrubs, i.e. *Crotalaria* spp. which in turn are replaced by *Acacia* spp. 2–3 years after establishment. At the same time species from adjacent areas begin to volunteer on the mine waste. This change in species composition is considered to be a natural and positive response by the plant community to changing habitat conditions.

Volunteers on waste rock include the ground orchid, *Spathoglottis* spp. and *Caldcluvia celebica* and the trees *Ficus bougainvillei, F. indigofera, F. polyantha, Pipturis* sp. and *Parasponia andersonii.* On tailings volunteers have included *Pandanus* sp., *Macaranga* sp., *Terminalia brassii, Buchanania macrocarpa* and *Pipturus* sp.

No active revegetation effort is carried out on waterlogged tailings deposits. All such areas rapidly and naturally recolonize.

The future structure of the plant communities, the composition of the soil and the time frame over which significant changes will occur are not precisely known. Such information will only become available with regular monitoring. The research programme upon which revegetation at BCL has been based, however, increases our confidence of successfully fulfilling the original objectives, i.e. obtaining a self-sustaining, maintenance-free plant community.

Acknowledgements. All chemical analyses for this study were performed by the Analytical Service Department of Bougainville Copper Limited. The advice and assistance given in this respect by Mr. F. Grieshaber is gratefully acknowledged. Acknowledgement is given to Bougainville Copper Limited for permission to publish this work.

References

Adams F, Henderson JB (1962) Magnesium availability as affected by deficient and adequate levels of potassium and lime. Soil Sci Soc Am Proc 26:65–68

American Society of Agronomy, Soil Science Society of America (1982) Methods of soil analysis, 2nd edn. ASA, SSSA Madison, Wisc

Andrew CS, Pieters WHJ (1972) Foliar symptoms of mineral disorders in *Phaseolus atropurpureus*. CSIRO Div Trop Pastures Tech Pap 11

Andrew CS, Robins MF (1969) The effect of phosphorus on the growth and chemical composition of some tropical pasture legumes (1) Growth and critical percentages of phosphorus. Aust J Agric Res 20:665–674

APHA, AWWA, WPCF (1980) Standard methods for the examination of water and waste water. APHA, AWWA, WPCF, Washington, DC

Archer IM, Marshman NA (1984) Revegetation at Bougainville Copper. Min Mag 115:307–313

Archer IM, Marshman NA (1985) Revegetation of waste rock and tailings. Internal Report Bougainville Copper Ltd.

Barrow NJ (1978) Inorganic reactions of phosphorus, sulphur and molybdenum in soil. In: Andrew CS, Kamprath EJ (eds) Mineral nutrition of legumes in tropical and subtropical soils. CSIRO, Aust

Beckworth RS (1964) Sorbed phosphate at standard supernatant concentration as an estimate of the phosphate needs of soils. Aust J Exp Agric Animal Husb 5:52–58

Bohn HL, McNeal BC, O'Connor CA (1979) Soil chemistry. John Wiley, New York

Bradshaw AD, Chadwick MJ (1980) The restoration of land. Blackwell, Oxford

Bramfield SM, Cumming RW, David DJ, Williams CH (1983) The assessment of available manganese and aluminium status in acid soils under subterranean clover pastures of various ages. Aust J Exp Agric Animal Husb 121:192–200

Doll EC, Lucas RE (1983) Testing soils for potassium, calcium and magnesium. In: Walsh LM, Beaton JD (eds) Soil testing and plant analysis. Soil Sci Am Inc

Fox RL, Kamprath EJ (1974) Phosphate adsorption isotherms for P requirements of soils. Soil Sci Soc Am Proc 34:903–907

Graham ER, Fox RL (1971) Tropical soil potassium as related to labile pool and calcium exchange equilibria. Soil Sci 111:319–322

Greenland DJ (ed) (1981) Characterization of soils. Clarendon, Oxford

Hartley AC (1976) Tailings and waste revegetation at Bougainville. PNG Sci Soc Port Moresby PNG

Hartley AC (1979) Weathering studies of Bougainville mine tailings in lysimeters. Aust J Soil Res 17:355–360

Kamprath EJ (1970) Exchangeable aluminium as a criterion for liming leached mineral soils. Soil Sci Soc Am Proc 31:348–353

Kamprath EJ (1978) Lime in relation to Al toxicity in tropical soils. In: Andrew CS, Kamprath EJ (eds) Mineral nutrition of legumes in tropical and subtropical soils. CSIRO Aust

Kittrick JA, Fanning DS, Hossner LR (eds) (1982) Acid sulfate weathering. SSSA Spec Publ 10 SSSA Madison, Wisc
Lindsay WL (1979) Chemical equilibrium in soils. Wiley & Sons New York
Mosser F (1942) Calcium nutrition at respective pH levels. Soil Sci Soc Am Proc 3:339–343
Robson AD (1978) Mineral nutrients limiting nitrogen fixation in legumes. In: Andrew CS, Kamprath EJ (eds) Mineral nutrition of legumes in tropical and subtropical soils. CSIRO Aust
Robson AD, Reuther DJ (1981) Diagnosis of copper deficiency and toxicity. In: Lonergan JF Robson AP, Graham RD (eds) Copper in soils and plants. Academic Press, Australia
Smeck NE (1972) Phosphorus; an indicator of pedogenetic weathering processes. Soil Sci 115:199–206
Stumm W, Morgan JJ (1980) Aquatic chemistry, 2nd edn. Wiley New York
Thompson JA (1980) Production and quality control of legume inoculants. In: Bergerson FJ (ed) Methods for evaluating biological nitrogen fixation. John Wiley & Sons, New York Chichester, pp 489–533
Tisdale SL, Nelson WC (1975) Soil fertility and fertilizers. MacMillan, New York
Uehara G, G Gillman (1981) The mineralogy chemistry and physics of tropical soils with variable change clays. Westview Trop Agric Ser 4
Vincent JM (1970) A manual for the practical study of the root-nodule bacteria. IBP handbook. Blackwell, Oxford

Biological Engineering of Marine Tailings Beds

DEREK V. ELLIS and LAURA A. TAYLOR[1]

1 Introduction

At coastal and marine mines, tailings are discharged to the sea, often in enormous amounts, and they form beds similar in many ways to the deposits naturally settling there (see Chap. 6). We should expect these tailings beds to be colonised by marine organisms natural to the area in ways comparable to the biological colonisation of mine waste on land. However, compared to the enormous literature about reclamation processes on land, there is virtually no published information on the reclamation of submerged and shoreline mine tailings. This chapter collates what little information is available, largely from limited distribution reports. We also raise issues about reclaiming marine-discharged tailings beds to biologically productive ecosystems. This is what we mean by the term "biological engineering" in our context.

This chapter should be read in conjunction with Ellis (1988) which provides basic environmental data about mines discharging tailings to the sea.

1.1 Mine Waste Reclamation

Reclamation is the process by which a disturbed ecosystem is brought back by human action to a productive condition, for example, as farmland, forest, recreational areas or for traditional subsistence by original inhabitants. Reclamation is especially important at large open pit or strip mine sites, where much land is disturbed. In many countries such mines are required to ensure appropriate reclamation; it is implicit that the other conflicting demands on the land resource were only temporarily postponed or interrupted by the mining operations, not eliminated forever.

At coastal mines and marine mining sites, large amounts of tailings may be discharged and deposited over great areas (Ellis 1988). For example at Island Copper Mine in northern Vancouver Island, British Columbia, Canada, the concentrator mill discharges approx. 40,000 t day^{-1} of tailings to an adjacent fiord. From 1970 to 1981, 106,000,000 t of tailings had accumulated in the fiord receiving system. Approximately 50 k^2 of fiord bed in Rupert and Holberg Inlets have received substantial deposits, which by 1982 after 11 years deposition were spread in a layer averaging 15 m deep (max. depth 35 m). An extensive area was covered

[1]Department of Biology, University of Victoria, B.C. V8W 2Y2 Canada

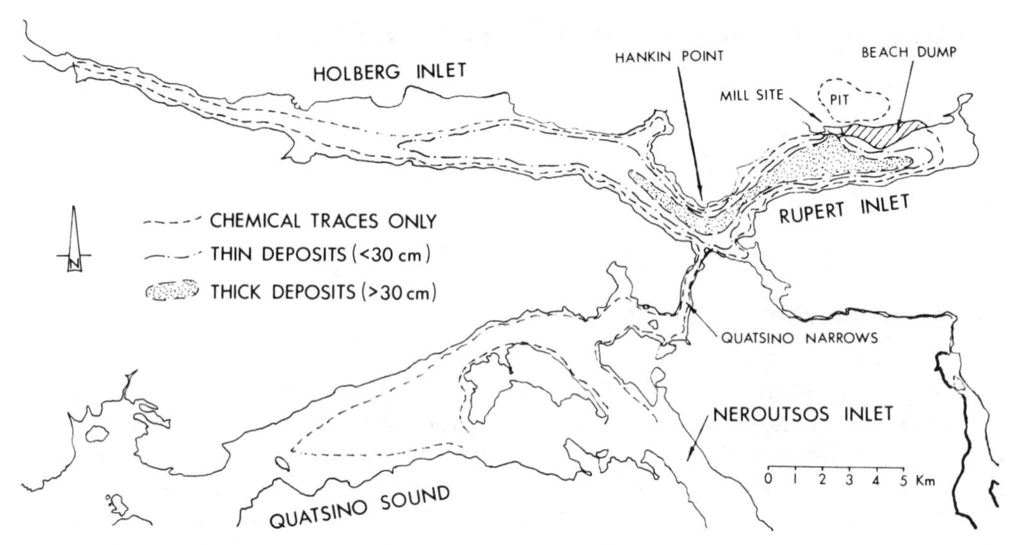

Fig. 1. Distribution of tailings deposits from the Island Copper mine in 1984

by less than 30 cm of tailings (Fig. 1). An additional area had received traces of tailings only detectable chemically. The total area affected had almost reached its present limits by 1973 after 2 years discharge, and has not extended greatly since then as deposits thicken rather than spread further.

There are other resource uses in the Island Copper receiving area. The fiord supports substantial stocks of Pacific salmon and Dungeness crab, both of which are important fishery species. It is not inevitable that a mine discharging tailings to a site with such biological production and commercial harvests will cause substantial, and permanent losses of the fishery resources. A coastal mine at a site with fishery resource values should meet the cost of ensuring return of submarine tailings beds to a productive marine ecosystem (not necessarily the same as previously) as inland mines meet similar reclamation charges. Such a requirement will require continuous planning and reclaim operations during the lifetime of a coastal or marine mine, as it does for a mine on land.

1.2 Biological Enhancement of the Seabed

The bulk of attention to enhancement of marine ecosystems has been by means of artificial reefs (D'Itri, 1985). These are hard structures created by dumping materials in shallow water, e.g. car bodies, truck tires or specially engineered interlocking concrete anchors. These reefs provide a habitat for growing organisms, and if they are built in shallow water where plant growth can occur, they become miniature ecosystems and important fishery sites. They grow attached plants (marine algae), and in the tropics corals. Consequent from this primary biological production, there follow plant-eating fish, and the range of animal herbivores, carnivores and particle eaters, which make a productive ecosystem.

Experience with artificial reefs provides little information about the enhancement of unconsolidated deposits: such as natural oceanic and riverine silts, clays, sands and their anthropogenic analogues of mine tailings. However, they can be considered part of the overall approach to underwater reclamation of tailings beds, since artificial reefs potentially can be built on unconsolidated tailings, once compacted, and at a depth sufficiently shallow for plant growth to occur.

The biological enhancement of unconsolidated deposits has generally been considered by aquaculturalists, rather than engineers. The considerations have largely been what species will grow on what kind of substrate; sand, silt, clay or some mixture? Accordingly, attention has been directed to selecting appropriate species, and then to considerations of enhancement. Such species may include shellfish, such as shrimps and clams, and groundfish such as sole and plaice. It may be necessary to consider such beds as nursery areas, raising only young to a stage at which they migrate seeking conditions more favourable to their increasing size. In some cases the beds have been modified to provide wide but shallow depressions, which appear to attract some species, and facilitate their growth (Mottet 1985). It may also be possible to use established intertidal aquacultural practices in shallow beds, such as planting stick inserts on which oysters or mussels may grow. Alternatively, the area may be divided by weirs into series of pools, which can be used for different life stages of a species, or for several species simultaneously as in on-land fish farms.

Mine tailings in shallow water present additional problems for reclamation, especially those arising from pollution. Shallow water beds by definition are within the surface belt of illumination in which all marine plant growth occurs, hence the belt of most biological production. Generally, the productive zone lies at less than 20-m depth on the continental shelf, but may extend to 30 m in clear water around oceanic islands. If problems such as contamination or acute poisoning should occur, then the risk of substantial impact is higher in illuminated zones than it would be were the tailings deposited in the dark at less productive but greater depths. Furthermore, if construction requirements for discharging to shallow water require a causeway to be built, from the tailings or its coarsest sand fraction, there will also be risk of increased leaching of toxins from rainwater and surface acid drainage. Acid drainage is not as serious a problem in the marine environment as on land due to the almost unlimited buffering capacity of seawater at the interface with the tailings, and restricted vertical movement of pore water (or solute diffusion) in deposits.

This leaching potential for trace metals from tailings deposits presents a further complication. There is a large body of literature on the chemistry of the topic (see other chapters in this vol.). In a biological context, the important components are whether leached trace metals can be present in concentrations toxic to organisms, whether they bioaccumulate in the tissues or guts of organisms to levels whereby those organisms contaminate their predators, or whether trace metals biomagnify. Biomagnification means that toxins progress up an ecosystem food chain increasing in concentration as they go, so that top predators including ourselves or fishery species may be poisoned. The risk becomes more serious at each level of predation if biomagnification occurs.

The well-known reduced abundance and diversity of organisms on "polluted" sediments is generally believed due to toxicity. But this effect may also be due to changed particle-size arrays or avoidance by drifting larvae of settling from the water column onto deposits chemically repelling them in some way (Menzie 1984).

In general, the literature on enhancement of seabeds, both engineering and biological, is relevant to the reclamation of marine tailings deposits. Reclamation is a special form of enhancement.

2 Available Information

2.1 Artificial Reefs

Artificial reefs are believed to have started when fishermen threw rocks overboard to supplement the few productive reefs on barren coasts. This was happening in Japan in the 1600s (Mottet 1985). So significant was the practice that in one area 100,000 40–50-cm rocks were dumped during the 1870s. Reefs were barely starting at that time in the USA. They appeared there in the mid-20th century and were built largely in the interests of recreational fishermen (Stone 1985). Initially, reefs were of wood and trees, but progressively scrap was introduced as a construction material: cars, tires, ships, demolition rubble. This introduced a serious complication: the needs for scrap disposal could introduce pollutants at the reef sites. Hence, reef construction now may require appropriate approval by environmental as well as fishery authorities.

Assessments were introduced to determine how effective such reefs were in meeting their stated objective of improving fisheries. In general, it appears that yields in the fishery zone around a reef can be increased 10–20 times, and that test reefs when built next to natural reefs can generate an amount and diversity of fish similar to the natural reef fauna. Their success indicates that they should be considered as a part of tailings bed reclamation.

There have been a number of trends in reef utilisation which need to be considered if they are to be included in any program of reclaiming marine mine tailings.

2.1.1 Materials and Construction

Scrap materials are still frequently used, of which tires from cars and trucks are the most common, due to their abundance and resistance to dissolution. Their resistance is desirable, since the target life for a reef must be measured in tens of years [30 years according to Grove and Sonu (1985)]. Tires need processing into stable structures, which will not move or break up. They need ballasting (with concrete), holes at any upper surface (to allow air or other gases to escape) and to be tied together (for example with tire threads which can be expected to have as long a life as the tires). Tires may need compacting to increase density. The largest such reef contains 2,000,000 tires. They are believed not to leach toxins, hence are pollutant free. Certainly, reefs built from them attract, shelter and produce fish.

In Japan and elsewhere, fishery and engineering research has produced many structures built from concrete, steel, fibre-reinforced plastic and other materials designed to meet criteria of stability, break resistance and ease of placement. They also meet the biological needs of the fish or other species targeted for enhancement. There are three basic designs: (1) platforms, (2) chambers and (3) blocks.

Platforms are almost two-dimensional and provide horizontal surfaces to which marine algae can attach. These are particularly important in areas such as Japan where there is a significant algal industry for food or processing.

Chambers are three-dimensional structures which provide shelter for fish as well as support for their food stocks. Fish and other organisms will use the inner space of chambers and crevices, as well as the outer space. The higher the chamber is built, the more likelihood of enhancing larger fish needing detectable reference structures in otherwise featureless seabeds. In contrast, the higher the structure, the less stable.

Blocks are the base material for building breakwaters, and other devices such as submerged reefs, serving to reduce wave and current energy, while incidentally providing a biological habitat.

At one site, near a coal-fired power plant, the particulate combustion by-products have been processed into resistant blocks which are used as reef materials. These were tested for toxin leaching chemically and by bioassay and were determined to be harmless. Materials strength appeared to increase with immersion, and the reefs built were getting fish occupants within 3 weeks of construction (Woodhead et al. 1982, 1985). Comparable heavy, seawater-resistant blocks can possibly be built from mine tailings.

Reef design should also facilitate harvesting, and where necessary maintenance operations. For example, a reef could be accessible to divers if necessary for shellfish gathering, or cleaning of fouling organisms competing with targeted species. Reefs should not interfere with fishing operations or boat navigation, although this is more a matter of placement than design.

2.1.2 The Concept of Targeting

Different species of fish and invertebrate shellfish have different demands on their habitat in terms of shelter and food supply, and in terms of depth, current speeds and adjacent substrate. The skill of building artificial reefs is now at a stage where particular species should be targeted for enhancement by an artificial reef project. The species selected will in large part be determined by the needs of local populations, whether subsistent, recreational or commercial fishermen. In general terms, the target species must be fish or marine invertebrates which require or are attracted to reefs, either as residents or on migration during their life-cycle habitat changes. The reef must maximise whatever their needs are, for example accessible inner space for shelter from predators, or attachment space for plant food. Marine algae could also be the targeted species, or even shellfish such as crustacean shrimp or molluscan snails. The reefs must also be detectable by mobile species, hence they should be placed in ways to lead fish from one reef to another. They must be siltation free, so that surfaces do not become covered in layers of silt which reduce their attractability to larval stages requiring clean settlement space. There are obviously a number of structural features which must be combined in ways matching species needs and environmental forces, hence targeting allows setting design priorities.

As mentioned previously, placement of the reefs is also a consideration if only to avoid hazards to navigation. One factor determining placement is the nature of the substrate at the site. This can range from bedrock, for which the reef provides supplementary, three-dimensional features, to unconsolidated deposits ranging from coarse sand to fine clays. The compactibility of deposit beds varies and this will determine the degree to which heavy structures placed or dropped will sink in, and hence lose reef value even if stability is improved. In general, sands, provided they are not actively moving under fast currents, will support reefs, but silts and clays may be too soft. A rule of thumb test for supporting strength is the Mathews hand test (Mathews 1985). If a diver's hand can be pushed into the sediment to beyond the wrist, the substrate will not bear a heavy structure sufficiently well.

Placement should also be such that the zone of fishing around the reef should just touch that of adjacent reefs. This maximises the fishery area, and relates to the design feature of leading fish from one reef to another.

It may be necessary to seed a newly built reef with young or adults of the targeted species to ensure that these forms gain a foot-hold on the structure before competitors arrive naturally.

Incorporation of artificial reefs into an underwater reclamation plan is not simply a matter of dumping scrap, but requires considerable social and biological planning, and engineering design.

2.2 Colonisation Experiments

In recent years a number of mines have investigated whether tailings will support regrowing benthos. There are two methods of approach: experimental tests and field surveys. The former involves either laboratory or field experiments with natural or artificial substrates (Cairns, 1982), and the latter involves surveys of biological growth on tailings beds. Descriptions of the former follow in this section. Seabed surveys are described in Section 2.3.

2.2.1 The Proposed Quartz Hill Molybdenum Mine, Alaska (see also Ellis 1988)

Smothering and recolonisation experiments using removable plastic containers in implanted trays were undertaken in the expected tailings receiving area in 1982 and 1983 (VTN, 1983). It was found that much benthos could survive smothering by up to 8 cm of bulk-sample tailings for periods of up to 90 days, but that few organisms burrowed into the tailings from below during that time. The tailings implants were colonised by species similar to those of control implants over periods from 6 to 25 months. The speed of colonisation varied with season and depth. Tailings recolonisation took 1.7 times longer than the controls. Recovery was faster in spring than other times of year. Recovery was also faster at shallower depths (95 and 150 m) than deeper (210 m).

Winiecki and Burrell (1985) investigated recolonisation of defaunated sediment trays in one of the possible fiord basins for Quartz Hill tailings as part of the pre-operational baseline surveys by the mine. They found that through 78 weeks of their experimental implants the number of organisms, number of species and total weight of organisms (biomass) increased. They concluded that equilibrium conditions had not been reached at 78 weeks based on best-fitting curves being straight lines. Nevertheless, by that time with 90 taxa and >3,000 organisms per sample unit (=0.1 m^2), a diverse and abundant array of organisms had colonised (cf. numbers and variety of organisms near Island Copper Mine, Tables 1 and 2). All biological parameters measured were affected by seasonal changes, e.g. higher organism densities in spring and lower in winter, and lower species diversities in spring and higher in winter. A limited number of the colonising species became abundant with time, and some subsequently declined, in ways that suggested they might be useful indicators of the stage of ecological succession.

There is a very substantial body of literature on recolonisation processes of sediments, both polluted and unpolluted (although not mine-derived). Example reports have been given by McCall (1977) and Arntz and Rumohr (1982).

2.2.2 Island Copper Mine, Canada (see also Ellis 1988)

In 1982 one of the authors (LT) started a 12-month series of artificial substrate colonisation experiments in Rupert Inlet, the fiord that receives tailings from Island Copper Mine (see Fig. 2 for location of experiments). The objectives of this research were: (1) to describe initial colonisation of undisturbed tailings; (2) to determine whether benthic colonisation was adversely affected by a tailing substrate and; (3) to outline the effects of season on benthic colonisation, and to

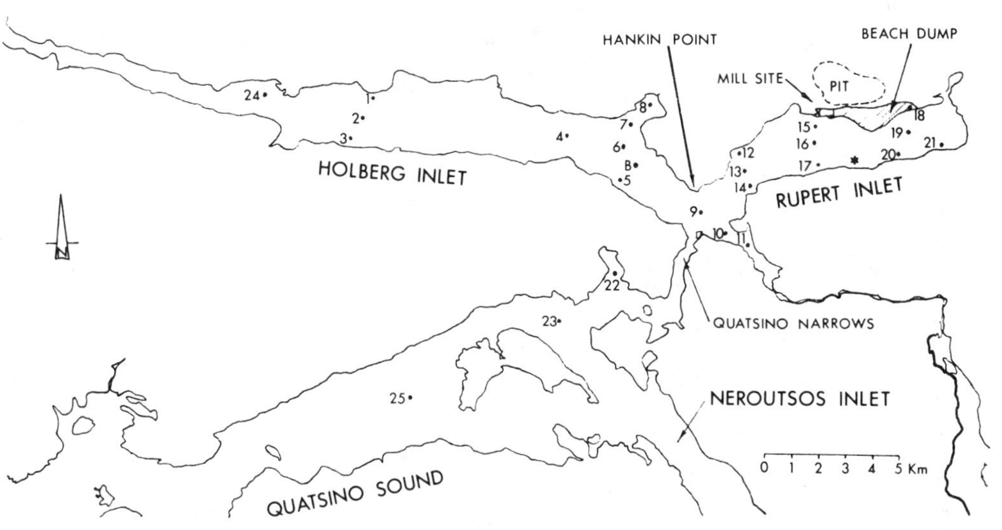

Fig. 2. Seabed sampling stations at the receiving area for Island Copper mine tailings. *Recolonisation studies experimental area

Table 1. Numbers of animals per square metre burrowed into seabed deposits at routinely sampled stations near Island Copper Mine[a]

	1	2_L	3	4_L	5_L	6_L	7	8	9_H	10_L	11	12_L	13_H	14_H	15_H	16_H	17_H	18	19_H	20_L	21	22	23	24	25	B_H
1970[b]	2380	4600	3540	2180	3080	4400		-	-	-	-	1620	2180	-	-	-	-	-	2160	2760	-	-	-	-	-	-
1971	6540	1040	960	560	1760	5660	4040	3740	8320	3880	4240	1680	5480	1320	3340	-	8200	5640	4020	3880	8900	5660	3520	-	-	-
1972[c]	2880	380	900	700	1260	1860	1060	1320	2800	2280	3820	980	160	1740	60	-	360	1820	500	1340	2840	1480	1520	-	-	-
1973[c]	2500	1380	1040	3580	4900	7500	2900	1700	5400	2780	6800	1500	1060	1500	240	-	260	1820	1820	2180	5260	2860	3980	560	-	-
1974	1420	2460	1420	1380	2460	3380	1080	1640	3000	2240	1280	1140	940	2320	3040	6840	4140	2120	2620	880	2960	1680	2620	1380	-	-
1975	540	1340	1400	3140	1060	960	760	1400	1200	1720	1240	260	1040	3280	740	-	-	860	1620	400	1640	1220	1940	1630	-	-
1977	1491	733	427	2479	5220	5698	1638	1256	1358	2741	1931	1472	217	3996	127	618	453	1275	1447	478	3779	1071	1243	631	-	6577
1978	1965	855	1397	1403	2456	3795	995	1365	2150	823	1206	1461	434	1563	172	0	6	670	778	976	4325	651	1040	708	-	5390
1979	3106	950	778	2692	3598	7361	1276	3074	2079	1518	1288	1729	147	3534	364	26	172	2092	1059	1155	11647	1614	1257	453	708	16431
1980	3936	1091	778	389	2194	11775	5505	3349	5141	4152	4095	2354	6	4580	989	759	1595	-	2028	1218	16010	1773	1320	466	293	12890
1981	3967	612	96	510	1977	5230	4108	3483	3814	1824	6940	3617	1084	3043	6	96	1340	-	1225	1569	10219	3074	3521	982	1359	-
1982	1767	1378	829	3470	2328	3719	2526	7023	274	4050	4803	1250	38	1340	0	421	1040	-	2736	2137	7036	4600	1614	1569	2028	1556
1983	4561	1639	619	2545	2245	5384	7406	5396	1397	6621	19391	2526	32	4784	32	1359	1920	-	2577	3680	10859	8656	4682	4734	3126	1148
1984	3132	1617	1454	3802	3457	1396	8133	2602	8393	11954	12030	4235	836	6557	26	848	1001	-	1863	2838	6780	7195	3049	3068	2756	842

[a]Stations were sampled in September each year except where shown. Tailings discharge started after the 1971 collections.
[b]October.
[c]December.
H = heavy tailings (>30 cm); L = light tailings (< 30 cm).

Table 2. Variety of species burrowed into seabed deposits at routinely sampled stations near Island Copper Mine[a]

	1	2_L	3	4_L	5_L	6_L	7	8	9_H	10_L	11	12_L	13_H	14_H	15_H	16_H	17_H	18	19_H	20_L	21	22	23	24	25	B_H
1977	24	18	16	16	20	17	23	35	12	29	27	18	15	16	4	4	4	23	17	18	20	25	44	20	–	13
1978	30	13	16	12	17	25	27	29	13	19	22	23	7	16	5	0	1	21	13	21	25	31	27	19	–	21
1979	32	13	15	18	23	15	29	32	13	27	27	25	8	23	4	3	9	28	10	23	27	42	47	19	29	17
1980	31	22	17	15	17	18	33	38	22	35	40	32	1	28	9	8	12	–	13	27	31	40	38	18	15	18
1981	22	17	9	12	24	23	40	37	10	25	42	35	9	17	1	3	16	–	13	29	61	44	74	36	57	–
1982	21	25	16	19	23	25	43	46	5	38	46	33	4	19	0	5	11	–	20	32	53	64	49	30	59	11
1983	41	27	10	23	24	23	55	43	13	37	56	32	4	30	2	2	6	–	15	37	50	65	83	45	54	12
1984	42	27	27	30	37	32	52	35	26	60	53	39	12	37	2	4	4	–	17	34	58	68	82	38	51	10

[a]Figures are the number of species counted in triplicate samples.
L = Light tailings (< 30 cm); H = heavy tailings (> 30 cm).

identify the seasons when this colonisation is most susceptible to further perturbation from tailing discharge.

Two types of artificial substrates were used in the field experiments; mine tailings from Island Copper and, as a control, fine marble sand (Imperial brand, Imperial Limestone Co., Vancouver, B.C.) of similar particle size and range to the tailings. The control substrate contained low levels of heavy metals and was devoid of the chemicals added to Island Copper's tailings. Heavy metal levels in the control approximated those present in natural sediments in the Island Copper area. Once every 2 months experimental units, consisting of two tailing and two control artificial substrates, were placed on the bottom of Rupert Inlet at 12–13 m depth. Units were left in place for 1,2,4,8 and 12 months. The timing is diagrammed in Fig. 3.

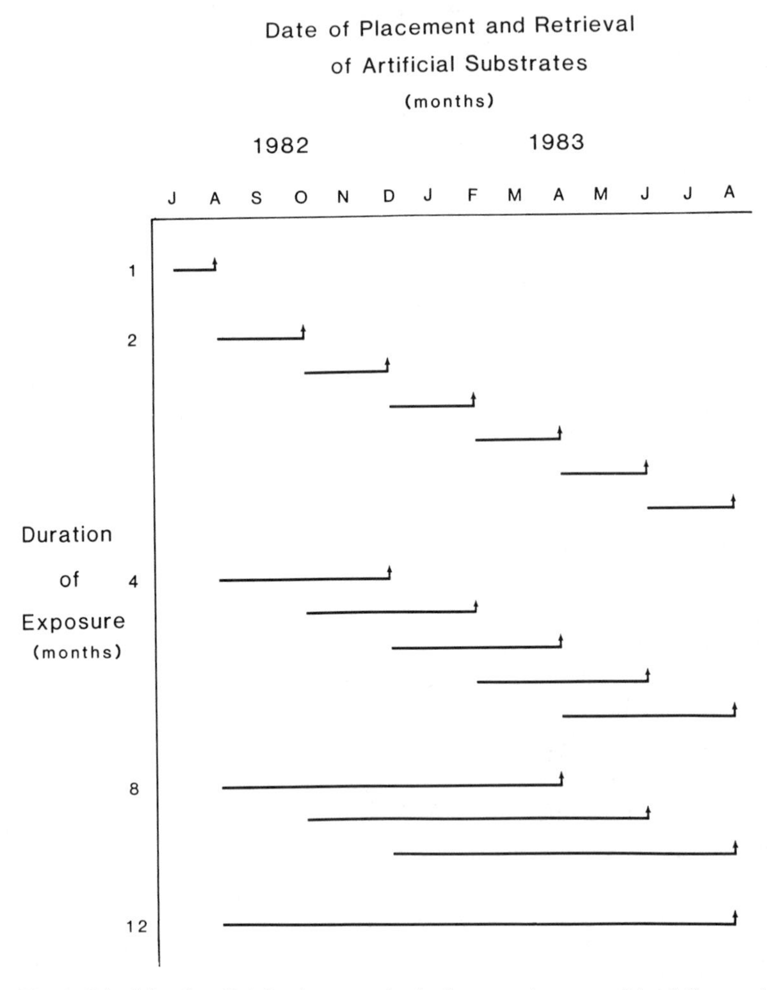

Fig. 3. Schedule of artificial substrate colonisation experiments at Island Copper Mine. Each *line* represents one experimental unit (two tailing and two control artificial substrates). *Arrows* indicate month of retrieval

In both tailing and control substrates the number of individuals and taxa generally increased with length of substrate exposure, i.e. with time. However, the rate of increase varied seasonally. During winter, numbers remained stationary or declined slightly, while comparatively sharp increases occurred in the spring, summer and fall (Fig. 4). This seasonality indicates that the time required for a tailing bed community to reach equilibrium will depend on the season in which it initially becomes available for colonisation. Hence, the time to equilibrium varies.

Preliminary analysis suggests that the numbers of individuals and taxa in tailing artificial substrates were still rising after 12 months of exposure. It appears that tailing communities in Rupert Inlet require more than 12 months to attain equilibrium conditions.

Tailing and control substrates were colonised by similar arrays of taxa. However, in every experimental unit deployed, regardless of season or length of exposure, tailing substrates supported fewer numbers of individuals and fewer taxa than their controls. This indicates that similar biological communities were developing on the two substrates, but development was occurring at a slower rate in tailings than in controls. Since the two substrates differed primarily in chemical

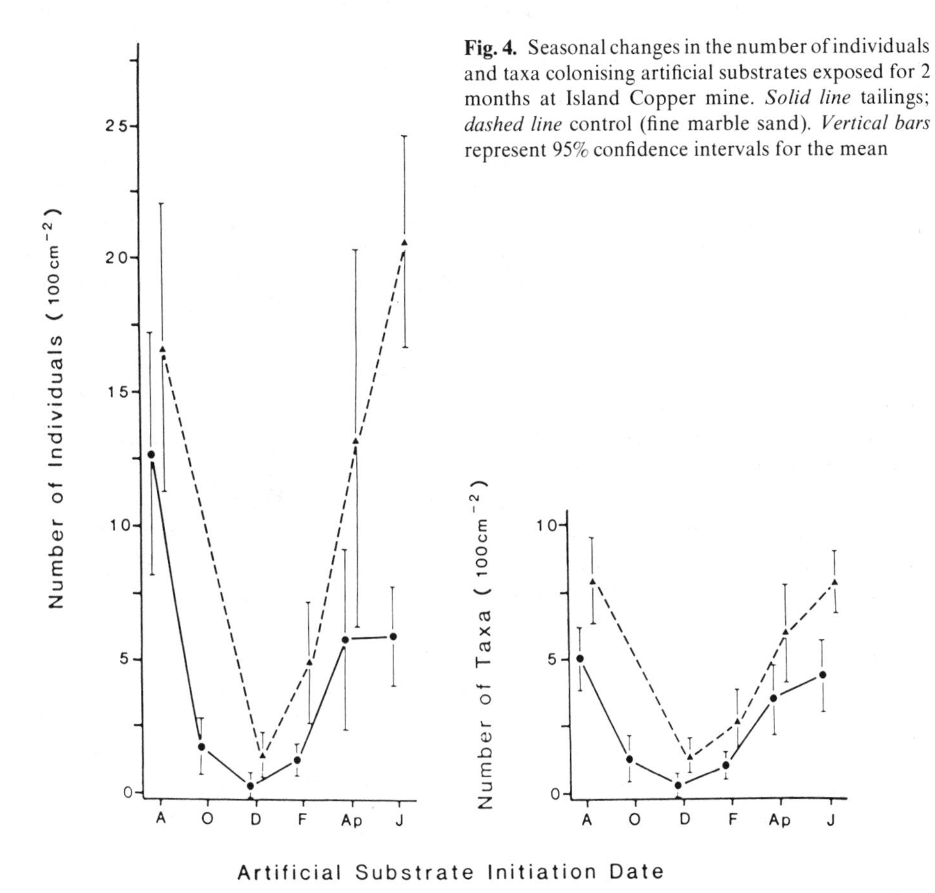

Fig. 4. Seasonal changes in the number of individuals and taxa colonising artificial substrates exposed for 2 months at Island Copper mine. *Solid line* tailings; *dashed line* control (fine marble sand). *Vertical bars* represent 95% confidence intervals for the mean

composition, the results suggest that some chemical component of the tailings adversely affected benthic colonisation.

Given that tailings depress the rate of benthic colonisation, there are two possible paths that a developing tailing bed community may follow. The tailing community may continue to develop until it reaches the same equilibrium state as would be found on natural substrates in the area. However, the time needed to reach this equilibrium would be longer on tailings than on natural substrate. Alternatively, tailing community development may stop at an earlier stage in succession, until the factor responsible for this is removed from the system. Examples of such removal are blanketing of tailing beds by natural deposits or removal of the offending component by dissolution. Long-term studies on tailing colonisation are needed to establish which of the above paths tailing colonisation follows. Despite the depressed rate of benthic colonisation on tailings, it must be emphasized that tailing artificial substrates were colonised by benthic organisms and supported diverse benthic communities.

The number of individuals and taxa colonising artificial substrates fluctuated seasonally (Fig. 4). Numbers were high in the fall, declined over the winter and rose again in the spring and summer. The winter decline probably reflects a corresponding decline in the rate of reproduction and growth of benthic organisms.

As recolonisation in winter is slow, impacts from tailing discharge on immigrant colonisers (larvae and adults) would be limited during that season. It is in the late summer and early fall that tailing colonisation is most susceptible to perturbations from tailing discharge.

Tailing discharge at Island Copper is currently ongoing (Ellis 1988) and the mine monitors nearby benthic communities annually (see Sect. 2.3.1). Species presence-absence data from the artificial substrates can be compared with similar data from the annual benthic surveys in 1982–83, using the ordination technique of principal components analysis (Gauch 1982). Figure 5 displays the results. Each sample is positioned on the basis of the similarity of its species composition to that of all other samples.

The distribution of artificial substrates in Fig. 5 demonstrates a gradient of increasing community development from left to right on the diagram. That is, in substrates initiated concurrently, the distance along axis 1 generally increases as exposure time increases. In addition, experimental substrates usually are situated lower on axis 1 than their controls. Axis 2 clearly separates the majority of artificial substrates from the benthic survey samples. Ordination axes are do not necessarily represent specific environmental parameters. Here, axis 2 appears to integrate several differences between the fiord seabed and the artificial substrates, e.g. sediment particle size, location in Rupert Inlet and community age.

Island Copper's benthic stations close to the tailing outfall or in the main path of the tailing turbidity current (group A; stations 13,15,16; Fig. 2) are grouped close to artificial substrates that showed little development (tailing substrates exposed for 2 or 4 months over the winter). This indicates that these highly impacted sites are being held in the early stages of colonisation, probably by constant disturbance from tailing deposition. Stations further away from the outfall and turbidity current (group B; stations 12, 14, 17, 19, 20, 21; Fig. 2) are distributed further along

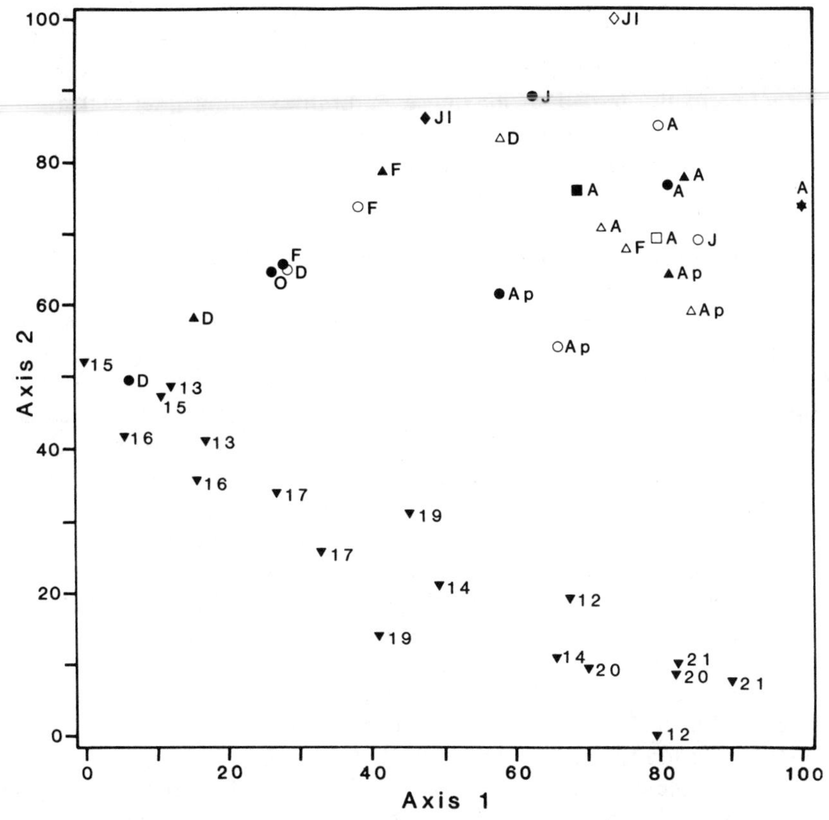

Fig. 5. Non-centred principal components analysis of the taxonomic assemblages inhabiting artificial substrates and Island Copper's 1982 and 1983 benthic samples from Stations 12–21 (see Fig. 2). *Numbers* indicate benthic survey station. *Letters* refer to the initiation month of artificial substate experiments: *closed symbols* tailings; *open symbols* controls; *diamonds* 1 month of exposure; *circles* 2 months; *triangles* 4 months; *squares* 8 months; *star* 12 months

axis 1 than group A stations. This suggests that group B stations are at later stages of colonisation and experience less impact than are group A stations. A continuum of the degree of impact is also evident within the two groups.

2.2.3 Shoreline Tailings – Vegetation Transplants

Tailings deposited on shorelines or engineered into a causeway carrying the tailings line to deeper water can be subjected to conventional reclamation procedures, i.e. vegetation seeding. One of the authors (DE) has seen seeded vegetation (mangroves) at a tailings line causeway at the Marcopper Copper mine in the Philippines (Ellis, Table 1). Mangroves appear to be the best choice for such transplant experiments in the tropics since their natural reseeding mechanism consists of germinating seedlings evolutionarily adapted to salt-water implanting between high and mid-tide levels on shore (Snedaker and Snedaker 1984).

At Island Copper Mine, Canada, part of the shoreline waste dump has been experimentally graded to maximise naturally seeding intertidal algal growth immediately below sites of conventional land reclamation (Buchanan 1982). Shoreline landforms can be shaped for biological growth.

2.3 Biological Surveys of Stabilised Tailings Beds

At a number of mine sites there have been biological surveys on stabilised tailings beds. Conclusions can be drawn about the general pattern and timing of tailings recolonisation. These sites fall into two groups. At one group, such as Island Copper Mine, the mine continues to operate, and the sites surveyed included those receiving no tailings, very light or heavy deposits. At another group of sites, such as the Kitsault molybdenum mine, sampling has followed closure of the mine.

Current theory maintains that unconsolidated seabeds subject to natural or other catastrophes, e.g. slumping, or other obliteration, recolonise by a successional series (Rhoads and Germano 1982). The early stages are represented by small pioneering species, largely polychaete worms or crustacea, with fast breeding responses to just tolerable conditions. Later stages comprise slower breeding and deeper-burrowing species which bioperturbate and hence modify and condition the sediment substrate. This stage will also include polychaete and crustacean species, but different species than the first colonisers, plus a diverse fauna of molluscs, echinoderms, etc. In the early stages productivity, i.e. growth rates and turnover (breeding), can be high. Seabed-feeding of larger species which represent fishery resources may actually benefit from the high productivity of a recolonising seabed, as farmers benefit from the productive but unnatural early successional stages of domestic crops that they induce on ploughed land.

2.3.1 Island Copper Mine, Canada

Tables 1 and 2 show numbers and diversity of benthos at sampling stations arrayed as in Fig. 2. These benthos are burrowing (infaunal) forms and are collected from 0.052 m^2 of seabed by means of a Ponar sampler (Ellis and Jones 1980) digging to approximately 10-cm depth. The counts are of animals retained after screening the samples through a 0.6-mm mesh. There have been various ways of analysing the data to demonstrate the extent of differences between sampling stations with tailings and without (Anon 1985), although no state of the art statistical analysis (which is needed) has yet been attempted on the enormous data bank of 23–26 stations over 15 years. The mine surveys have collected many thousands of specimens representing tens of species per sample (three samples per station, each survey for averaging purposes). Procedures have been improved at intervals, especially in 1977, and the data from that year on are more reliable than previously.

At stations with heavy tailings (>30 cm) the benthos were often reduced to virtually zero both in terms of numbers of animals and variety of species (Tables

1 and 2; Figs. 1 and 2). However, on occasions, the following year's survey showed numbers and diversity of species considerably recovered, e.g. Station 13 from 1980 to 1981, and Station 16 from 1982 to 1983, when numbers rose from 6 and 421 m^2 respectively, to 1084 to 1359. These recoveries were not to levels of benthos supported at stations without tailings which approximate 1,000–10,000 m^2. In general, heavy tailings stations (Stations 9, 13, 14, 15, 16, 17, 18 and B) appear to have less benthos than other stations, but this needs a rigorous statistical analysis to confirm. There were occasions when numbers and diversities on heavy tailings were high, with several thousand animals per square metre representing 15–40 species, e.g. Station 14 in 1984 and Station B 1977 to 1980.

We conclude that benthos can exist in considerable numbers on tailings, even heavy deposits, but that they are subject to repeat obliterations, presumably by changes in location of the tailings turbidity current, from slumps and erosion of current levees, and surges of suspended tailings over the seabed. Benthic reductions of this type are known to occur naturally (Levings et al. 1983; see Station 3 in 1981). Recolonisation back to ambient numbers of several thousand per square metre may take 1 or more years.

Ellis and Heim (1985) compared some of these infaunal numbers in Table 1 with numbers of large interface living species (epifauna) surveyed from submersible. They found that there could be substantial reductions of these large epifaunal species and changes in their diversity, while the tailings at adjacent, routinely monitored stations supported an abundant burrowing fauna.

Jones and Ellis (1976) had earlier shown that at least one species of polychaete worm could settle and grow on tailings, but at a rate less than normal. At least one of the many species of such polychaetes were able to build their protective cases from tailings (Fig. 6). At least one species of large clam, *Humiliaria kennerlyii*, has been collected in Rupert Inlet only in tailings, in shallow beds of coarsest (sand) fractions. Given appropriate underwater conditions, apparently seabed stability and low or zero toxicity, the tailings bed can be colonised by substantial numbers of living marine organisms within a year of obliteration.

2.3.2 Kitsault Molybdenum Mine, Canada (Ellis 1988)

Two surveys of benthos on tailings and natural sediments nearby have been made during mine operations and after closure (Kathman et al. 1983, 1984). The first was in 1982 a few weeks prior to closure, and the second 1 year later, in 1983. In 1982 after 18 months operations, benthos at tailings-impacted sites were distinguishable by few species, but 1 year later after closure the tailings-impacted sites could not be clearly differentiated from the unimpacted sites, but these unimpacted sites had shown a diversity reduction (apparently not mine related).

Similar benthic surveys had been made in 1974 prior to the mine re-opening to determine whether previous mine operations terminating 2 years previously (1967–1972) had affected the benthos in ways still detectable (Ellis and Jones 1975). Benthos near the mouth of the creek, which received tailings and discharged them to the sea, were only separable by one of several characteristics indicating environmental stress. These sites had benthos in as large numbers and in as high

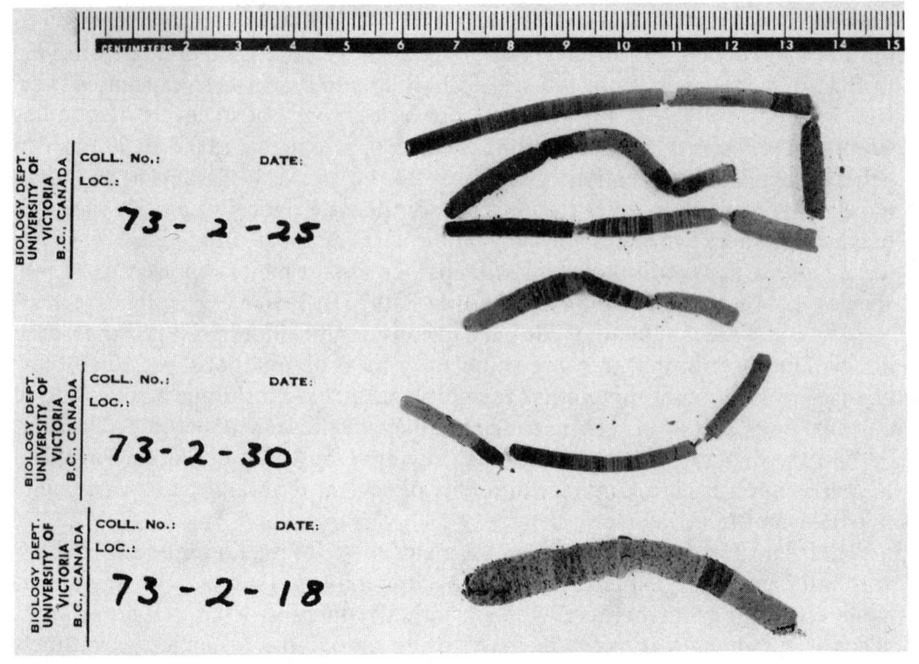

Fig. 6. Tubes of a polychaete worms showing bands of tailings (*white*) and natural deposits (*dark*) (Reproduced from Jones 1984)

weights as elsewhere, but fewer species present in large numbers (three compared to seven species). If impact had occurred from 1967–1972, it had disappeared in the 2-year period since mine closure.

2.3.3 Brynnor Iron Mine, Canada (Ellis 1988, Table 1)

At the Brynnor iron mine, tailings were discharged to the beach, close to the loading dock, until 1968 when the mine was closed. The beach has areas of haematite sand (spills and dust from the loading wharf) and tailings clays (Levings 1975). In 1975 the magnetite sand supported a diverse community of organisms including eelgrass and the edible littleneck clam (*Venerupis japonica*). Fewer species had colonised the tailings, but they included the edible soft-shell clam (*Mya arenaria*) known to tolerate harsh environments, and some polychaete worms. Levings suggested that the compactness of the tailings or interface turbidity prevented colonisation. It was not necessarily toxicity of the tailings. Sub-tidal surveys reported eelgrass beds, many small invertebrate animals and good catches of rock crab. Pacific oysters were abundant on interspersed suitable habitat, and the area was a registered oyster lease.

3 Discussion

3.1 Summaries of Available Information

3.1.1 Recolonisation of the Seabed

Recolonisation is the process by which an ecosystem receives new organisms. On seabeds, the organisms arrive either as larvae settling from the water column above, or as mobile animals, flatfish and crabs for example, actively swimming or crawling in. Also, some types of worms and other invertebrates may burrow their way in through the sediments.

Recolonisation of seabeds occurs continuously and naturally, but especially each spring in temperate regions after the winter period of natural mortalities, slow growth, slow sexual maturation and reduced breeding. This recolonisation is also the process by which seabeds recover biologically from naturally occurring catastrophes such as slumping, and from man-made impacts of dumping or spilling masses of particulates.

It appears from extensive recolonisation studies (for references see Sect. 2.2.1) that where there is complete obliteration of existing species, successful recolonisation can start immediately when conditions have stabilised. It depends of course on whatever larvae are present at the time. The more larvae (as in spring), the quicker the recolonisation. The process varies seasonally, but within a single year (or even a single summer) an early successional stage of a recovering seabed community can be established.

Experimental investigations to date have not extended much beyond 1 year. However, the alternative investigative technique of seabed surveys confirms that within a year extensive mine tailings beds can be recolonised and support many organisms. In at least one case (Kitsault) after 2 years, species associations were present which showed few differences from those on unimpacted adjacent seabeds.

It is to be expected that known latitudinal differences in larval biology will affect recolonisation. In tropical regions larvae may be relatively few but stable in numbers through the season. In polar regions larvae are rare in winter, but abundant in summer. In temperate regions larvae vary considerably in seasonal abundance and availability.

3.1.2 Tailings and Fisheries

Fisheries continue at sites of marine tailings deposition. Periodic fishing of salmon, crab and prawn stocks is well documented at Island Copper Mine, Canada. One of us (DE) has seen subsistent fishing in progress at both Marcopper and Bougainville mines in the Philippines and Papua New Guinea respectively. In these two cases, there was active fishing around the tailings plume, which appeared to have attracted some species of fish.

In contrast, placement of the discharge system can cause fisheries losses. Tailings discharged to rivers for gravity flow to the sea can block movements of fish which must complete their life cycles by river migrations. Although direct turbidity effects on free-ranging fish at marine mine sites have not been well investigated, the well-documented susceptibility of coral reefs to siltation (Gomez et al. 1981; Brown and Howard 1985) should be considered at tropical mine sites. Coral reef fisheries can be important (Huntsman et al. 1982).

There is no known case of fish toxicity arising from biomagnification through marine tailings disposal. Nevertheless, the potential is there, and merits appropriate investigation at sites of intensive local resource use. Toxicity should particularly be checked at smelters and refineries, which are more likely to affect the chemical composition of the waste discharge than a concentrator mill.

3.1.3 Special Concerns of Subsistent Fisheries

At some mine sites, the local population is dependent on fisheries for their own subsistence, or for local marketing, or it has a strong cultural dependence on marine resources for social and religious needs. Such situations can be obvious in tropical regions, but may also occur elsewhere, e.g. claims of native Indians in western Canada.

Any country controlling mining can be expected to have priorities for resolving potential resource conflicts between local citizens and a mine development. The various resolutions include monetary compensation (continuing or a single payment), or relocation of the people concerned. There is also potential to restore or substitute productive ecosystems, both during operations and after mine closure. These can be at a subsistent level, or in the form of aquacultural developments. For example, at Island Copper, the mine is partner in a local salmon enhancement program on the river from which their process water is taken. It is possible to seek balances between complementary compensatory procedures.

3.2 Discharge and Reclamation Objectives

3.2.1 Deep, Shallow or Shoreline Discharge?

Tailings have been discharged to the sea at various depths, and patterns of impact with depth are now apparent from the collated case histories.

Shoreline discharges cause extreme surface turbidity, which can reduce primary biological production. A prograding shoreline can obliterate productive intertidal and nearshore fishery resources. In contrast, shoreline tailings are accessible for engineering and reclamation. Conventional reclamation seeding (mangroves) has been tried, and appears to proceed well, as does natural seeding in tailings. However, shorelines are subject to wave erosion, hence resuspension and redistribution of the materials and leached toxins by longshore drift. Tailings above surface may generate acid waste leaching problems.

Shallow water discharges within the euphotic (plant growth) zone will also impact stretches of the biologically most productive depths. Where a causeway has been built to discharge to a selected shallow water point, problems similar to those of shoreline discharge occur. There have been plans to dyke and backfill embayments for tailings disposal, converting shallow water to land (with attendant acid waste and wave erosion problems.

Deep-water discharge places tailings below the euphotic zone and constrains impact to less productive depths. At sites where such depths have been close to shore, e.g. northern fiords, Atlas Mine (The Philippines), deep discharge, with turbidity current flow to greater depths, has been the preferred option. In silled fiords, there is an additional advantage that the tailings are constrained within a pit, although the inevitable reduction in fiord depth may need consideration. Does the pit have the capacity to take the mine's lifetime of tailings, and will infilling raise the seabed into the productive euphotic zone? Deep-water discharge needs to consider whether local upwelling as at Island Copper will upwell some fraction of the tailings to shallow water and the surface.

3.2.2 Flat Seabed or Artificial Reefs?

Tailings beds are inevitably as flat as natural seabeds formed from deposits of river sediments and plankton. The development of artificial reef technology permits engineering and placement of reefs on these flat beds. Although several mining operators have discussed such plans with one of us (DE), there appears to have been no application to date. In one case the mine-derived initiative was stopped due to local objections that use of worn haul-truck tires would be polluting. This could have been checked by bioassays.

3.2.3 Target Species?

Artificial reef technology is now at a level where it is routine procedure to select target species for enhancement. The necessary engineering can then be designed with the biological needs of these species in mind, e.g. their food supply and the primary production on which it is based, physical shelter and protection from predators, attraction of roving species. Fishery procedures must also influence reef design and placement. Targeted species range from marine algae through shellfish such as crustacea and bivalve molluscs, to resident and migratory fish. Knowledge of local species, their potential abundance and habitat needs, and their value as resource species is needed for good engineering in any proposal for fishery enhancement.

3.3 Biological Engineering

We use the term "biological engineering" to denote reclamation of tailings beds to productive ecosystems. By reclamation we mean enhancement of natural recolonisation processes, so that the return to biological productivity is both speeded up and directed.

Major engineering considerations in the marine environment are water depth, and exposure to storms and waves. Accordingly, since these engineering parameters will vary with coastal location, we summarise information in terms of mines discharging to open coasts, shallow embayments and fiords. For each habitat we consider whether there is deep as well as shallow water nearby, taking the separation point between deep and shallow as the depth at which primary biological (plant) production is limited, i.e. approximately 20–30 m. The potential for resuspension and upwelling is also a factor in determining suitable water depth.

An important factor underlying whether biological engineering should be allowed is whether the discharged tailing stream is toxic or not, and whether the deposited tailings will leach toxins. Such toxicity can be determined by appropriate bioassays (White 1984), and can be minimised by process controls. If there is residual toxicity, reclamation procedures should take into account the extent to which chemicals will contaminate organisms and biomagnify. There may be circumstances under which reclamation should not be encouraged, and others in which residual toxicity after deposition may be the prime determinant of the reclamation objectives and procedures. It is of course also necessary to be sure that any engineered structures, such as artificial reefs, are not built from toxic materials which might contaminate the reclaimed ecosystem.

3.3.1 Open Coasts

Open coasts have high energy shorelines, and discharging tailings to such coastlines exposes the pipelines, any causeways and other structures to storm and wave damage. At the Jordan River mine in Canada (Ellis 1988) the tailings line broke twice in 2 years of operation, and tailings deposited closer to shore than intended. This resulted in slight copper bioaccumulation by intertidal clams dispersed down the line of longshore drift some 20 km over 4 years (Ellis and Popham 1983).

Unless discharged beyond the depth with potential for resuspension by storm waves, tailings beds will be unstable and liable to redispersion. It is conceivable that widespread dispersion of a thin tailings blanket may be a preferred disposal option over a smaller but thicker tailings bed at depth or perched on a shallow sloping shelf. In either case, engineering should be designed for the option chosen, and not allow unintended dispersal patterns to occur, either by default of specific objectives or by accident.

The physical conditions on open coasts are sufficiently severe that we can see little potential there for biological engineering of tailings beds. The prime consideration if tailings disposal is considered in such a habitat is to minimise resource losses by well-chosen disposal objectives and accident-proof engineering.

3.3.2 Shallow Embayments

Embayments are coastal indentations with extensive shallow water, and are relatively protected by their configuration from wave action. Many types of biological engineering are physically practical, but the determinants of which particular balance is chosen should be biological and social. Established enhancement procedures such as artificial reefs, ponding and aquaculture are all possible. Shallow water and shoreline vegetation (algae, mangroves, marine grasses) can be seeded or transplanted as are their analogous forms in conventional land reclamation.

Presently, untried possibilities could include shaping of the tailings beds to provide irregular contouring by changing the pipeline discharge point, or dumping tailings from a scow. Such shaping of deposits must consider stability after discharge ceases, and erosion potential for resuspending and redistributing the tailings to a geomorphological equilibrium. Also, it might be possible to use mine waste products in a beneficial way, e.g. building artificial reef blocks from compacted consolidated tailings, or haul-truck tires bound together.

3.3.3 Fiords

Fiords are deep narrow submerged valleys, and they provide reasonably accessible deep water close to shore. In silled fiords, e.g. traditionally defined fiords with a terminal glacial moraine at its mouth or somewhere along its length, there is a deep pit, in which tailings can be constrained.

For mine tailings to be discharged to a fiord, there is a basic set of calculations needed. These are to determine the volume of tailings that will be discharged during the lifetime of the mine, the degree to which tailings will compact when deposited and hence their final volume, and the volume of the fiord pit receiving the tailings. From these data, figures can be derived for the final depth of the tailings bed, and particularly whether the infilling will raise the seabed into biologically productive shallow water. Oceanographic data are needed to determine whether active upwelling may force tailings to the surface, and whether the amount upwelled is significant.

Reclamation procedures such as artificial reefs for fish and shellfish production may be useable in deep fiords, provided there is reason to predict that species present in deep water can be enhanced in amounts which are significant for both production and harvesting.

A presently untried concept is that the fiord seabed should be deliberately raised into shallow water, so that enhancement can utilise the greater biological productivity of the illuminated zone.

Other possibilities with or without raising the seabed to the productive zone include attempting to enhance productivity by adding fertiliser in some form as is done in reclamation practice on land. The skill in fertilisation is to add nutrients required by the colonising species; the long-term objective is to enhance the natural soil-forming processes. The equivalent formation of "soil" in the seabed, i.e. organically rich deposits, may be enhanceable also. It would be needed particularly in tropical oceanic areas with their low marine production, and may be beneficial even in high rainfall coastlines where river outflow and seasonal phytoplankton blooms naturally contribute organically-loaded deposits.

Such river outflows provide a natural capping to tailings beds. These caps might also be enhanceable if needed for any reason, e.g. enhancing stability or minimising leaching of toxic compounds. On final closure the grinding mills of the concentrator plant could process waste rock to spread a cap of non-metal-bearing sand or a grind mix with good compaction qualities.

4 Final Comment

Marine tailings beds can be considered as prospective reclamation sites in the same way as mine waste dumps on land. The objective should be to return the reclaimed area to biological productivity, but not necessarily producing the same crop as before. In the sea, reclamation is particularly important due to the almost inevitable impact of the tailings on local fisheries, and the dependence of some residents on the fisheries for subsistence, commerce, recreation or spiritual needs. Reclamation should be targeted at particular fisheries species. It should not be a generalised strategy of enhancing by whatever means are handy or cheap. Particularly, it should not be a convenient way of getting rid of mine scrap or wastes.

At the present time, knowledge of fishery enhancement in both biological and engineering terms in such that innovative procedures for reclaiming tailing beds are waiting for trials. These encompass:

1. Developing a set of artificial reefs to enhance reef fisheries;
2. Shaping and fertilising a tailings bed to enhance ground fish or shellfish fisheries;
3. Capping a tailings bed to enhance stability or to minimise leaching;
4. Raising a fiord seabed into shallow water so that biological productivity is enhanced by primary production.

These separate procedures could be combined in various ways at specific sites. They must, however, be properly monitored for continuous quality control of the manipulated area so that the proposed objective is reached within the intended time.

Acknowledgements. We are very grateful for cooperation from many agencies and people. The management and environmental staff of Island Copper Mine, Bougainville Copper Mine and US Borax have provided one or both of us opportunities for research, or to review data, and also gave permission to publish proprietary information. We have benefited from discussions and provision of reprints by many people too numerous to mention by name. We thank them collectively. Funding for academic aspects of our work has come from the Natural Sciences and Engineering Research Council of Canada (LT), the Graduate Research Engineering and Technology Awards of British Columbia (LT) and the University of Victoria Faculty Research Fund (DE). Tom Gore prepared photographs from our negatives, Leonard Craig drafted Figs. 1 and 2 and Katharine Ellis typed the manuscript. Figure 6 is reproduced from an M.Sc. thesis, University of Victoria, by permission of A.A. Jones.

References

Anon (1985) Annu Environ Assess Rep, Island Copper Mine, 2 vols

Arntz WE, Rumohr H (1982) An experimental study of macrobenthic colonization and succession, and the importance of seasonal variation in temperate latitudes. J Exp Mar Biol Ecol 64(1):17–46

Brown BE, Howard LS (1985) Assessing the effects of "stress" on corals. Adv Mar Biol 22:1–63

Buchanan RJ (1982) A government inquiry into mine waste disposal at Island Copper Mine. In: Ellis DV (ed) Marine tailings disposal, chap 7. Ann Arbor Sci Press, pp 255–270

Cairns J, Jr (ed) (1982) Artificial substrates. Ann Arbor Science Press

D'Itri FM (1985) Artificial reefs, marine and freshwater applications. Lewis, Chelsea, Mich

Ellis DV (1988) Case histories of coastal and marine mines. In: Salomons W, Förstner M. (Ed.) Chemistry and biology in solid waste. Springer, Berlin Heidelberg New York, pp 73–100

Ellis DV, Heim C (1985) Submersible surveys of benthos near a turbidity cloud. Mar Pollut Bull 16(5):197–203

Ellis DV, Jones AA (1975) Alice Arm project, review of benthic infauna data. Univ Victoria Rep

Ellis DV, Jones AA (1980) The Ponar grab as a marine pollution monitoring sampler. Can Res Tech Sect June/July, pp 23–25

Ellis DV, Popham JD (1983) Accidental formation and subsequent disappearance of a contaminated beach. In: McLachlan A, Erasmus T (eds) Sandy beaches as ecosystems. Proc Ist Int Symp Sandy beaches. Junk, The Hague

Gauch HG, Jr (1982) Multivariate analysis in community ecology. Cambridge Univ Press, 298 pp

Gomez ED, Birkeland CE, Buddemeier RW, Johannes RE, Marsh JA, Jr, Tsuda RT (eds) (1981) The reef and man. Proc 4th Int Coral Reef Symp. Mar Sci Centre, Univ Philippines, 2 vols

Grove RS, Sonu CJ (1985) Fishing reef planning in Japan. In: D'Itri FM (ed) Artificial reefs, marine and freshwater applications, Chap 6. Levis, Chelsea, pp 187–252

Huntsman GR, Nicholson WR and Fox WW Jr (eds) (1982) The biological bases for reef fishery management. NOAA Tech Memo NMFS-SEFC-80

Jones AA (1984) Effects of mine tailing on benthic infaunal composition in a British Columbia inlet; with special reference to sampling, instrumentation, and the biology of *Ammotrypane aulogaster* (polychaeta, Opheliidae). M Sci Thesis, Univ Victoria

Jones AA, Ellis DV (1976) Sub-obliterative effects of mine-tailing on marine infaunal benthos. II: Effect on the productivity of *Ammotrypane aulogaster*. Water Air Soil Pollut 5:299–307

Kathman RD, Brinkhurst RO, Woods RE, Jeffries DC (1983) Benthic studies in Alice Arm and Hastings, Arm., B.C., in relation to mine tailings dispersal. Can Tech Reph Hydrog Ocean Sci 22:30 pp

Kathman RD, Brinkhurst RO, Woods RE, Cross SF (1984) Benthic studies in Alice Arm, B.C., following cessation of mine tailings disposal. Can Tech Rep Hydrog Ocean Sci 37:57 pp

Levings CD (1975) Biological reconnaissance of an abandoned mine tailings disposal area in Toquart Bay, Barkley Sound, B.C. Fish Res Bd Can Mss Rep Ser 1362:6 pp

Levings CD, Foreman RE, Tunnicliffe VJ (1983) Review of the benthos of the Strait of Georgia and Contiqueus Fiords Can. J. Fish. Aquat Sci 40:1120–1141

Mathews H (1985) Physical and geological aspects of artificial reef site selection. In: D'Itri FM (1985) Artificial reefs, marine and freshwater applications, chap 4. Lewis, Chelsea, Mich, pp 141–148

McCall PL (1977) Community patterns and adaptive strategies of the infaunal benthos of Long Island Sound. J Mar Res 35(2):221–166

Menzie CA (1984) Diminishment of recruitment: a hypothesis concerning impact on benthic communities. Mar Pollut Bul 15(4):127–128

Mottet MG (1985) Enhancement of the marine environment for fisheries and aquaculture in Japan. In: D'Itri FM (ed) Artificial reefs, marine and freshwater applications, Chap 2. Lewis, Chelsea, Mich, pp 13–112

Rhoads DC, Germano JD (1982) Characterization of organism-sediment relations using sediment profile imaging: An efficient method of remote ecological monitoring of the seafloor (Remots TM System). Mar Ecol Progr Ser 8:115–182

Snedaker SC, Snedaker JG (eds) (1984) The mangrove ecosystem: Research methods. UNESCO monographs on oceanographic methodology 8, 251 pp

Stone RB (1985) History of artificial reef use in the United States. In: D'Itri FM (ed) Artificial reefs, marine and freshwater applications, Chap 10 Lewis, Chelsea, Mich, pp 3–12

VTN Environmental Sciences (1983) Benthic smothering and recolonization studies in Boca de Quadra. Rep US Borax, 50 pp

White HH (ed) (1984) Concepts in marine pollution measurements. Univ Maryland Press, College Park, 743 pp

Winiecki CL, Burrell DC (1985) Benthic community development and seasonal variations in an Alaskan fiord. In: Gibbs PE (ed) Proc 19th Eur Mar Biol Symp. Cambridge Univ Press, pp 299–309

Woodhead PMJ, Parker JH, Duedall IW (1982) The coal-waste artificial reef program (C-WARP): A new resource potential for fishing reef construction. Mar Fish Rev 44(6–7):16–23

Woodhead PMJ, Parker JH, Duedall IW (1985) The use of fly ash from coal combustion for artificial reefs construction. In: D'Itri FM (ed) Artificial reefs, marine and freshwater applications, chap 8. Lewis, Chelsea, Mich, pp 265–292

Reclamation of Pyritic Mine Spoil Using Contaminated Dredged Material

J.W. SIMMERS[1], G.S. WILHELM[2], J.M. MARQUENIE[3], R.G. RHETT[1], and S.H. KAY[1]

1 Introduction

The Ottawa, Illinois area strip mine reclamation project was initiated in 1978 as a Productive Uses Project (PUP) of the US Army Engineer Waterways Experiment Station (WES) Dredged Material Research Program (DMRP). It was designed to demonstrate a potential productive use of dredged material, i.e., the large-scale reclamation of pyritic mine spoil or overburden resulting from area strip mining activities. This productive use of dredged material may be a viable future disposal alternative. The area strip mine reclamation project described herein was created to demonstrate the feasibility of using a cover of dewatered dredged material to reclaim surface mine spoil deposits.

The initial objective of the reclamation effort was the abatement of erosion and acid drainage from exposed pyritic overburdens. This overburden contains excessive amounts of iron pyrite (FeS) which, when oxidized and wetted, runs profusely off steep-sided spoil ridges and produces water high in sulfuric acid. The slow rates at which these spoil areas revegetate and their barren extensiveness foreshadow endless decades of acid runoff if efforts are not made to ameliorate their condition.

In this demonstration, the topography of the spoil ridges was recontoured and dredged material was used as a substrate for vegetational development. Placement of the dredged material over the pyritic mine spoil buffered both the acid runoff and limited the infiltration of water as it allowed the establishment of a dense growth of perennial, graminoid vegetation. A control area not treated with dredged material, though leveled and seeded similarly, remained essentially barren.

During the study, certain physicochemical parameters of the test plots within the demonstration project were monitored as were changes in the colonial ecosystem as it became established. The dredged material used in this demonstration project contained low levels of PCBs (polychlorinated biphenyls), PAHs (polyaromatic hydrocarbons), and toxic heavy metal contaminants, so bioassay and biomonitoring techniques were applied to assess the degree to which contaminants were available and mobile, not only in the initial stages of revegetation, but also as the vegetation developed and the dredged material aged.

[1]US Army Engineer Waterways Experiment Station, Vicksburg, Mississippi 39180, USA
[2]The Morton Arboretum, Lisle, Illinois 60532, USA
[3]Division of Technology for Society, TNO, 1785 AJ Den Helder, The Netherlands

2 Methods and Materials

The construction of the demonstration site has been described in detail by Perrier et al. (1980) and Harrison and van Luik (1980) in WES technical reports. The following is a synopsis of those reports and a description of methodologies used to monitor floristic development and contaminant mobility.

2.1 Construction

The demonstration site was created amid parallel ridges of unvegetated pyritic overburden left after mining stopped in the early 1930s. These eroding ridges of mine spoil were leveled to form a gently sloping plateau (Fig. 1). Four test plots were constructed (a control plot and three treatment plots).

The treated plots received a 0.9-m-thick cover of dewatered dredged material obtained from the Alsip disposal facility of the Metropolitan Sanitary District of Chicago. The plots are shown in Fig. 2. They measure about 24 × 55 m and are isolated from each other and from the surrounding mine spoil by 0.6-m-high, beveled containment dikes. The dikes' integrity is maintained with an inert

Fig. 1. Aerial view of demonstration reclamation site and the pyritic overburden from strip mining at Ottawa, Illinois. Plot 4 is on the *left*: the untreated Plot 1 is on the *right*. Note the Parshall flumes in the *upper left-hand corner* of each plot

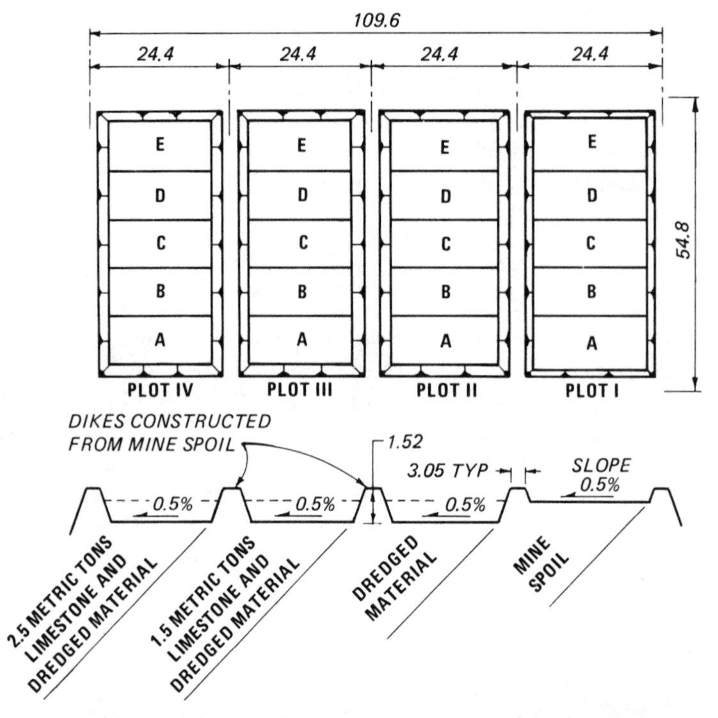

Fig. 2. Site plan and profile views of demonstration reclamation plots I-IV, including subplots *A-E*, at Ottawa, Illinois (Perrier et al. 1980)

waterproof shroud. Each plot was designed to effect a diagonal slope of about 0.5%, the lowest portion of which was designed to drain water volumes exceeding the field capacity of the substrate through a Parshall flume. In recent years, differential settling of the dredged material has compromised the original slope contours, so currently there are shallow low areas where water accumulates rather than runs off.

The control plot, designated Plot 1, consisted of leveled, untreated mine spoil.

Plot 2, located just south of Plot 1, was constructed of dredged material overlying untreated mine spoil.

In Plot 3, about 1.5 t (11 t ha^{-1}) of crushed agricultural limestone were incorporated into the top 15 cm of mine spoil before the layer of dredged material was applied.

Plot 4 was treated similarly to Plot 3, except that about 2.5 t (17 t ha^{-1}) of limestone were applied.

2.2 Soil Water and Runoff Water Sampling

Surface water was monitored for 1 year and soil water was monitored for 2 years. Soil water monitoring was accomplished by means of pressure-vacuum soil water samplers. These were installed in duplicate at 0.6-m depth in the control plot and at three depths (0.6, 0.9, and 1.5 m) and two locations (25 and 35 m) in each of the treated plots. Parshall flumes constructed of fiberglass-reinforced plastic were used to gauge runoff. The flumes were placed in the lowest, southwest corner of each plot, and runoff was collected during selected storm events.

Standard procedures for laboratory analysis of water samples were followed by Harrison and van Luik (1980) and are summarized here. Each sample collected was drawn into an acid-washed container and placed immediately in a refrigerated chest and held at approximately 4° C. Water sample pH was determined within 2 h of sample withdrawal. Following the pH determination, samples were divided and preservatives added according to the methods applied. All samples, both with and without preservatives added, were transported to the laboratory in refrigerated chests. Water samples without preservatives added were held for determination of acidity, chloride, and sulfate, while water samples with preservatives added were analyzed for heavy metals and nutrients.

2.3 Vegetation

After construction in 1978, all four plots were seeded with five types of perennial grass and one legume as shown in Table 1. Rates of application of pure live seed varied between 17 and 22 kg ha^{-1}, with a total seed application of 112 kg ha^{-1} per plot. After seeding, wheat straw mulch was placed on each plot at a rate of 4.5 t ha^{-1}. The mulch was sprayed with an asphalt emulsion to form a binder.

In July of the same year, a survey of the vegetation was conducted to record initial flushes in annual and biennial development (Perrier et al. 1980). In 1979 another survey was conducted which noted a shift toward the establishment of the seeded perennials. A regularly scheduled vascular plant survey was begun in 1980 and continued each spring through 1985. In 1982 summer and fall surveys were begun as well.

The goal of the survey was to gather only such data as necessary to describe in general terms the species composition of the plots so that notable changes in composition over time might be recorded. It was decided that the simplest approach was to make estimates of the percent phytomass (standing crop) of the

Table 1. Plant species seeded and their rates of application at the Ottawa, Illinois strip mine reclamation demonstration site in 1978

Seed mixture Grasses	Application rate (kg ha^{-1})
Kentucky Blue Grass (*Poa pratensis*)	17
Kentucky 31 Tall Fescua (*Festuca arundinacea*)	22
Lincoln Smooth Brome (*Bromus inermis*)	17
Blackwell Switch Grass (*Panicum virgatum*)	22
Perennial Rye (*Lolium perenne*)	17
Legume	
Bird's Foot Refoil (*Lotus corniculatus*)	17
Total	112

individual plant species within each plot. The estimates were made in increments of 5% inasmuch as finer increments could not meaningfully be discerned. Scattered individuals or plants with negligible phytomass were recorded as present only in trace amounts (ranging from a single individual to hundreds), but always were perceived as comprising less than 3% of the phytomass. Estimates were made by the same recorder in order to maintain a perceptual uniformity throughout the survey.

In an effort to simplify the estimates, as well as to make them more accurate, each demonstration reclamation plot was divided into five subplots (Fig. 2 A-E). Estimates were made on the plants in each subplot relative only to other plants in that subplot. A plant, for example, might have been perceived to comprise 15% of the phytomass in one subplot and then remained undetected altogether in one or more of the remaining subplots. Phytomass estimates for each plant species in the five subplots were averaged so that their mean values are expressed in terms of their importance relative to all other plants in the plot as a whole on a scale of 1 to 100 (Table 2).

Table 2. Summary of the major vascular plant species and their relative importance (expressed as percentage), and changes in species composition from year to year in spring in each of the four demonstration reclamation plots at Ottawa, Illinois

Plot 1		1980	1981	1982	1983	1984	1985
Smooth Brome		9	1	2	1	8	13
Tall Fescue		57	17	4	7	30	10
Yellow Sweet Clover		1	23	5	20	–	–
Switch Grass		31	56	79	64	50	57
Wild Black Cherry				1	1	1	8
Coefficients of similarity:	1980	100	67	44	46	44	42
	1981		100	65	66	58	54
	1982			100	89	75	70
	1983				100	87	71
	1984					100	83
Total number of species:		7	11	20	19	20	26
Plot 2		1980	1981	1982	1983	1984	1985
Smooth Brome		77	90	91	94	89	94
Field Thistle		–	–	1	1	4	2
Tall Fescue		9	1	1	1	–	–
Coefficients of similarity:	1980	100	79	72	62	57	52
	1981		100	80	62	60	59
	1982			100	72	73	64
	1983				100	88	69
	1984					100	73
Total number of species:		31	37	38	37	36	27
Plot 3		1980	1981	1982	1983	1984	1985
Smooth Brome		85	90	84	91	88	93
Grass-leaved Goldenrod		–	1	6	1	2	
Bird's Foot Trefoil		1	1	2	2	1	3

Table 2. *(continued)*

Plot 3		1980	1981	1982	1983	1984	1985
Kentucky Blue Grass		1	1	1	2	1	1
Tall Goldenrod		1	1	3	–	–	–
Tall Fescue		2	1	1	1		
Common Ragweed		2	1	1	–	–	–
Switch Grass		–	–	–	–	2	1
Orchard Grass		2	1	–	–		
Coefficients of similarity:	1980	100	87	65	55	49	44
	1981		100	75	58	59	49
	1982			100	69	68	61
	1983				100	77	67
	1984					100	73
Total number of species:		23	30	39	39	34	27

Plot 4		1980	1981	1982	1983	1984	1985
Smooth Brome		74	67	50	76	56	71
Grass-leaved Goldenrod		1	2	18	6	14	5
Tall Fescue		16	5	3	1	1	–
Tall Goldenrod		1	2	6	–	7	2
Kentucky Blue Grass		–	11	7	3	2	3
Bird's Foot Trefoil		1	1	3	1	2	3
Reed Canary Grass			1	1	4	4	–
Field Thistle		–	–	4	2	1	2
Evening Primrose		–	1	1	2	1	3
Hairy Chess		1	5				
Downy Brome		1	5				
Annual Sunflower		1	1	1	–	1	–
Common Ragweed		1	1	1	–	–	1
Switch Grass		–	–	–	1	1	2
Blue Vervain				–	–	1	2
Yellow Sweet Clover		1	1	–	–		
Coefficients of similarity:	1980	100	84	71	60	54	59
	1981		100	78	71	62	72
	1982			100	78	75	79
	1983				100	79	73
	1984					100	81
Total number of species:		28	34	40	42	43	46

For the comparison of species composition changes from year to year, coefficients of similarity were calculated using the standard relationship:

$$2C/A+B,$$

where C equals the number of species in common, A represents the number of species in one plot, and B represents the number of species in the compared plot. Botanical nomenclature, species concepts, and common names are those employed by Swink and Wilhelm (1979).

2.4 Contaminant Mobility

The initial chemical analysis of the dredged material used in the demonstration project indicated the presence of some toxic heavy metals; therefore, a decision was made to follow the contaminant mobility within the developing ecosystem. Studies utilizing bioassay and biomonitoring techniques were conducted to evaluate toxic heavy metal mobility and the possible presence of organic contaminants such as PCBs and PAHs.

The first studies, initiated in 1979, were designed to address the potential toxic metal uptake by the dominant plants in the developing, successional ecosystem. Samples of Kentucky 31 Tall Fescue and Lincoln Smooth Brome were collected at the field demonstration site. The plots of the site were designated I-IV and subdivided into subplots A-E (Fig. 2). In each subplot, the most abundant stands of Smooth Brome and Fescue were identified and collections were made. Each sample consisted of the total biomass, predominantly one of the species above, that could be encompassed by a square sampler (28.7 × 28.7 cm) made from a folded carpenter's ruler. The plants were clipped within 5 cm of the ground, and each sample was placed, along with an acetate label, into a plastic bag. It was secured with a twist tie and placed on ice in an insulated cooler chest for shipment to the Waterways Experiment Station. Each collected sample was photographed prior to collection. Voucher specimens of Brome, Fescue, and other plants present at the site were collected for reference and inclusion in the WES Research Herbarium.

At the WES, the extraneous plants were separated from the samples by comparison with the herbarium reference material. Plant material was washed with distilled water and oven dried at 70° C. Subsequently, the plant materials were ground into a coarse powder, digested with nitric acid, and analyzed for arsenic (As), cadmium (Cd), chromium (Cr), copper (Cu), lead (Pb), nickel (Ni), zinc (Zn), and mercury (Hg). All methods used are described in Simmers et al. (1980).

Concurrent with the native plant biomonitoring effort, the WES plant bioassay, utilizing Yellow Nutsedge (*Cyperus esculentus*), was applied in the field at Ottawa. This test (Folsom et al. 1981) has been used to describe potential plant uptake of heavy metals from sediments and soils. The 7.6-liter inner bucket of the "double bucket" apparatus (Fig. 3) was embedded in the site in each plot. The bucket was filled with dredged material collected from the plot. The Nutsedge was allowed to grow in each embedded bucket for 45 days prior to harvest and analysis. Results of the field plant bioassay were compared to colonizing plant data and to the Nutsedge data base developed from collections in naturally occurring areas in the Great Lakes region (Simmers et al. 1980).

Plant studies were augmented by the application of an earthworm bioassay (Simmers et al. 1985). This study was designed to aid in determining the major routes of contaminant mobility. The earthworm (*Eisenia foetida*) was used as an indicator of bioavailability of metals in the leaf litter, the surface layer of dredged material (30 cm), and the deep layer of dredged material near the mine spoil (100 cm). These data further clarify the contaminant mobility aspects of the reclamation technique in relation to management of large-scale disposals.

The earthworms were obtained from WES cultures and randomized into subsamples of about 150 worms (approximately 20 g wet wt.) and placed in

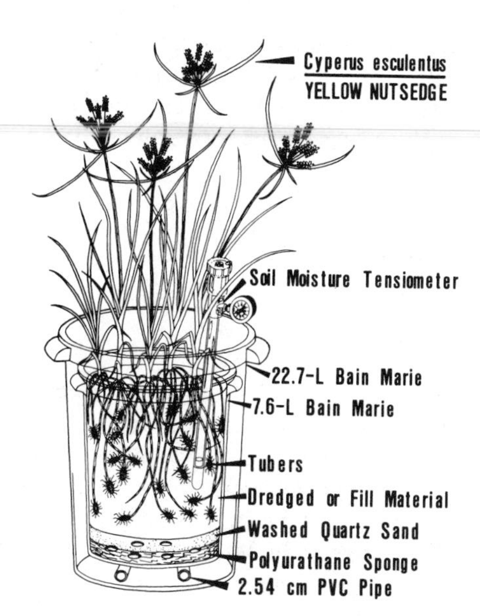

Fig. 3. Plant bioassay double bucket apparatus designed by the US Army Engineer Waterways Experiment Station (Folsom et al. 1981)

replicated composite 7.6-liter samples of dredged material from 30- and 100-cm depths as well as the leaf litter. Exposure was allowed for 28 days prior to the recovery of the worms for analysis. Temperature in the growth chamber used for laboratory bioassay testing was maintained at 15° C with low light conditions. The lower 3–5-cm section of the laboratory test containers was kept saturated with water, while the top was kept dry. This allowed the worms to find their optimum moisture conditions in the test materials.

Earthworms also were exposed in the field in 7.6-liter buckets as in the plant field bioassay. After recovery, the earthworms were transported to the laboratory in small amounts of their respective media. Both field-exposed and laboratory-exposed earthworms were purged for 48 h at 10° C on moist filter paper prior to chemical analysis.

Each tissue sample was homogenized using a PT 10/35 Brinkman Homogenizer (Sybron Corp, Westbury, N. Y.) equipped with a PTA 20S titanium blade. The resulting homogenate was divided into two subsamples (5 g for heavy metals analysis and 10 g for hydrocarbon analysis) and wet-digested.

Tissues and substrates were analyzed for heavy metals using atomic absorption spectroscopy and US Environmental Protection Agency (USEPA) sample digestion procedures (Delfino and Enderson 1978). Test samples were analyzed for Cd, Cu, Cr, Pb, Ni, and Zn using a Perkin-Elmer Model 2100 Heated Graphite Atomizer and a Perkin-Elmer Model 5000 Atomic Absorption Spectrophotometer. Mercury was determined using a Perkin-Elmer Model 503 Atomic Absorption unit using the cold vapor technique. Arsenic was determined using a Perkin-Elmer Model 305 Atomic Absorption unit with a MHS-10 hydride generator.

Techniques for hydrocarbon analyses of earthworm tissues and substrates also followed recommended EPA procedures (USEPA 1982). Organic compounds in sediments were extracted using a hexane-acetone solvent, and those in tissues were extracted with 4% sodium hydroxide. PCBs were determined following Soxhlet extraction and silica gel cleanup by means of electron capture gas chromatography. The PAH fraction was separated by silica gel chromatography and subjected to capillary gas chromatography.

3 Results

3.1 Quality of Soil Water

During the first 2 years after site construction, soil water was sampled at three depths in the treated plots and at one depth in the control plot. The three depths chosen for the pressure-vacuum samplers placed in the treatment plots assured that the samples represented the dredged material, the dredged material/mine spoil interface, and the mine spoil. The data summarized in Table 3 indicate that the pH of the dredged material near the surface dropped slightly with the establishment of vegetation and then remained relatively constant through the monitoring period. Also of interest, however, is the trend shown by the pH measurements at the 1.5-m depth, where the soil water was collected below the dredged material/mine spoil interface. Here, the soil water became less acidic.

Table 3. Soil water pH in control and treatment plots at the Ottawa, Illinois demonstration reclamation site (After Harrison and van Luik 1980)[a]

Plot	Meter depth	Nov 77	Jun 78	Sampling date Nov 78	May 79	Sep 79
				pH		
1: Untreated mine spoil	0.6	5.4	3.8	3.5	3.7	3.4
2: No lime added	0.6	8.0	6.7	7.7	6.5	6.7
	0.9	8.0	6.6	8.0	6.5	6.7
	1.5[a]	5.4	5.8	4.8	5.0	6.4
3: 11 t ha^{-1} lime	0.6	8.0	6.9	7.5	6.5	6.7
	0.9	7.9	6.7	6.8	5.5	–
	1.5[a]	2.5	2.9	2.6	5.6	5.7
4: 17 t ha^{-1} lime	0.6	8.0	6.8	7.7	6.5	6.7
	0.9	5.7	6.6	7.1	6.6	6.8
	1.5[a]	4.6	3.9	4.0	6.5	6.7

[a]The 1.5-m depth in Plots 2–4 is below the applied dredged material and in the original mine spoil

3.2 Quality of Runoff Water

The pH and acidity of the runoff collected during a storm event are tabulated in Table 4. The data demonstrate the physical effectiveness of the use of dredged material in pyritic mine spoil restoration insofar as its capacity to reduce acid runoff. Extensive analyses of the runoff for heavy metals were conducted. No

Table 4. Selected runoff water quality values collected during a storm event April 10, 1978 (After Harrison and van Luik 1980)

	Control plot (1)		Treated plots (1, 2, and 3)	
	High value	Mean value	High value	Mean value
pH	3.34	3.52	7.15	7.28
Acidity, mg l^{-1}	1340	582	20	16
PO_4, mg l^{-1}	0.05[a]	0.05[a]	0.06	0.05[a]
TKN, mg l^{-1}	0.85	0.65	1.80	0.95
NH_3, mg l^{-1}	0.42	0.33	0.25	0.22
NO_3, mg l^{-1}	3.0	2.2	0.32	0.32
SO_4, mg l^{-1}	1500	633	300	131
Fe, mg l^{-1}	3.66	1.22	1.34	0.50
Cl, mg l^{-1}	1.0[a]	1.0[a]	6.0	1.2

[a] Below detection limits.

movement of the toxic metals present in the dredged material was noted in the runoff water, and levels of As, Cd, Cu, Pb, Cr, Hg, and Ni were below detection limits of 0.02–0.05 mg l^{-1}.

3.3 Vegetation

During the year in which the plots were planted (1978), the floristic composition of the three plots consisted of a vigorous flush of ubiquitous annuals: Common Ragweed (*Ambrosia artemisiifolia* var. *elatior*), Beggar's Ticks (*Bidens polylepis* and *B. frondosa*), Black Mustard (*Brassica nigra*), and Heartsease (*Polygonum lapathifolium*). Of the species planted in the treated plots, only Perennial Rye was present to any significant extent. Switch Grass and Kentucky Blue Grass were not at all evident, though they later became established in each plot. Tall Fescue, Smooth Brome, and Bird's Foot Trefoil appeared only sporadically, fully overwhelmed by the annuals mentioned above.

By 1979 it was apparent in the treated plots that two of the planted perennial grasses (Smooth Brome and Tall Fescue) were becoming well established. The annuals were much less evident. Remaining particularly in the low areas, were stands of Common Burdock (*Arctium minus*), a biennial, and Beggar's Ticks (*Bidens polylepis*), an annual. The untreated plot continued to contain only scattered clumps of Smooth Brome and Tall Fescue in its drier, eastern two subplots; the other three subplots were barren.

By 1980, the three treated plots had become nearly completely vegetated by a mix of Smooth Brome and Tall Fescue. The untreated plot remained largely bare of vegetation, though in areas where plants had been growing, Switch Grass was becoming dominant. In 1985, the species composition of all the plots changed by an average of 49%±8. Even the dominant species, especially in Plot 4, changed rather dramatically over the 5-year period.

Table 2 is a summary of the major vascular plant species in each of the four demonstration plots as perceived in the spring of the year. Each species is represented by a modified importance value expressed as a percent. The table also shows changes in species composition from year to year, expressed as changes in similarity coefficients. Each succeeding year is increasingly different from 1980, the year in which systematic vegetation sampling began. The total number of species in each plot for each year is shown at the bottom of each plot table.

Most of the species present were apparent only in amounts comprising less than 3% or so of the total phytomass; however, in Plot 2, 1980, two species (Tall Fescue and Smooth Brome) together comprised 86% of the total phytomass. The other 14% was constituted by an aggregate of another 29 species. Each of these minor species, while individually interpretable perhaps as insignificant, may in fact play significant roles in the physicochemical evolution of the developing soil horizons.

3.4 Contaminant Mobility

The results of heavy metal analysis for Plots 2, 3, and 4 of the field plant bioassay study conducted in 1981 are compared in Table 5 with the analysis of the dredged material reported by Perrier et al. (1980) and the results of a field survey of *Cyperus* growing in naturally occurring wetlands around the Great Lakes by Simmers et al. (1980). The limited colonization of the control plot (Plot 1) by plants was not within the scope of this contaminant mobility and succession study. These results indicate that in spite of the relatively high substrate metal concentrations (e.g., Cd), the plant uptake, as demonstrated by *Cyperus*, was within the ranges of the naturally occurring plants. The mean concentration in the leaves of the bioassay plants was slightly elevated, but Cd was relatively unavailable to plants. Lead and Ni were not accumulated to any notable extent.

Heavy metal contents of the Fescue and Smooth Brome were determined shortly after the experimental plots were constructed. Statistical analysis indicated no significant trends in the variation among all subplots of the treated plots. An analysis of variance showed no significant differences to indicate any contaminant

Table 5. Contaminants ($\mu g\ g^{-1}$) in dredged material and in bioassay plant tissues, 1981

contaminant	Dredged material[a] (n = 4) Mean SD	Field plant bioassay (n = 16) Mean[c]	Range	Field Survey[b] (n = 31) Mean[c]	Range[c]
As	–			d	
Cd	13.0± 0.6	0.81	0.56± 1.37	0.51	d± 1.78
Cr	165.6± 4.9	d		2.35	d± 10.30
Cu	113.0± 13.2	12.26	9.65± 19.20	6.53	d± 26.96
Pb	507.0± 7.0	0.93	0.69± 1.15	5.22	d± 43.56
Ni	53.2± 3.5	d		3.29	d± 12.20
Zn	1123.0± 53.0	84.2	63.3 ± 106.0	74.8	8.6± 237.1
Hg	0.9± 0.3	–		d	

[a]From Perrier et al. (1980).
[b]From Simmers et al. (1980).
[c]d = Detection limit.

concentration gradients at the site, either along the slopes or across the plots. Based on these results, the Smooth Brome and Fescue data are shown as means for the dredged material portion of the field site (Table 6). Observed changes in floristic composition suggested potential changes in rhizosphere environment, so the plant heavy metal concentrations were evaluated again in 1983. Interestingly, over the 4 year period, Fescue had declined to the point where sufficient phytomass for replicated sampling was unavailable. For the Brome, however, bioavailability of at least Cu, Pb, and Ni appeared to have changed. Neither Brome nor Fescue contained a high Cd level; the Fescue did appear, however, to have accumulated Pb and Ni to levels that may be of possible concern, though there are no applicable FDA action levels (USFDA, undated). At the time the data were collected, Brome was responsible for comparatively little toxic metal uptake. Had the Fescue remained a major part of the ecosystem, management procedures might have been necessary to retard movement of heavy metals into other levels of the ecosystem.

Table 6. Mean contaminant levels ($\mu g\ g^{-1}$) in two dominant plant species in the treated demonstration reclamation plots 2–4 at Ottawa, Illinois

Contaminant	Substrate[a]	Festuca[b] (n = 9) 1979	Bromus[b] (n = 16) 1979	Bromus[b] (n = 14) 1983
As	–	d	d	d
Cd	13.0± 0.6	1.05± 0.73	0.78± 0.57	0.94± 0.25
Cr	165.6± 4.9	5.45± 1.53	2.94± 0.59	1.72± 0.39
Cu	113.0± 13.2	5.11± 1.65	7.67± 1.09	15.94± 2.24
Pb	507.0± 7.0	12.17± 2.4	3.25± 1.46	3.80± 0.80
Ni	53.2± 3.5	5.15± 2.14	3.21± 0.93	1.64± 0.46
Zn	1123.0± 53.0	72.43± 13.46	55.49± 12.35	75.54± 14.70
Hg	0.9± 0.3	d	d	d

[a]From Perrier et al. (1980), n = 4.
[b]d = Detection limit.

A comparison of the earthworm tissue levels of metals from the three test media is shown in Table 7. Smooth Brome and Fescue were collected and analyzed earlier and are the source of the thick leaf litter or duff layer on the now grass-dominated plots. Cadmium appears to be available from the leaf litter layer, while Cu and Ni appear to be more bioavailable in the dredged material. Lead is apparently equally available in all three media.

The field screening bioassay for PCBs and PAHs was an attempt to apply current analytical capabilities not previously available to an older site where a major contaminant may have been overlooked. These data are shown in Table 8. PCB congeners were present in low levels in the earthworm tissues, but they may have been present originally in higher levels had earthworms been exposed at the disposal site in Alsip where the material was dewatered. Similarly, PAHs were also present in earthworm tissues.

Table 7. Contaminant concentrations in earthworms, leaf litter, and substances from treated demonstration reclamation plots 2–4 at Ottawa, Illinois ($\mu g\ g^{-1}$ dry w.) 1983

Test material	Cd	Cu	Pb	Ni
Background earthworm	3.67±0.51	9.55± 1.00	1.50± 0.65	2.00±0.77
Earthworm	14.07±5.37	9.17± 1.56	2.17± 0.46	1.87±0.46
leaf litter	3.27±0.73	15.66± 1.65	–	5.89±0.20
Earthworm	9.03±0.89	25.83± 4.20	2.87± 0.69	5.23±0.61
30-cm depth[a]	10.00±0.50	127.00± 8.60	620.00±69.90	51.50±3.20
Earthworm	8.23±0.21	25.37± 1.03	5.27± 2.00	5.33±0.31
100-cm depth[a]	9.18±1.63	116.7 ±10.30	585.00±22.80	50.10±2.16

[a]Dredged material.

Table 8. Concentrations of selected PCBs and PAHs (ash-free dry wt.) in experimental worms exposed to dredged material for 28 days at the Ottawa, Illinois reclamation demonstration site in 1983

PCBs ($\mu g\ kg^{-1}$)	Worm tissue concentration
2,5,2′,5′ Tetrachlorobiphenyl	53.0
2,4,5,2′,5 Pentachlorobiphenyl	87.0
2,4,5,2′,4′,5′ Hexachlorobiphenyl	270.0
2,3,4,2′,4′,5′ Hexachlorobiphenyl	300.0
2,3,4,5,2′,4′,5′ Heptachlorobiphenyl	210.0
PAHs ($\mu g\ g^{-1}$)	
Anthrene	0.10
Benzo (b) flouranthene	0.18
Benzo (k) flouranthene	0.14
Benzo (g,h,i) perylene	1.90
Benzo (a) pyrene	0.39
Benzo (e) pyrene	0.46
Chrysene	0.10
Dibenzo (a,i) anthracene	0.71
Dibenzo (a,i) pyrene	0.77
Perylene	0.14
Phenanthrene	0.12
Pyrene	0.11
Triphenylene	0.11

4 Discussion

4.1 Soil Water and Runoff

The basic objective of the demonstration project was accomplished as demonstrated by the soil water and runoff data collected. The regrading of pyritic mine spoil and application of a dewatered dredged material cover halted acid runoff. The pH at the dredged material/mine spoil interface remained near neutral, and vegetation was established. From a simple restoration viewpoint, the project should be considered successful and serve as an example for pyritic surface mine spoil restoration. There was no indication in the physical data to recommend the application of lime at the interface; it appears from a soil water pH view, that the dredged material alone was satisfactory. The lime rate, however, may have had an impact on the successional patterns in the plant communities developing on the site.

At this point, the restoration portion of the project is complete. The following discussion will address the ecosystems which are becoming established and the movements in the ecosystem of contaminants from this particular dredged material.

4.2 Vegetation

The inevitable conclusion concerning the vegetation is that from the first year until the present, it has continued to change, and to succeed beyond what it was the year before. Changing soil horizons and the development of a rhizosphere consisting of a slowly changing root-soil chemistry are evidently changing the physicochemical character of the dredged material.

Once a homogeneous mixture of barge canal sediment, the dredged material is becoming increasingly stratified and chemically diverse. It is becoming clear that one of the challenges facing practitioners who attempt to restore any kind of land is to come to understand that pedogenesis and plant succession will be ongoing factors and that their impact on developing ecosystems will have to be addressed. It will be necessary to understand what factors control the kinds of plant succession which can occur under a given set of conditions. At Ottawa it is likely that, if perennial grasses had not been planted, the succession would more rapidly have been under the influence of woody plant growth, e.g., ashes, box elders, elms, cherries, cottonwoods, or some mixture of them. In this case this would likely have had an effect on rhizosphere development and contaminant movement patterns quite different from what we see today. An inevitable vegetational succession will occur on each site and progress according to the initial availability of seeds and diaspores. For this reason, it is apparent that a “no management” plan must be regarded as an active management decision.

4.3 Contaminant Mobility

The incidental presence of toxic metals in the dredged material at the Ottawa demonstration site appeared to have little direct effect on the plant community. If there was an effect, it is likely that macroecological factors involving early successional competition has obfuscated it. Both of the dominant grasses in the demonstration plots contained negligible tissue levels of heavy metals when analyzed. Results from the application of the field version of the WES plant bioassay suggest that there is no unusual movement of contaminants from this dredged material into the vegetation. The only notable situation that could be contaminant-mediated was the decrease of the Fescue at the site concurrently with relatively elevated Pb tissue levels. At this time, this can only be considered a coincidence and not a selection of successional species by contaminant influence. Other than the Pb levels in Fescue (Table 6), dominant plants and bioassay plants did not indicate contaminant levels beyond what is occurring in reference populations found in the Great Lakes area.

Earthworm bioassay procedures conducted in the growth chamber and screening tests conducted in the field did indicate areas of potential concern. Table 7 indicates a potential route of bioaccumulation of Cd by the earthworm in a laboratory setting. The presence of earthworm species which feed in the leaf litter zone could allow a significant flow of Cd from dredged material through the earthworm to predators. This route of contaminant movement has not been verified in the field at the Ottawa demonstration site.

The presence of PCB congeners and PAHs in the test earthworms is probably quite typical for older dredged material disposal sites. It is suspected that at the time of original dredging many organic contaminants were present. The levels accumulated by the earthworm biomonitors may well reflect PCB and PAH levels below detection limits in the dredged material at the present time.

5 Summary and Conclusion

Dewatered dredged material can be used effectively in the reclamation of pyritic mine spoil. The demonstration project described here showed that a cover of dredged material stopped acid runoff while allowing speedy revegetation.

The vegetation now established at the site is changing in composition at a steady rate; pedogenesis is under way. How changes in contaminant movement will be expressed (considering pedogenesis and rhizosphere development) is the object of continued research, as is research into how management protocols can be used to control these processes to produce positive results.

Although some toxic metals, PCBs, and PAHs are present, there do not now appear to be any potential routes of contaminant uptake other than the movement of Cd from leaf litter as indicated by the bioassay earthworms. The presence of low levels of contaminants in dredged material need not eliminate it from consideration for a productive use; however, it is necessary to conduct the appropriate bioassay monitoring procedures and be prepared to implement appropriate management strategies consistent with the productive use.

Acknowledgments. This research was conducted as part of the US Army Corps of Engineers Dredging Operations Technical Support Program (DOTS), under the auspices of the United States/Netherlands Memorandum of Understanding concerning dredging and related technology.

References

Delfino JJ, Enderson RE (1978) Comparative study outline methods of analysis of total metal in sludge. Water Sewage Works 125 (RN) R32–34;47–48
Folsom BL, Jr, Lee CR, Bates DJ (1981) Influence of disposal environment on availability and plant uptake of heavy metals in dredged material. US Army Eng Waterways Exp Stn, Tech Rep EL-81-12, Vicksburg, Miss
Harrison W, Luik A van (1980) Suitability of dredged material for reclamation of surface-mined land. US Army Eng Waterways Exp Stn, Tech Rep EL-80-7, Vicksburg, Miss
Perrier ER, Llopis JL, Spaine PA (1980) Area strip mine reclamation using dredged material: a field demonstration. US Army Eng Waterways Exp Stn, Tech Rep EL-80-4, Vicksburg, Miss
Simmers JW, Folsom BL, Jr, Lee CR, Bates DJ (1980) Field survey of heavy metal uptake by naturally occurring saltwater and freshwater marsh plants. US Army Eng Waterways Exp Stn, Tech Rep EL-80-5, Vicksburg, Miss
Simmers JW, Marquenie JM, Rhett RG (1985) Bioavailability of heavy metals in relation to potential use of dredged material for large scale acid mine spoil restoration. In: Lekkas T (ed) Heavy metals in the environment. Int Conf Proc, Edinburgh, UK, pp 214–216
Swink F, Wilhelm G (1979) Plants of the Chicago region. The Morton Arboretum, Lisle, Ill
(USEPA) (1982) Test methods for evaluating solid waste, 2nd edn. SW-846
USFDA (undated) Action levels for poisonous or deleterious substances in human food and animal feed. Washington, DC HFF-326

Part III Management

Towards a Long-Term Balance Between Economics and Environmental Protection

MALTE FABER and GERHARD WAGENHALS[1]

1 Introduction

In contrast to the other contributions to this volume, our chapter concentrates on an economic approach to the problems of mine wastes and dredged materials. We stress the existence of trade-offs between economics and environmental protection, which may be interpreted in terms of the unifying notion of entropy. The aim of this chapter is to show:

1. That there is a strong presumption that the market allocation of depletable resources is intertemporally suboptimal resulting in prices which are too low to allow for sufficient protection of the environment and of the needs of future generations;
2. How the concept of entropy may serve to make way towards a reconciliation of economics (or, more exactly, between the consumption of goods) and environmental protection in the case of mine tailings and dredged material.

It may interest the reader that our approach is also in contrast to the other treatments of environmental and resource problems in economic literature, for in many respects economics has lost its biophysical foundation (Faber 1985, p. 316). We use the notion of entropy to come to grips with this aspect and to try to develop a physically-oriented production theory. With our natural science and economic approach we hope to further at the same time interdisciplinary research between economics and the natural sciences which seems to us expedient if not necessary to contribute to the solution of the complex and intricate environmental and resource problems (cf. Faber and Proops 1985, p. 604).

Section 1 deals with the role of resource prices in market economies, suggesting that they often may be too low compared to prices which result in an intertemporal optimum.

Based on a multisector model of the economy, Section 2 derives fundamental qualitative equations dealing with the interrelations between mine wastes, entropy and economic variables like mining costs or energy requirements for extraction. Section 3 illustrates our main arguments and Section 4 checks their predictive capabilities, relying on data of two of the world's major mineral-producing countries. The final section presents a critique of our approach and summarizes the main results of our study.

[1]Alfred Weber-Institut, University of Heidelberg, Grabengasse 14, D-6900 Heidelberg, FRG

2 The Role of Resource Prices in Market Economies

Consider the model of an ideal market economy with private ownership. If utility-maximizing consumers and profit-maximizing producers take market prices as given, an equilibrium price system exists. An equilibrium price system is a vector of prices which makes the actions of all economic agents compatible with the total resources available. The allocation of resources corresponding to an equilibrium price system is optimal in the sense that within certain institutional, technological and behavioural constraints there exists no other feasible allocation of the factors of production (including resources) of an economy where everyone is at least as well off and at least one agent is strictly better off (Debreu 1959; Malinvaud 1985). An "allocation" of factors of production in an economy is a very complex concept. Allocation entails the information (1) how much of each of the factors of production is used to produce a good, and (2) how much of each good is produced. The most important institutional condition for the existence of an equilibrium price system and optimal allocation of resources is the existence of a complete set of markets for all goods. This includes:

1. A complete set of futures markets; and
2. A complete set of markets for environmental pollution.

In the following we shall show that conditions (1) and (2) are not fulfilled in reality and that therefore prices in market economies may not provide an optimal allocation of resources.

2.1 Non-Existence of a Complete Set of Futures Markets

Minerals and metals do not regenerate themselves naturally, i.e. they are not renewable. Ceteris paribus, in the course of time the extraction of such resources becomes more and more expensive due to the declining ore concentration of the deposit and ensuing higher energy requirements in mining. Figures 1 and 2 illustrate this fact for a hypothetical deposit.

At the latest in the long term, when a metal's concentration in the earth's crust becomes too small, substitution takes place and the resource can no longer be extracted economically. Economic agents have to switch to alternative materials.

The existence of a complete set of futures markets for all time periods is a condition for the existence of an efficient market solution. These markets could conceivably consider the effects of declining concentration (Fig. 1) and the corresponding increase in energy requirements (Fig. 2). In reality, however, if futures markets for metals and minerals exist at all, contracts are traded only for metals available in less than 2 years. Futures markets extending for decades into the future do not exist. Thus, futures prices quoted in reality do not necessarily allow for the fact that metals are not renewable in the long run.

Summing up, missing futures markets extending to periods where the exhaustibility of resources may be felt significantly represent the first reason why the market system may fail to achieve an intertemporally optimal allocation of resources.

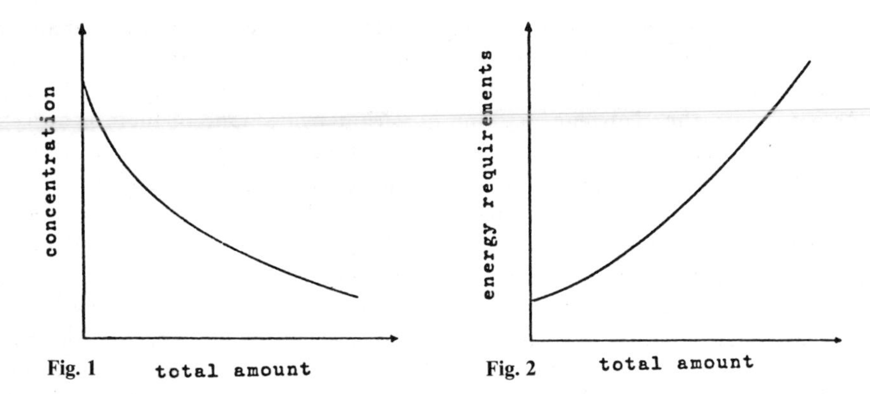

Fig. 1. Concentration depending on total amount extracted
Fig. 2. Energy requirements to extract, mine and mill one unit of a resource depending on total amount extracted

2.2 Non-Existence of a Complete Set of Markets for Environmental Pollution

Extracting, mining and milling of raw materials as well as their use in fabricating and their disposal cause environmental damage. For example, conversion of concentrates may involve operations of roasting with the emission of sulphur dioxide gases. If they are not eliminated from the atmosphere their noxious character may cause production externalities and may seriously affect living beings. Runoffs from solid waste piles, both as sediments and as solutions containing metal salts, also may pose serious problems. Emissions trading, as practiced partially in the USA, may serve as an example for a market for environmental pollution in reality (Rehbinder and Sprenger 1985). But typically none or only insufficient markets of this type are available. Therefore, if externalities due to environmental pollution exist, price-taking profit-maximizing behaviour does not lead to an optimal allocation of resources.

To sum up, externalities due to environmental pollution in extracting, mining and milling represent the second reason why markets may fail to achieve an optimal allocation of resources.

2.3 Consequences

Our considerations suggest that a competitive price system does not lead to an optimal intertemporal allocation of resources. The right price of a depletable resource should cover intertemporal opportunity costs due to their eventual exhaustion and opportunity costs due to externalities, like potential environmental damages because of mine wastes. Because the price system alone does not provide an efficient solution to the problem of intertemporal allocation, raw materials are "too cheap" and thus they are not used economically.

If exhaustible resources were priced "correctly", i.e. if their price included the components described above, then, for example, the safe deposition of mine wastes and dredged materials and recycling measures could be managed and financed by firms. There would be no necessity for state authorities to intervene. In the absence of special markets (1) and (2), which make the market allocation optimal, the state has to intervene. This can be done in various ways. One of them is a tax and subsidy policy which generates a welfare improvement for at least one agent without hurting other agents. However, the form of this policy depends sensitively on the technology used by the producers and on the preferences of the consumers. Since the required intervention depends on more accurate information than currently available, it is not a priori certain whether public policy will be able to improve matters. In each case, the pros and cons of each measure have to be assessed carefully. In qualitative terms, however, it may be said that the market prices of minerals and metals are too low in comparison to the prices which are intertemporally optimal. Thus, intervention should tend to increase mineral and metal prices to allow an intensification of environmental protection and to provide for future generations. This should also include the deposition of mine wastes which are currently not yet exploitable with a profit. We will turn to this point in the next section.

Up to now, we showed that there is no presumption that the market allocation of extractive resources is intertemporally optimal due to missing future markets and externalities resulting from environmental pollution. From now on we mainly concentrate on these two effects, because they are most important for the management of mine wastes and dredged materials.

The basic trade-off between economics and environmental protection can be illustrated as follows. Extracting a mineral from the environment produces a concentrate and a waste material. Extraction decreases the concentration of the resource in the environment (see Fig. 1). Labour and capital inputs required to extract one unit of a resource depend on its concentration, which declines with increased accumulated extraction of a resource (see Fig. 2). Thus, extraction costs increase in the course of time ("intertemporal economic effect") and simultaneously the burden on the environment increases ("environmental effect"). The increase of the latter is due to two causes. (1) Decreasing concentration makes it necessary to mine increasing amounts to extract one unit of a metal. (2) In a growing economy increasing amounts of many ores are extracted and in turn used for the production of increased amounts of consumption goods. These in turn will increase the amount of waste and thus in general environmental pollution.

In the rest of this chapter, the notion of "entropy", well known from thermodynamics, will be used to relate these two effects. According to Boltzmann's interpretation, entropy can be used as a measure of "disorder". Both the mining of a resource from a site with high grade ore as well as its final disposal into the environment, e.g. in the form of mine wastes, generate a continuous increase of our planet's entropy. Thus, there is an interdependence between entropy and the concentration of a resource, on the one hand, and between entropy and potential environmental pollution due to mine wastes and dredged materials, on the other (Faber et al. 1983b, Faber 1985). Both effects have an influence on costs and therefore on the profitability of a mining enterprise.

The following section will deal with these relationships in some detail theoretically, in the fourth section we will apply them to specific cases.

3 Entropy, Environment and Resources

This section describes the basic relations of a resource and an environmental model including in particular the deposition of mine wastes and recycling.

First, we turn to the modelling of resource extraction relying on the relationship between concentration, entropy and energy use. Then we deal with entropy as a highly aggregated measure of pollution. Finally, we sketch the main features of a resource and environmental model including mine wastes relying on the notion of entropy as a unifying concept.

3.1 Entropy and Concentration of a Resource

First, we discuss the relation between entropy and the concentration of a binary mixture of molecules. Consider a physical mixture of n1 atoms of type 1 and n2 atoms of type 2. Let

$$n := n1 + n2 \tag{1}$$

denote the total number of atoms in a given site. Then,

$$x := n1/n \tag{2}$$

defines the concentration of type 1 atoms (e.g. a metal) in the site. According to Kittel and Krömer (1984, p. 312), the change in entropy due to unmixing n1 atoms of the metal and n2 atoms of type 2 sums up to

$$\Delta\sigma = nB(x \log x + (1-x)\log(1-x)), \tag{3}$$

where B denotes Boltzmann's constant ($B = 1.38066 \times 10^{-23}$ JK^{-1}). $\Delta\sigma$ is negative, because log x and log (1–x) are negative. Since we deal with a physical mixture of atoms, we can neglect enthalpy. Therefore, the change in Gibbs' free energy gives the minimum energy requirement of unmixing n1 atoms of the metal and n2 atoms of type 2:

$$\Delta\mu = -T.\Delta\sigma, \tag{4}$$

where T denotes absolute temperature in Kelvin (cf. Chapman, Roberts 1983, p. 91). Figure 3 presents a three-dimensional graph of Eq. (4), i.e. it shows the change in energy of unmixing two types of atoms depending on temperature T and on concentration x for a given number of atoms n in a site. The figure demonstrates clearly: the higher the mixture of the two types of atoms and the higher the absolute temperature, the higher the energy requirements to unmix them.

Up to now, our considerations referred to changes in entropy and energy for n atoms in a site. Now we concentrate on total changes in entropy and energy *per mole* of one resource (e.g. resource 1, which is in our illustration the metal) in a site. First, we have to divide Eq. (3) by n1. This gives $\Delta\sigma/n1$, i.e. the change in entropy per atom in the metal. Secondly, it is a corollary of Avogadro's law that the number of molecules in a mole is the same for all substances. This number, called Avogadro's constant, is approximately $N = 6.02205 \times 10^{23}$ mol^{-1}. Multiplying

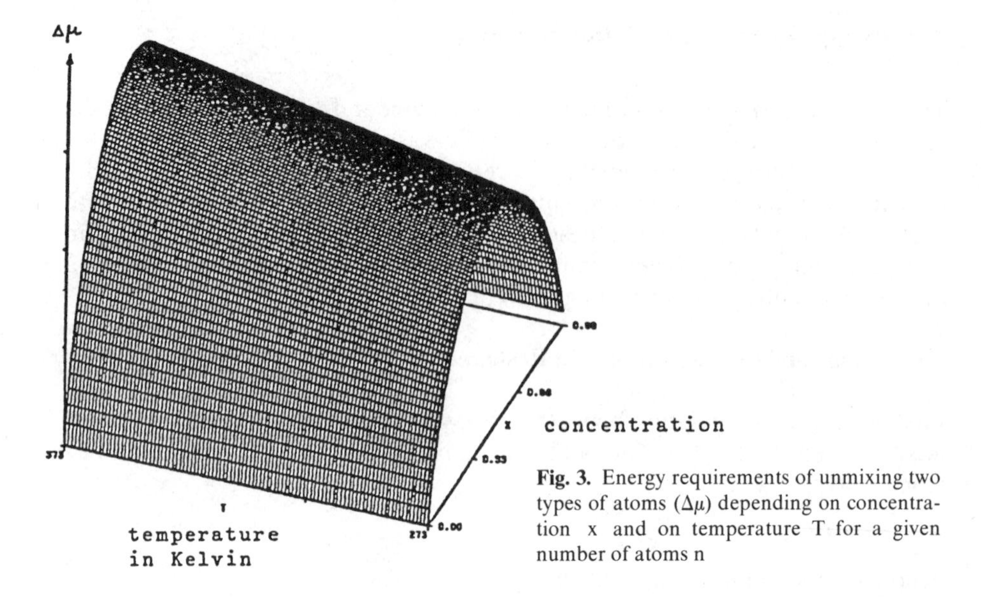

Fig. 3. Energy requirements of unmixing two types of atoms ($\Delta\mu$) depending on concentration x and on temperature T for a given number of atoms n

$\Delta\sigma/n1$ with Avogadro's constant finally gives the change in entropy per mole of resource 1:

$$\Delta S_{n1} = (N \cdot \Delta\sigma)/n1. \tag{5}$$

Inserting Eqs. (2) and (3) in Eq. (5) gives

$$\Delta S_{n1} = NB(\log(x) + \frac{1-x}{x} \log(1-x)). \tag{6}$$

Using the formula for Gibbs' free energy again, the minimum energy requirement to extract 1 mol of a resource is:

$$\Delta U_{n1} = -T\Delta S_{n1}. \tag{7}$$

Faber et al. (1983b, p. 103) present a different derivation of this formula. Illustrating Eq. (7), Fig. 4 shows the energy requirements to extract 1 mol of a resource depending on concentration x and as a function of temperature T. The figure demonstrates that the lower the concentration and the higher the temperature, the higher is the energy required to extract 1 mol of resource 1.

This illustrates the interrelations between concentration, entropy, energy requirements and therefore also energy costs of extracting 1 mol of a resource.

3.2 Entropy and Pollution

Mining, milling and metallurgical treatment of minerals entails production of a sizable amount of wastes. Tailings disposal constitutes a severe problem for the mining industry. Based on Kümmel (1980), this section shows how to construct a pollution function using entropy as an aggregate measure of environmental pollution.

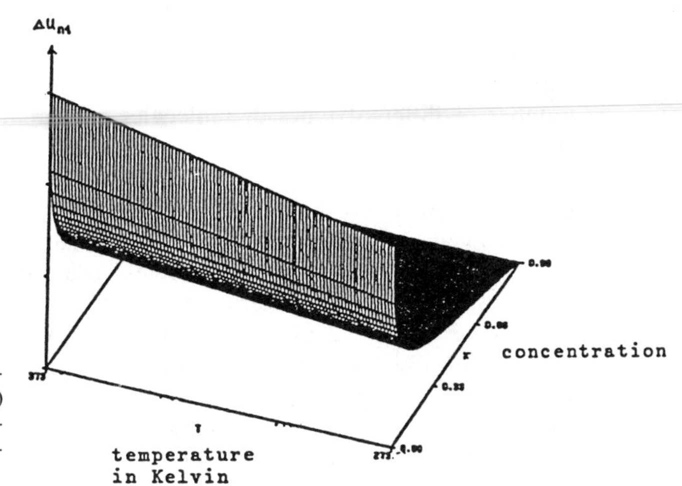

Fig. 4. Energy requirements per mole (ΔU_{n1}) depending on concentration x and on temperature T

For a given space of an economic system V and for a given increase in entropy per unit time dS/dt:

$$U := (dS/dt)/V \tag{8}$$

defines a total measure of pollution. The dimension of U is J K^{-1} per unit of volume. Assuming the existence of m different types of molecules contributing to environmental pollution, for each sort of molecule i (i = 1,..., m), a corresponding degree of environmental pollution U_i may be constructed (for details see Kümmel 1980, pp. 25–26). Given specific critical upper limits of pollution U_{ci}, which are determined by society and which must not be exceeded, and given natural purification rates U_{oi} for each type of molecule i,

$$P_i(U_i) = \left(1 + \exp\left(\frac{U_i - U_{ci}}{U_{oi}}\right)\right)^{-1} \tag{9}$$

defines a sensible measure of the welfare loss due to pollution with type i molecules. Welfare loss increases (or economic welfare decreases) with growth in pollution and with declining margins between natural purification rates and critical upper limits of pollution.

A measure of total welfare loss due to environmental pollution depends on the welfare loss due to each type of molecule and in general has to allow for synergetic effects. In general, a mapping

$$L = L(p_1(U_1), p_2(U_2), \ldots, p_m(U_m)) \tag{10}$$

may be used as a measure of welfare loss due to total pollution.

There are two important differences between the entropy approach to concentration and costs and the entropy approach to pollution:

1. The welfare loss function L defined in Eq. (10) is generally not a continuous function of environmental pollution due to different types of molecules, while

concentration and extraction costs principally depend on energy input continuously;

2. Ore extraction in a mine usually concentrates on a few elements only, while m, the number of different types of molecules contributing to environmental pollution, is very large.

Both points make it very difficult to apply the entropy approach immediately to practical problems of pollution. Of course, it would be desirable to compare instead directly the development of environmental protection measures in the field of mine wastes and dredged materials with improvements of techniques of deposition and the corresponding costs. However, data are not available to achieve this goal systematically. Sousa's (1981) comprehensive survey of capital expenditures made by the US copper industry to comply with environmental legislation, especially the Clean Air Act and its amendments, does not include general data on the relationship between costs and the "best available technology economically achievable" standard, let alone specific data for mine wastes and dredged materials on a site-specific basis. Mining enterprises consider these data as highly confidential.

Available evidence suggests that the actual behaviour of decision makers is sometimes irrational from an economic point of view. For instance, in the last decade costs considered to be just tolerable to remove and dispose dredged materials were 5–10 Deutsche mark/Dutch Guilders per cubic metre in Hamburg and Rotterdam harbour. Now, it seems that available alternatives for ponding techniques as presently applied in Hamburg, e.g. disposal on spoil heaps or thermal treatment for construction material, both after separation and dewatering will cost approx. ten-fold, without any corresponding improvement of the state of technique (and with questionable results with respect to environmental compatibility). At the same time, the Rotterdam alternative of disposing contaminated dredged materials on an artificial island, a widely accepted solution, will cost the same or even less than with the techniques used previously for this material (inland sites or dumping at sea). (This example was supplied to us by Ulrich Förstner.)

Thus, publicly available data does not allow the direct use of the entropy approach to pollution. It would be necessary to carry through an extensive field study on mine wastes or dredged materials in a particular region. In order to gain insight into the development of abatement costs due to environmental protection, it would not be sufficient to consider only a point in time or a relatively short period, but it would be necessary to take into account a time period of several years. A corresponding field study of water pollution and the costs of water treatment plants for the German federal state of Baden-Württemberg from 1971 to 1979 was done by Faber et al. (1983a). A similar effort for mine wastes or dredged materials is highly desirable, but beyond the scope of this Chapter.

In summary, missing data do not allow us to systematically analyze the cost developments due to anti-pollution measures on the basis of the entropy approach. Additionally, the two difficulties mentioned above prohibit an immediate empirical implementation of Eq. (10). The entropy approach and in particular Eq. (10), however, can serve as an aid in considering the welfare loss due to pollution. This will be shown in the following section, where we will develop a model which can be applied more directly to the problems of mine wastes and dredged materials.

3.3 Resources and Environment

This section sketches a resource and environmental model and outlines its main results regarding intertemporally optimal prices and waste disposal.

Assume the existence of a production possibility set which can be described by a set of I linear production processes R_i $(i=1, ..., I)$. In general, each process uses capital, labour, resources and materials (as well as flows from the environment) to produce a desired good.

In each period of time, production process R_i uses l_i units of labour, k_i units of capital, r_i units of resource and m_i units of a material, yielding at the beginning of the next period, one unit of a desired product, and as "by-products" $(1-c_i)k_i$ units of capital as well as s_i units of waste. In an obvious notation, after Bliss (1975, p. 80), we may write:

$R_i: l_i$ labour $+ k_i$ capital $+ r_i$ resource $+ m_i$ units of a material $\rightarrow 1$ desired product $+ (1-c_i)k_i$ capital $+ s_i$ waste $(i=1,.., I)$.

Space restrictions do not allow one to analyze problems of the environmental impact of mine tailings and dredged material in terms of this general approach. Thus, we have to simplify the above model drastically, hoping not to lose too much of its empirical content. For details, the reader is referred to Faber et al. (1983b) and Faber (1985).

Consider the following simple model economy: The final target of the economy is the production of a consumption good. There are two ways to produce this good. The first production process uses only labour as an input, the second process additionally uses a certain type of machine, which may be vaguely described as a capital good. The consumption good, as well as the inputs of capital and labour are assumed to be completely homogeneous and divisible. Capital has to be produced using a resource and it depreciates exponentially with rate c_i. As a by-product every production process emits also waste, which pollutes the environment. The waste can be treated and transformed into purified waste which is not harmful to the environment. For reasons of simplicity we consider only one resource which has either to be extracted or has to be won from mine wastes and/or dredged materials.

Formally, the model can be described as follows:

$R_1: l_1$ labour $\rightarrow 1$ consumption good $+ s_1$ waste;
$R_2: l_2$ labour $+ k_2$ capital $\rightarrow 1$ consumption good $+ (1-c_2)k_2$ capital $+ s_2$ waste;
$R_3: l_3$ labour $+ r_3$ resource $\rightarrow 1$ capital $+ s_3$ waste;
$R_4: l_4$ labour $+ 1$ waste to be treated $\rightarrow 1$ purified waste $+ s_4$ waste;
$R_5: l_5(x)$ labour $+ 1/x$ extraction material $\rightarrow 1$ resource $+ (1-x)/x$ waste;
$R_6: l_6(x^*)$ labour $+ 1/x^*$ waste material $\rightarrow 1$ resource $+ (1-x^*)/x^*$ waste.

Thus, the production possibilities of our economy can be described just by $I=6$ production processes. Each of the first two processes (R_1 and R_2) may be used to fabricate a homogeneous consumption good. R_3 produces the capital good. R_4 describes a waste management process. The last two processes describe the possibilities to produce the resource. Contrary to the first four processes, where the assumption of fixed labour coefficients appears to be realistic, now labour input requirements depend on the degree of concentration of the resource in the extracted material itself (i.e. on x in R_5), or in the waste material (i.e. x^* in R_6).

Assume that minerals are extracted from sites according to their concentration and assume that the amount of labour used for extraction, mining and milling is proportional to the level of energy which is necessary to extract one unit of the resource. Then, according to Eq. (7), describing the change in Gibbs' free energy, labour (energy) input requirements rise inversely with concentration and are proportional to the cumulated quantities of the resource already extracted in the past.

Hypothesize the existence of a social welfare function, which gives us the "social utility" depending positively on the economy's consumption possibilities and negatively on the degree of environmental pollution as measured by the function L derived in the last section (see Eq. 10), for example. As explained in Section 2.3 great difficulties exist in applying Eq. (10) to empirical problems. For this reason it is necessary to use instead of Eq. (10) other welfare measures to evaluate the damage of environmental pollution (Mäler 1985). Since they are also difficult to implement empirically, it is often expedient to incorporate the value judgements of society concerning the tolerable levels of pollution in form of restrictions. This implies, for example, that deposited dredged materials have to fulfill certain requirements concerning concentrations of certain substances.

Maximizing the correspondingly altered social welfare function subject to technological, resource and environmental constraints gives shadow prices for all goods. In contrast to the market price of a good, the corresponding shadow price is not determined on the market by the fundamentals of supply and demand, but it is computed such that it mirrors the value "society" gives to this good. Thus, in contrast to the market price, the shadow price also takes care of the two effects mentioned in Section 1, namely the "intertemporal economic effect" and the "environmental effect". Chapters 7 and 8 of Faber et al. (1983b) present a detailed derivation. They show that the shadow price of one unit of the resource consists of three terms:

1. The first expression comprises direct labour costs, which depend on concentration x.
2. The second term includes a sum of prices referring to future periods up to the economy's planning horizon. These prices may be interpreted as intertemporal opportunity costs resulting because a unit of the resource utilized in period t is unavailable for disposition in later periods. This is the "intertemporal economic effect" mentioned in Section 1.3.
3. The third term accounts for environmental pollution due to the use of the resource. This is the "environmental effect", also referred to in Section 1.3.

To briefly summarize: (1) shadow prices increase with declining concentration of a resource, (2) shadow prices increase with decreasing tolerable levels of pollution, and (3) shadow prices differ from market prices, thus indicating that the market allocation is not optimal.

Figure 5 presents graphs of typical functions describing the decline of concentration of a resource in a site and the corresponding development of the concentration in the emissions of waste. This concentration increases in the beginning in line with growing use of the resource, later it declines again. Let x^* be the concentration necessary for an efficient use of the resource when producing

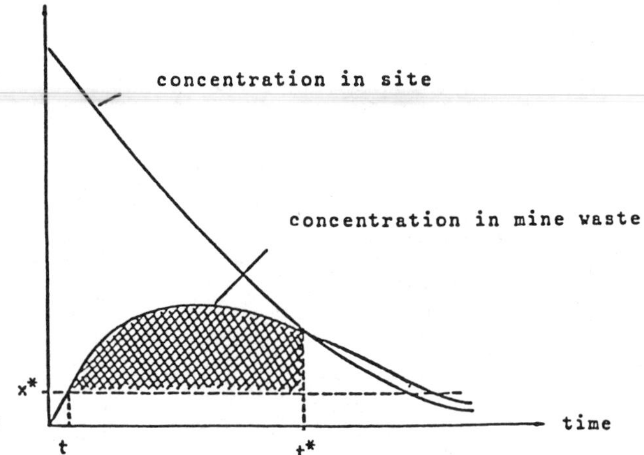

Fig. 5. Resource concentrations in site and mine waste

with the technique consisting of production processes R_2, R_3 and R_5. If mine waste was used for recycling purposes beginning in t* only, then the potential of the cross-hatched area between t and t* would not be used for recycling. This would have two unfavourable effects:

1. The resource would be disseminated in a diffusive form between t and t* and might even pollute the environment;
2. The possibility of developing an artificial site in form of a disposal site is not taken care of. We note that it may be economically efficient to use this disposal site for recycling purposes already before t*, because the natural sites with higher concentrations may have small volumes.

Summing up: Our considerations suggest that waste disposal should start early, first, to maintain a low level of pollution and therefore high environmental quality, and secondly, to reduce the increase in entropy by means of recycling and thus to decrease the energy requirements for extraction in the long run.

4 An Empirical Application

This section presents an empirical illustration of some micro- and macroeconomic aspects of the theory developed in the last section. Due to missing data we have to restrict ourselves to problems of resources.

Based on changes in concentration, measured by the average grade of exploited ore, we derive the minimum energy requirement to extract 1 mol of a resource according to the thermodynamic relations derived in the last section. Then, we check whether declining concentration, increasing entropy and growing energy requirements are reflected in costs and in market prices.

Our study deals with two cases: (1) a single copper mine in Chile (microeconomic perspective), and (2) total copper mine production of the United

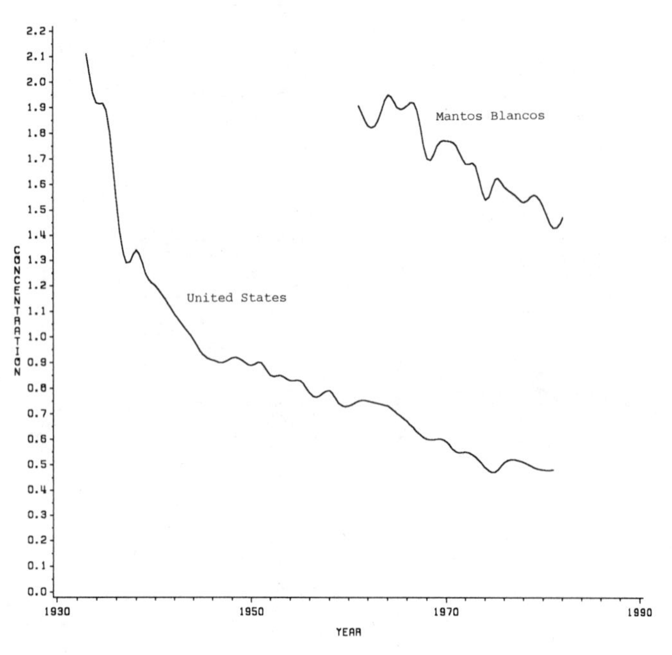

Fig. 6. Average annual grade of copper ores, United States (aggregated) and Mantos Blancos mine

States (macroeconomic perspective). Figure 6 shows the progressive lowering in average annual ore grades first, for all US copper mines together, and secondly, for the Empresa Mineria de Mantos Blancos, Chile's second largest mine producer of the red metal.

Both time paths of average ore grades (concentration) exhibit a long-term declining yield of copper metal from ores. The mineral occurrences in a single mine like Mantos Blancos, which went on stream only in 1961, are not very homogeneous due to irregular high-grade ore pockets. This fact is reflected by the greater variability of Mantos Blancos' ore grade as compared to the US average grade. The latter ore grade is derived from aggregation of all US copper mines, which generally consist of operations on large low-grade disseminated orebodies which are chiefly mined by open-pit or block-caving methods.

Given ore grades, we calculated the changes in entropy associated with changes in concentration, and thus the minimum changes in energy necessary to sort out the red metal according to Eq. (7) derived in Section 3.1. Figures 7 and 8 present the resulting indexes of minimum energy requirements, along with indexes of real unit costs and real market prices in the case of the Mantos Blancos mine (Fig. 7) and in the case of all US copper mines together (Fig. 8). Prices and costs are deflated by the US producer price index.

Consider first our microeconomic case, the Chilean Mantos Blancos mine. A glance at Fig. 7 shows that neither energy requirements, costs nor market prices vary in line with each other. From the data shown in this figure, we calculated correlation coefficients for the period 1961–1982. They amounted to only 18% for

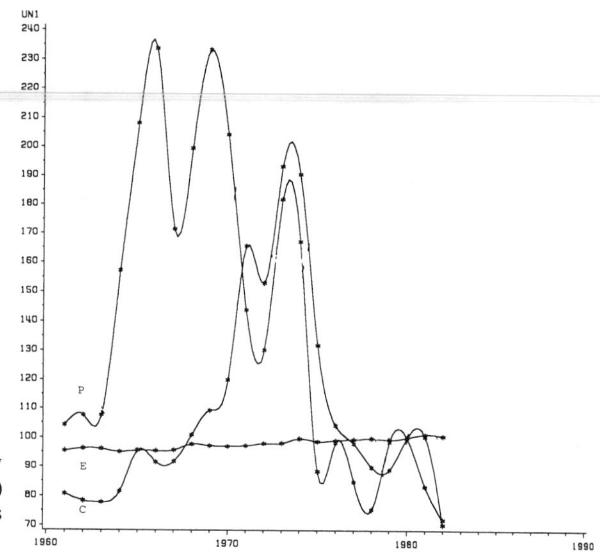

Fig. 7. Indexes of minimum energy requirements (E), real unit costs(C) and real market prices (P), Mantos Blancos, Chile, 1980 = 100

the relation between energy requirements and real unit costs and to just 26% for the relation between real unit costs and real prices.

Looking at the historical facts, this result is not very surprising: unit costs were very high in the early 1970s because of structural changes in the political regime of the country. While energy requirements in Mantos Blancos increased rather steadily, real unit costs varied considerably due to non-economic factors. Neither costs nor prices reflect the eventual exhaustion of the red metal extracted at the Mantos Blancos mine. Of course, the London Metal Exchange copper price is a market price determined by the fundamentals of the world copper industry. The share of Mantos Blancos' copper output is almost negligible when compared to the total world's production of primary copper. Thus, it cannot be expected that cost changes of a single mine directly influence the world copper price.

In contrast to this microanalysis, the macroanalysis of all US copper mines together looks quite different. Figure 8 shows that real costs and energy requirements vary in line by and large. Indeed, calculated from the data depicted in Fig. 8, we obtain a correlation coefficient of 96% for the relation between these two variables. Again, at first glance, the relation between market prices and costs seems to be relatively weak: the coefficient of correlation amounts to just 10%. But the reader should realize that we consider the last 2 decades, when real prices have been declining in spite of increased energy requirements and increasing real costs. In the long run it seems very probable that concentration will decline substantially (Prokop 1975), leading to higher energy inputs due to increasing entropy requirements to physically separate metals from ores, and eventually to increasing real prices (Slade 1982; Wagenhals 1985).

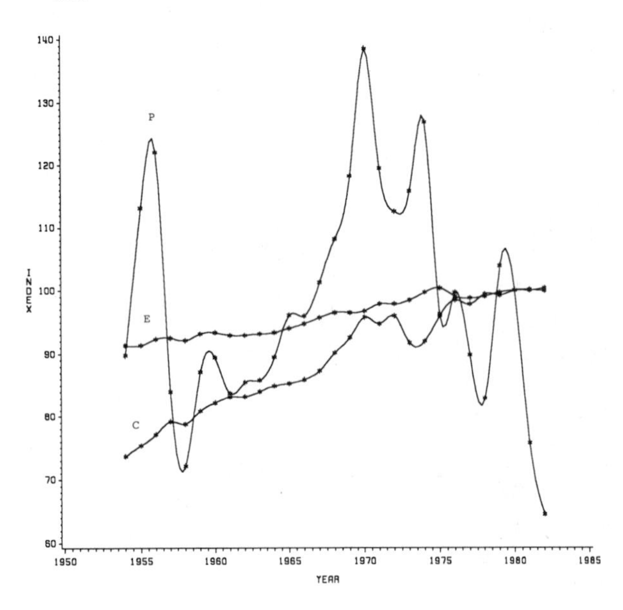

Fig. 8. Indexes of minimum energy requirements (E), real unit costs (C) and real market prices (P), all copper mines, United States, 1980 = 100

5 Critique and Conclusions

As mentioned in the introduction, the considerations in this chapter represent an attempt to capture certain aspects of the long-term balance between economics and environmental protection which are appealing in studying problems of mine wastes and dredged materials and which may not be reflected by other methods of economic analysis. However, even though our study provides convenient tools to assess the paths to an increased balance between economics and environmental protection, it has serious limitations which should be made explicit.

First, in Section 2, we have assumed that consumers and producers take resource prices as given. While this supposition is true for most world markets for minerals and metals, it is certainly wrong for some markets where market power prevails (e.g. platinum, mercury and molybdenum markets).

Secondly, as we have already pointed out in Section 3, our model is restricted to a set of six special linear production processes. This is a very high level of aggregation compared to real economies with millions of production processes where factors and products may be substitutable.

Thirdly, the formula for Gibbs' free energy (Eq. 4) and the minimum energy requirement to extract 1 mol (Eq. 7) gives the minimum thermodynamic limits. Our considerations implicitly assume that total energy needs vary in line with the minimum energy requirements. This is plausible from a macroeconomic point of view, but not necessarily from the microeconomic perspective of a single mine.

Furthermore, our model neither considers technological inventions or innovations nor political or social factors in a satisfactory manner. (See, however, Faber and Proops 1985, pp. 608–609.)

Additionally, due to the difficulty of collecting accurate data, our case studies in Section 4 had to be restricted to 2 decades and two countries. Information on ore grades and real costs is generally confidential, most data referring to Chile were only made personally available to one of the authors of this Chapter due to the friendly support of the Comision Chilena del Cobre.

Although we bear the above considerations in mind, we still feel that our analysis provides additional qualitative insights into the interrelated processes of resource extraction and environmental protection which in general has been neglected in economic studies up to now. Several results emerge from this study:

1. From a microeconomic perspective, minimum energy requirements and real costs are correlated only moderately. Our case study of the Mantos Blancos mine in Chile illustrates this claim. Site-specific variables heavily influence costs on the microlevel.
2. From a macroeconomic point of view, the coefficient of correlation between minimum energy requirements and actual costs may be considerable. This claim is exemplified by our case study of all US copper mines together.
3. These two results suggests that the notion of entropy (Eq. 3) and the corresponding notion of minimum energy requirement (Eq. 7) are not apt instruments to analyze processes on the level of a single mine, i.e. on a microeconomic level, but that they may serve as powerful aids in considering macroeconomic consequences of trade-offs between economics and the environment.
4. While higher minimum energy requirements due to declining ore grades are reflected in real costs to some extent, the correlation between energy requirements and real prices has been very weak in the 2 decades our examples refer to. Twenty years in history amount, however, to less than a generation. During this relatively short period, thermodynamic minimum energy requirements summed up to just some 1 to 5% of the total energy used in concentrating, smelting and refining metals. Our feeling is that in the long run concentration will decline strongly and that thus resource prices will eventually increase in line with increasing future energy needs even if potential technological improvements are taken into account.
5. In the short run the weak correlation between minimum energy requirements and real prices corroborates our hypotheses derived in Sections 2 and 3 where we showed that the market's allocation of exhaustible resources is not intertemporally optimal. Indeed, at least in the last few years, not only the market prices of many minerals and metals have been too low to allow for a sufficient provision for:
   1. The intertemporal economic effect mentioned in the introduction and explained in Sections 2.1 and 2.3, as well as
   2. The environmental effect referred to in the Introduction and clarified in Sections 2.2 and 3.3.
6. Between 1971 and 1980 the share of investments in environmental protections in total investments amounted to some 4% in the Federal Republic of Germany, to 7.7% in the United States and to 14% in Japan (Schenkel 1982). According to our own estimates, future costs of environmental protection measures will increase considerably. Until the end of this decade their share in total investments may even double in the OECD countries. For Japan we expect an increase in costs of water protection up to 100 billion US dollars.

To cover (part of) these costs, an increase in metal and mineral prices is a necessary condition. It is mandatory to take care of the needs of future generations and to reduce environmental pollution including a lowering of the environmental impact of mine wastes and dredged materials. Whether the intertemporal allocation mechanism of the market will reach this normative target without state intervention is still an open question.

Acknowledgement. We are grateful to Friedrich Breyer, Ulrich Förstner, Ingo Pellengahr and John Proops for their comments.

References

Bliss CJ (1975) Capital theory and the distribution on income. Elsevier/North Holland, Amsterdam Oxford
Chapman PF, Roberts F (1983) Metal resources and energy. Butterworths, London Boston
Debreu G (1959) Theory of value: an axiomatic analysis of economic equilibrium. John Wiley & Sons, New York
Faber M (1985) A biophysical approach to the economy: entropy, environment and resources. In: Gool W van, Bruggink J (eds) Energy and time in economic and physical sciences. Elsevier/North Holland, Amsterdam, pp 315–337
Faber M, Proops LR (1985) Interdisciplinary research between economists and physical scientists: retrospect and prospect. Kyklos 38:599–616
Faber M, Niemes H, Stephan, G (1983a) Umweltschutz und Input-Output-Analyse. Mit zwei Fallstudien aus der Wassergütewirtschaft. Mohr/Siebek, Tübingen
Faber M, Niemes H, Stephan, G (1983b) Entropie, Umweltschutz und Rohstoffverbrauch. (Lecture notes in economics and mathematical systems, vol 214). Springer, Berlin Heidelberg New York (English translation appeared in 1987: Entropy, Resources and Environment, Springer)
Kittel C, Krömer, H (1984) Physik der Wärme. 2. Aufl. Oldenbourg, München Wien
Kümmel R (1980) Growth dynamics of the energy dependent economy. Hain Königstein
Mäler K (1985) Welfare economics and the environment. In: Kneese AV, Sweeny JL (eds) Handbook of natural resource and energy economics, vol 1. Elsevier/North Holland, Amsterdam New York Oxford, pp 3–60
Malinvaud E (1985) Lectures on microeconomic theory, rev edn. Elsevier/North Holland, Amsterdam
Metallgesellschaft AG (ed), Metal statistics. Frankfurt am Main, publ. annually
Prokop FW (1975) The future significance of large copper and nickel deposits. Bornträger, Berlin
Rehbinder E, Sprenger RV (1985) The emissions trading policy in the United States of America: an evaluation of its advantages and disadvantages and analysis of its applicability in the Federal Republic of Germany. US Environ Protect Ag, Wahington, DC
Schenkel W (1982) Umweltschutz als Quelle für Innovation. Technologieforum Berlin. Umweltschutzamt, Berlin
Slade M (1982) Trends in natural-resource commodity prices: an analysis of the time domain. J Environ Econ Manag 9:122–137
Sousa LJ (1981) The U.S. copper industry: problems, issues, and outlook. US Dep Interior, Bur Mines, Washington, DC
US Bureau of Mines (ed). Minerals yearbook. Dep Interior, Washington, DC
Wagenhals G (1984) The world copper market. (Lecture notes in economics and mathematical systems, vol 233). Springer, Berlin Heidelberg New York
Wagenhals G (1985) Copper. In: Donges J (ed) The economics of deep-sea mining. Springer, Berlin Heidelberg New York Tokyo, pp 112–203

Rotterdam Dredged Material: Approach to Handling

J.P.J. NIJSSEN[1]

1 Introduction

In the Rhine estuarine region in which the city of Rotterdam is located, access for shipping must always be guaranteed.

This necessitates continuous dredging of the fairways and harbours and every year approximately 23 million m^3 of sediment must be removed by maintenance dredging work.

After World War II the port of Rotterdam was greatly extended (Europoort, Maasvlakte). This also resulted in a considerable increase in the amount of dredging work required.

The depth of the harbours and the fairways in the lower estuarine area is maintained by the Municipality of Rotterdam and the state (Ministry of Transport and Public Works).

In the 1970s understanding of the contaminants present in the dredged material and of their effects increased. The contamination of the dredged material is caused by the discharge of industrial and domestic effluents into the surface water.

Up to the 1970s the way in which the dredged material was handled was mainly determined by the method of dredging used, the costs and its possible uses. The silt from the western harbours and fairways (see Fig. 1), was discharged into the sea (13 million m^3). The remaining dredged material (10 million m^3), was mainly used for raising sites for the extension of the city or for agricultural purposes (see Fig. 2). In recent years problems have arisen in relation to the disposal of the dredged material on land sites, while at the same time discharge into the sea has been limited as much as possible. This has resulted in a lack of disposal sites in the vicinity of Rotterdam. Due to the contamination of the dredged material, local authorities did not permit the discharge of dredged material within their boundaries. Moreover, it was advised that recently discharged dredged material should not be used for a number agricultural purposes.

In 1975, taking into consideration the necessity for guaranteed access to the port, on the initiative of the municipal authorities of Rotterdam, the Provincial Council of South Holland set up the Steering Committee "Disposal of Dredged Material" (SCDD). The task of the SCDD was to develop a policy to solve the problems relating to the discharge of dredged material from the estuarine region.

[1]Municipality of Rotterdam, P.O. Box 6633, 3002 AP Rotterdam, The Netherlands

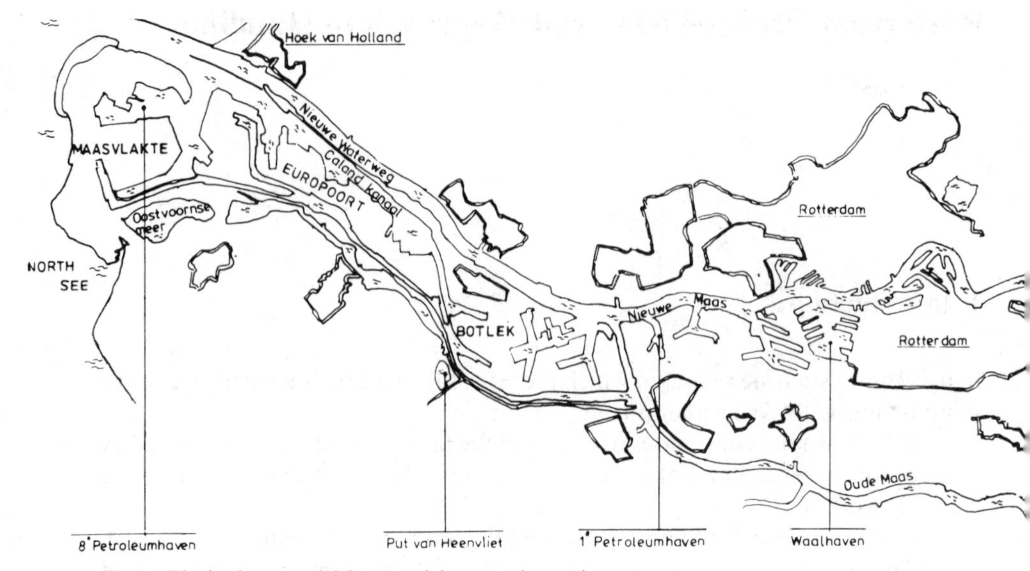

Fig. 1. The harbours and fairways of the estuarine region

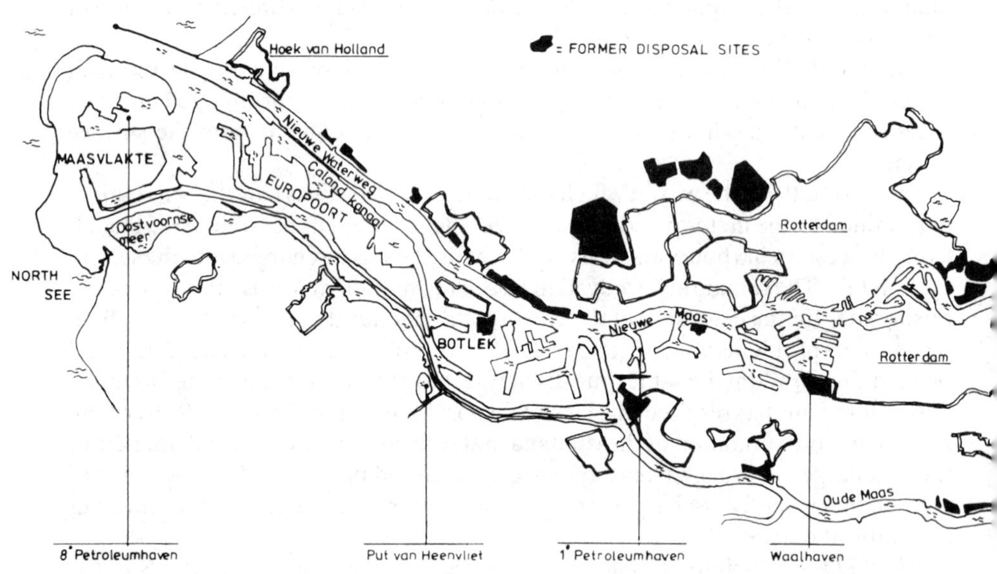

Fig. 2. The location of the former disposal sites for dredged material in and around the Municipality of Rotterdam

The public authorities involved were represented in the SCDD, these being the Province of South Holland, the Public Authority of Rijnmond, the Municipality of Rotterdam and the Ministry of Transport and Public Works and in addition, from 1979, the Province of South Gelderland. This chapter deals with the formulation of the policy since 1975. The policy carried out is greatly influenced by the laws relating to the environment. Simultaneously with the increase in the awareness that the dredged material was contaminated, more stringent laws relating to environmental practice were developed.

The discharge and handling of dredged material is governed by one or more environmental laws. In recent years the policy with regard to the discharge and processing of dredged material has also been closely followed by those with direct or indirect interests in this, such as committees concerned with the protection of nature, local authorities, industries, etc. The influence exerted by these pressure groups on the policy developed by means of their use of public meetings, publications and objections will also be considered in the following sections. Meanwhile, as will be shown, the processing of dredged material from the estuarine regions for useful applications or its discharge can be guaranteed for a number of decennia.

2 Qualitative Aspects of the Dredged Material

In the estuarine area both fluvial and marine sediments are deposited. Together with the suspended material, the Rhine and Maas carry contaminants into the harbours and fairways.

In comparison, the marine sediments are relatively clean in origin. Thus, the marine sediment has a positive influence on the quality of the dredged material. The influence of the mixing of marine sediment on the total quality of the sediment is illustrated in Fig. 3.

The quality of the dredged material is monitored by means of periodic sampling programmes. The sampling programmes were started in 1972 and repeated in 1974, 1977, 1981 and 1984.

The list of parameters to be determined during the sampling campaigns was arrived at during the process of the work. The choice of parameters is based largely on lists of substances which are considered damaging or potentially damaging to the aquatic environment. For the catchment areas of the Rhine and Maas lists of Rhine chemical behaviour were compiled within the framework of the "International Commission for the Protection of the Rhine Against Pollutants" (the IRC), and the almost identical European Community guideline 76/464/EEC. The physico-chemical analysis techniques also originated in a similar way, during their practical application, and have been standardised in order to ensure comparability of sediment samples, on the understanding that they will be kept in line with the NEN regulations (Netherlands Norm). A general indication of the quality of the dredged material is given in Table 1. Here, the quality of the dredged material for three typical harbours of the estuarine region is given for 1979, 1981 and 1984

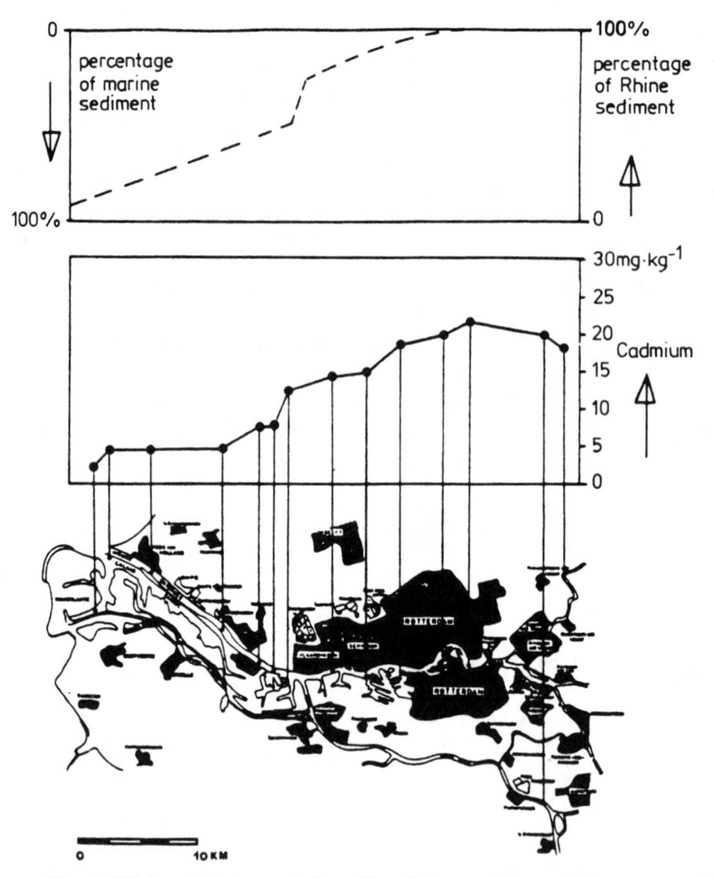

Fig. 3. Mixing of the sea and river silt and the resulting cadmium content (van Leeuwen et al. 1983)

(Delft Hydraulics Laboratory 1980, Wegman and Hofstee 1981, Gemeentewerken Rotterdam et al. 1982, 1984).

In recent years (from 1979), the quality of the Rhine water has improved in almost all considerations (Dijkzeul 1982). There has been a steep decline in the heavy metal contaminant content; the organic pollutant content has also declined, albeit less steeply. These improvements can be attributed, in part, to cleansing upstream and in part to the economic recession.

The improvements in quality mentioned are certainly not proportional to the improvement in the sediment in the estuarine area. This difference is caused by a complex of factors such as the mixing of marine sediments, localised discharges, hydrological and hydraulic differences and so forth. An indication that, in spite of these factors, the quality of the sediment is improving can been obtained by comparing the data relating to quality in the various harbours in different years.

In order to make such a comparison possible the extrapolated value for a fraction 50%< 16 μm for metals is given. The extrapolation is obtained by setting out

Table 1. An impression of the average quality of dredged material in the years 1979, 1981 and 1984 for the Waalhaven, the Botlekhaven and Europoort, whereby the heavy metals have been extrapolated to a fraction smaller than 16 μm (*)

Location / Parameter		Waalhaven			Botlek			Europoort		
		1979	1981	1984	1979	1981	1984	1979	1981	1984
% 16 μm		33.4	42.2	43.7	42.4	43.3	31.5	31.4	40.4	39.6
% Organic matter		9.2	8.6	8.2	4.3	7.2	5.8	4.6	5.3	6
% $CaCO_3$		12.7	12.1	29.8	16	15	16.5	16.3	18.9	18.11
Zn*	($mg\ kg^{-1}$)	1614	1081	1016	904	759	1079	326	259	256
Cu*	($mg\ kg^{-1}$)	203	156	142	141	119	137	41	41	39
Cr*	($mg\ kg^{-1}$)	417	249	187	336	161	173	169	79	97
Pb*	($mg\ kg^{-1}$)	320	320	233	221	229	240	84	95	80
Cd*	($mg\ kg^{-1}$)	21	18	13	16	11	12	3.6	4	3
Ni*	($mg\ kg^{-1}$)	66	54	44	43	48	63	29	30	24
As*	($mg\ kg^{-1}$)	38	25	13	24	24	38	20	17	13
Hg*	($mg\ kg^{-1}$)	3.7	4.1	3.3	7.0	4.6	3.8	1.0	0.9	0.8
EOCL	($mg\ kg^{-1}$)	1.65	9.4	8.2	9.9	12.9	4.6	0.8	4	1.9
Oil	($mg\ kg^{-1}$)	1695	2500	2320	3500	17.49	1537	894	637	520
PCBs Aroclor 1242	($mg\ kg^{-1}$)	-	0.40	0.13	-	0.22	0.05	-	0.07	0.02
Aroclor 1248	($mg\ kg^{-1}$)	-	0.70	0.35	-	0.33	0.22	-	0.09	0.11
Aroclor 1254	($mg\ kg^{-1}$)	-	0.70	0.36	-	0.31	0.21	-	0.11	0.12
Aroclor 1260	($mg\ kg^{-1}$)	0.02	0.34	0.25	-	0.25	0.11	-	0.06	0.07
HCB	($mg\ kg^{-1}$)		0.12	0.03	0.08	0.07	0.03	n.a.	0.01	0.01

– indicates no data; n.a. not analysed.

each separate observation in a harbour graphically, against the calculated fraction 16 μm. The calculated fraction % 16 μm is determined by the formula:

$$\%\ 16\ \mu m\ (\text{calculated}) = \frac{\%\ 16\ \mu m\ (\text{determined})}{100\% - \%\ CaCO3 - \%\ \text{organic matter}} \times 100\%$$

If it is assumed that there is a linear relation between the fraction (%) 16 μm (Delft Hydraulics Laboratory 1979) and the absolute value per parameter, the value at 50% 16 μm can be extrapolated.

In Fig. 4 the amount of maintenance dredging work carried out in the area managed by the Ministry of Transport and Public Works and the Municipality of Rotterdam in recent years is shown graphically.

From this figure it appears that there are great variations in the total amount of dredging work executed from year to year. To a large extent these differences are caused by variations in the amounts of silt brought in by the Rhine and Maas rivers and by the sea. The average annual volume of maintenance dredging work carried out amounts to some 23 million m^3 (13 million m^3 of which is handled by the state and 10 million m^3 by Rotterdam).

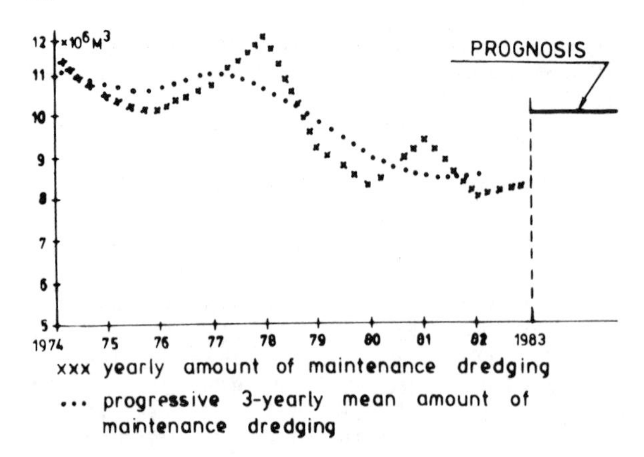

Fig. 4. The annual amount of maintenance dredging work carried out by the Ministry of Transport and Public Works and the Municipality of Rotterdam shown in m^3, with a density of 1,160 kg m^{3}

3 Legal Aspects Relating to Disposal of Dredged Material

Since 1976 dredged material has been considered to be waste matter. In the case of dredged material disposed of or processed on land the law relating to waste matter applies. The law governing the disposal of waste matter is characterised by the principal of suitability and the ICM criteria (Isolate, Control and Monitor). The suitability principle implies that at the time of the application for exemption from the waste material law, it must be shown that no other processing techniques are available. The design of the disposal site must take the ICM criteria into account.

The laws governing the pollution of surface water (FWA) and pollution of sea water (SWA) are applicable when disposal in open water (sea) or surface waters (lakes, ponds and clay pits, besides the major rivers) is under consideration. The FWA is also applicable to discharging into surface waters the water released from dredged material during disposal operations (FWA = Fresh Water Act/SWA = Sea Water Act)[2].

---

[2]In the disposal of dredged material with a layer thickness of 1.5 to 2 m a m^3 of dredged material has an initial density of approximately 1,160 kg m^{-3} (MKO 1985; van Tol et al. 1985). Under the influence of gravity and top loading (more layers) the density increases to approximately 1,400 to 1,500 kg m^{-3}. Water is driven out and the greatest amount is released onto the upper surface. This water is drained from the site and released into the surface water in its original state or after purification.

Legal regulations which must be taken into consideration during the disposal or processing of dredged material in the Netherlands are: the Town and Country Planning Act, the Chemical and Waste Products Act, the Protection of the Soil Act and, in certain cases, the Public Nuisance Act rather than the acts relating to the Disposal of Waste Matter, the protection of nature and the protection of the areas from which drinking water is obtained in South Holland. The law relating to Environmental Impact Statements has a separate place. The Town and Country Planning Act is applicable to the disposal of dredged material on land or in lakes, gravel pits, etc. The Town and Country Planning Act offers local authorities, including municipal authorities, the opportunity to prevent or delay the disposal of dredged material for a term of up to 10 years, only an Article 65 Procedure, giving ministerial consent, can overrule this appeal-delaying tactic.

On the basis of the concentration of contaminants found in the dredged material, with the exception of localised, very severely contaminated dredged material, dredged material does not fall under the Chemical and Waste Products Act. The other acts and regulations mentioned are only incidentally applicable.

The proposed law relating to reports on environmental impacts (EIR) will be treated further in Sections 5 and 8.4.

Amongst other things, the FWA is designed to preserve the function of the surface water by preventing its contamination. This act prohibits the discharge into surface water bodies of waste matter, contaminated matter or dangerous substances in any form whatsoever, without a permit. The granting of a permit depends on the degree to which the functions of the surface water in question can be protected by the setting of further regulations. To protect the functions of the surface water, norms, the so-called IMP (Indicative Long-Term Plan) values have been introduced. The values apply for a period of 5 years and after this may be adjusted if desired. This ensures that it is possible to satisfy the requirement in the setting of the norms and that they must be aimed at what is currently technically and economically feasible.

The aim of the SWA is the prevention of contamination of the sea and the maintenance of the functions of the sea. The law embodies the proscriptions of the treaty of Oslo (for the prevention of the pollution of the sea in consequence of the discharges from vessels and aeroplanes). In the law it is stated that it is forbidden to discharge a number of named substances or to take them on board with the intention of discharging them. No exemption from this proscription is possible. This proscription does not apply to substances which occur as trace elements in other substances.

In the second place it is forbidden to discharge waste matter, including dredged material, contaminated or dangerous substances which do not fall into the first category, or to take them on board with the intention of discharging them, without a permit.

Up to the present, no norms, guiding values or indications, which are specified in the FWA and aimed at the protection of the functions of the sea, have been developed.

4 Policy Framework

The lack of disposal sites for dredged material was the main reason for the establishment of a Managing Steering Committee Disposal of Dredged Material (SCDD) and an executive Coordinating Committee for the Disposal of Dredged material (CCDD).

The SCDD commissioned the CCDD to determine "where and under what threshold conditions can dredged material (in both the long and the short term) be disposed of".

In the spring of 1977 the CCDD issued a report relating to the short term. In this report the disposal sites are summarised on the basis of the amount of dredged material to be disposed of as well as on the basis of a number of limiting factors relating to the disposal options.

The limiting factors derived largely from the uncertainties relating to the environmental consequences of the discharge of contaminated dredged material. In addition, an acquisition group was set up which was charged with the task of entering into consultation with the municipal authorities involved, within whose boundaries the dredged material would have to be placed. The aim was to seek out the municipal bottlenecks and, if possible, assist in obtaining the sites which had been selected by the CCDD for disposal in the short term.

In fact, the acquisition group did not succeed in reaching agreement with the municipalities in question with regard to the acquisition of additional disposal sites.

The above gave reason for the setting up of a policy plan with the aim of finding a solution for the long, mid- and short-terms. It was also decided that the policy plan should also be based on an Environmental Impact Report. In September 1979, within the framework of a series of experiments of the Ministry of Public Health and Environmental Hygiene (see Sect. 5), the "Definitive Environmental Impact Report (D-EIR) Disposal of Dredged Material" was completed. In the EIR the environmental impacts of the three, at that time, most realistic disposal options were investigated. These possible disposal options are: disposal in the sea, disposal on land and disposal in deep ponds. In addition, attention is paid to alternative ways of handling the material.

In consultation with the Minister of Transport and Public Works, it was agreed that the taking of concrete measures by all the authorities involved, including the state, should depend on the content of the previously mentioned policy plan.

As a first step towards the policy plan, the "Provisional policy plan for the processing of dredged material" was issued at the end of 1980. In this an attempt was made to reach a balanced consideration, taking into account:

1. The experience of the various proposed methods of disposal which had been gained up to that time;
2. The environmental report which had been made;
3. The contribution of the various management sectors involved in the CCDD;
4. Previous statements on the processing of dredged material.

Emphasis was placed upon the provisional policy plan in the short term.

In consequence of the appearance of the provisional policy plan, a large number of reactions were received, which included suggestions of alternative disposal options. These reactions and the fact that since 1979 there had been very little progress in the use of new disposal sites, motivated a rather radical modification of the provisional policy plan. In the spring of 1982 the "Definitive policy plan for disposal of dredged material" was issued. In the definitive policy plan more attention was paid to the consideration of the possibilities in the long term (after 1985). In view of the importance of the D-EIR and the definitive policy plan they are given further consideration in later sections.

5 Final Report on the Environmental Impact of the Disposal of Dredged Material

N.V. Grontmij, Dwars Hederik and Verhey (D.H.V.) and Twijnstra and Gudde, in cooperation with the Ministry of Public Health and Environmental Matters, were commissioned by the SCDD to draw up the final report on the environmental impact of the disposal of dredged material (D-EIR).

The EIR was drawn up within the framework of the experimental environmental impact reports, which, amongst other things, meant that the content of the EIR had to be examined and evaluated by a panel of independent experts and that third parties should be given an opportunity to react to it.

The objective was "to indicate the anticipated environmental impact of the disposal of dredged material in the sea, on land and in lakes". In this it was assumed that "the maintenance of the depth of harbours and fairways in and around Rotterdam and Dordrecht" is intended to guarantee their accessibility and thereby the social and economic conditions and it will therefore be continued in the future.

In the D-EIR, disposal sites were not considered. The report is directed towards the question of what environmental impacts are to be expected when dredged material is disposed of at sea, on land or in lakes.

The D-EIR was completed before the policy plan and its provisions are encompassed by the policy plan. The D-EIR therefore has the character of a policy EIR. The disposal alternatives for the mid- and long term are central to the D-EIR.

No supplementary research was carried out for purposes of the D-EIR. The writers confined themselves to the available information and closed the D-EIR with an extended chapter "Gaps in knowledge and information".

The D-EIR is made up as follows: after the introduction and the formulation of the aim of the D-EIR come, dredged material: amount, composition, origin, classification and comparison with norms and normative guidelines.

After this, following a general introduction to the behaviour of the contaminants, the dispersal of contaminants in the sea, on land and in lakes is handled, leading to a quantification of the emissions per disposal site. In the final chapter consideration is given to the environmental impact of disposal at sea, on land and in lakes, alternative methods of handling, the costs and advantages and a risk analysis.

As stated earlier, the report ends with a chapter entitled "Gaps in knowledge and information", which is preceded by a chapter that goes more deeply into the application of the D-EIR.

The nature of the environmental impact caused by the disposal of dredged material, as described in the D-EIR is summarised in Table 2. With regard to the environmental impact of the disposal of dredged material in the sea, in the EIR it is concluded that in view of the open character of the food chain, the inability to control the dispersion of the contaminants is one of the most important disadvantages of disposing of contaminated dredged material in the sea. In the EIR the term "open character" is understood to mean: the large, uninterrupted geographical dispersion and also as the great species dependence/interdependence of the organisms in the marine environment. Moreover, it is also concluded that when disposal takes place in the sea considerable dilution occurs. It is, however, stated that even with such dilution the effects are still unpredictable.

When land sites are used, the dispersion of contaminants is dependent upon the local geochemical and geohydrological situation and intended use of the land after the completion of the discharge of dredged material. In the EIR it is concluded that the degree of impact on the quality of the soil and of the groundwater is also to a large extent dependent upon the choice of the location and on the intended use of the area. As is indicated little information relating to the broader ecological

Table 2. The nature of the environmental impact produced by the disposal of dredged material at sea, on land and in lakes, as distinguished in the D-EIR

	Disposal at sea	Disposal on land, filling of lakes	Disposal in lakes (partial filling)
Abiotic Environment			
Soil	Silt formation through sedimentation Contamination of the soil in the sedimentation areas	Changes in structure and physico-chemical composition	Contamination of the bottom
Groundwater	Not applicable	Seepage: flow to surface water Percolation: contamination of groundwater dependent on percolation velocity, pH, iron, permeability, etc.	Seepage: flow to surface water Percolation: mainly lateral flow from the water phase
Surface Water	Great turbidity caused by suspended silt particles Great dispersal of contaminants by sea currents	Seepage: eutrophication and possible salinisation. Chance of the release of heavy metals and substances foreign to the environment Great dispersal of contaminants by sea currents	Seepage: eutrophication and possible salinisation. Chance of the release of heavy metals and substances foreign to the environment Percolation: no important effects
Air	No important effects	Nuisance from smells during discharge of unripened dredged material	Nuisance from smells unnecessary, dependent upon the discharge techniques used
Biotic Environment			
Disposal site	Interference with existing ecosystems Changes in the soil structure resulting in changes in the composition and species of the flora and fauna Accumulation of contaminants in sedentary organisms	Interference with existing ecosystems Contamination of flora and local fauna	Interference with existing ecosystems due to turbidity
Environs of disposal site	Dispersion of contaminants with suspended silt particles and plankton as a result of sea currents	Minor impact on the environs	Outflow from lakes makes contamination of the receiving surface water possible

	Accumulation in the food chain of fishes and bottom-living organisms via plankton and bottom sediment		
Short term	See impact at disposal sites	Impediment to the growth of certain plants	See impact as disposal sites
Long term	Changes in the composition and species Chance of accumulation of contaminants in the food chain	Impoverishment of species and the composition Chance of accumulation in the food chain of local fauna	Impoverishment of species under the influence of eutrophication, soil structure and chance of accumulation of contaminants in the food cycle
Direct effects	See short term	See short term	See short term
Indirect effects	Nutritional cycle has an open structure resulting in accumulation in higher organisms	Nutritional cycle controllable Accumulation in higher organisms, controllable to a reasonable degree	Nutritional cycle with closed structure Accumulation in higher organisms limited
Human Environmental Health			
Foodstuffs	Chance of disappearance of interference with commercially important species of fish Contamination of fish or crustaceans	Contamination of agricultural and home-grown crops and meat and dairy products, in principle controllable	Not applicable
Various	Chance of negative impacts on beach recreation through silt formation in the water	No important effects. Depends mainly upon initial situation and intended function of the site (No paddling pools, allotments, etc.)	Both positive and negative impacts to be expected, depending on the initial situation and intended function
Spatial Pattern	Not applicable	Interference with or enrichment of existing land-use systems, depending upon location and intended use	No important impacts

impacts is available. When the EIR was being drawn up further research had only been carried out into the uptake of the contaminants present in dredged material by crops intended for consumption (Nijssen and Wijnen, 1985, Vreman and van Driel 1985). The expectation that if areas filled in by dredged material are not used for agricultural purposes, in principle, whereby a good measure of control over the environmental impacts can be gained, is based upon this investigation.

The potential uses made possible by this control, mentioned in the EIR are: industrial sites, recreational areas, depots for the supply of material for dike construction and the filling in of excavations.

For the disposal of the dredged material in lakes it is supposed that the environmental impact will be mainly of a local nature. When disposal takes place in sand pits, attention is drawn to the possibility of contamination of groundwater. A distinction is drawn between temporary and long-term effects. Temporary effects are caused by the (temporary) decline in the quality of the surface water (increase in turbidity; mixing of transport and consolidation water). The long-term effects are mainly related to the contamination of bottom-dwelling organisms and possibly to a larger degree to the quality of the groundwater. Here also, it is construed that little information relating to the effects on the bottom fauna is available. In the D-EIR, when treating the alternative a distinction is made between the reduction alternatives, the assimilation alternatives and the various disposal methods.

By reduction alternatives is understood: not dredging, reduction of the amount of dredged material and purification at source. It is assumed that both, not dredging and reduction of the dredged material, result only in a geographical displacement of the environmental problem. Only purification at the source can be accorded any value.

By assimilation alternatives is understood: temporary or permanent storage, purification of dredged material, use of dredged material as a raw material and combustion. Temporary storage can be meaningful in the case of calamities, whereby an amount of heavily contaminated dredged material must be stored to await a more suitable destination. Permanent storage can be considered when highly contaminated dredged material must be stored in a controlled way.

The purification of dredged material is one of the technical possibilities of which, in fact, the practical and economic execution remains limited to the desalinisation of dredged material. The use of the dredged material for the brickmaking industry is limited as the amount of dredged material available far exceeds the demand and the quality of the bricks is unfavourably affected. Lastly, combustion requires a very high energy input, while the contaminants (metals) in the remaining ash and in the fly ash form a residue.

In conclusion, a number of methods of disposal are distinguished in the EIR, which in fact correspond with the main divisions into disposal at sea, on land or in lakes. Of these, land reclamation and the formation of islands deserve particular mention (see Sect. 8.4).

The EIR finishes by mentioning gaps in knowledge and information. It is noted that, amongst other things, insight into the biological and, to a lesser extent, the abiotic effects resulting from the presence of contaminants in dredged material is lacking. The absence of norms for the quality of seawater, bottom and groundwater was also experienced as a lack in 1979!

6 Relevant Environmental Studies

6.1 Introduction

The D-EIR formed the general framework for the development of the policy. During and after the completion of the D-EIR, environmental studies which had a bearing on parts of this development were carried out.

The environmental studies were related to the dispersion of the contaminants occurring in the dredged material when disposing of it on land, the environmental impact of partial contamination (by the disposal of dredged material) in lakes and the demands which were made for the layout of former dredged material sites.

6.2 Disposal in Polders

After the publication of the D-EIR in 1980 an extensive geochemical/geohydrological investigation was completed. This was carried out in a disposal site the Broekpolder (Waterloopkundig Laboratorium 1980). The investigation, carried out by the Delft Hydraulics Laboratory, was commissioned by the Public Works Department of Rotterdam.

With the aid of geochemical investigation the extent to which contaminants occur in both the groundwater and the surface water in and around the Broekpolder was examined. The groundwater movements in and around this polder were determined from a geohydrological investigation; a mathematical model was used to calculate the past, present and future flow regimes.

A synthesis of the results of these studies led to a prognosis of the future movements of the contaminants with the groundwater in the direct environs of the former Broekpolder.

6.3 Disposal in Lakes

Since the completion of the D-EIR, two investigations have been carried out on the environmental impact of the partial filling of lakes: the Oostvoornse Meer, 1982 (MKO 1981 and Delft Hydraulics Laboratory 1980 and 1981) and the Heenvliet Pit in 1983 (Grontmij 1983 and Waterloopkundig Laboratorium 1983).

In both studies it is assumed that the surface water area becomes more shallow. The quality of the surface water is determined as a function of the degree of infilling, the quality of the dredged material disposed of and time. During consolidation the pore water is driven out upwards and this water contains contaminants.

The Delft Hydraulics Laboratory carried out model calculations for both studies, with the numerical chemical model CHARON for the Oostvoornse Meer and with the linked model CHARON/BLOOM for the Heenvliet Pit. To assist in the choice of input data and the calibration of the model, laboratory tests (including the determination of the physical/chemical constitution of the pore water) and field tests on a 50% scale (experimental discharge of Maas mouth and Waal Harbour silt in a pit in the 8th Petroleum Haven, see Fig. 1) were carried out.

On the basis of the model calculations and the experiments, it could be concluded that, disregarding the seasonal fluctuations, the content of almost all the parameters in the second and third year after the start of the disposal programme (disposal over 2 years in the Heenvliet Pit) increases until the pore water values are approached. (For these values see Table 3.)

Table 3. Average composition of pore water of dredged material from the eastern harbours

Parameter	Unit	Concentration
Ammonia	mg l^{-1}	82.5
Phosphate	mg l^{-1}	2.9
Chloride	mg l^{-1}	1,700–3,100
Cadmium	$\mu g\ l^{-1}$	0.91
Copper	$\mu g\ l^{-1}$	1.9
Zinc	$\mu g\ l^{-1}$	11.8
Lead	$\mu g\ l^{-1}$	6.0
Chrome	$\mu g\ l^{-1}$	10.0
Oil	$\mu g\ l^{-1}$	0.14
PCB[a]	$\mu g\ l^{-1}$	0.035
PCP	$\mu g\ l^{-1}$	0.011

[a]The sum of PCB isomers.

The main conclusions are:

1. The pH increases, so that due to the high ammonia concentrations a high concentration of ammoniac is attained;
2. The acid content is low in the second year; due, among other things, to the decreasing depth of the water, an improvement occurs thereafter;
3. The algal concentration remains low;
4. Organic contaminants, including specifically the PCBs, are absorbed by detritus and algae.

Neither of the two options, the partial filling of the Oostvoornse Meer and the Heenvliet Pit, were carried out. The reasons for this were diverse. In the case of the Heenvliet Pit, the council, in whose area the pit is located, decided against the project.

In the case of the Oostvoornse Meer, among others within the framework of the development of a large-scale site for the disposal of dredged material (see Sect. 8.4), the Municipality of Rotterdam considered that the partial filling of the Oostvoornse Meer was not desirable. The Oostvoornse Meer is a saline, stagnant, nutritially poor water body with a depth of approximately 40 m below sea level and as such is unique in the Netherlands.

6.4 Layout of Former Disposal Sites

At a later stage an investigation was carried out to determine in how far the decision to lay out dredged material disposal sites for the planned uses mentioned above and those intended for residential development could be considered a responsible action. The aims of the study were:

1. To identify and evaluate the potential risks to the biotic and abiotic environments;
2. On the basis of the risk evaluation, to present recommendations on the layout of the sites for their definitive use(s).

In order to determine what risks can occur during the layout and use of former dredged material disposal sites, i.e. risk identification, the potential means of transfer of contaminants from the soil into the surrounding environments are inventoried. The aspects studied, i.e. means of transfer, with the names of the participating or executing institutes, were:

1. Ground moisture and surface water (Delft Hydraulics Laboratory 1985);
2. Plastic drinking water pipelines (K.I.W.A., Gemeentewerken Rotterdam, D.W.L. 1986);
3. Flora (Gemeentewerken Rotterdam 1986);
4. Ingestion (L.H. Wageningen and Gemeentewerken Rotterdam 1986);
5. Building materials;
6. Atmosphere.

A potential measure is covering with 1.0 to 1.2 m of sand. For this reason the upward transport of contaminants in such a sand cover was also studied (Heidemij advice 1983). On the basis of the results of the risk identification and evaluation (Groenewegen and Nijssen 1984), it was specified that:

1. For residential use, covering by a layer of sand of 1.2 m (for gardens 0.9 m sand and 0.3 m humus) is to be recommended;
2. For industrial and recreational sites, with the exception of children's playgrounds and industrial housing, a cover of sand is not recommended;
3. After the completion of the layout, the councils, in whose territory such an area is located, must prepare a functional plan, with a view to continuing the monitoring of the effectiveness and efficiency of the measures taken.
4. To restrict the use of plastic pipes to PVC, as far as possible, there must be a careful choice of building material but the choice must not deviate from that customarily used in peat lands.
5. In the construction of underfloor spaces the aim should be not only the provision of a height of 0.5 m, but also adequate ventilation.

The above measures are valid for a fixed groundwater table of 1.0 to 1.2 below the surface level of the ground.

7 Handling of the Dredged Material from Harbours and Fairways in the Estuarine Area: Definitive Policy Plan

The definitive policy plan appeared in the spring of 1982. The purpose of the policy plan was to advance a socially responsible policy with regard to the handling of the material produced during maintenance dredging.

This was developed against the background of the need to maintain the required depth in the harbours and fairways now and in the future. The advancement of a policy directed towards the constructive use of the material and to a favourable balance of profit and loss was considered to be of lesser importance.

In the definitive policy plan a policy is formulated for the short and middle/long terms. This policy is partially based on a geographically orientated classification of the dredged material. The policy plan can be characterised as a general management aid. The plan has no legal or judicial basis, it gives the view of the SCDD with regard to the policy to be executed by the authorities involved with the handling of the dredged material in the short, mid- and long terms.

The classification, as introduced in the policy plan, is mainly geographical, associated with the east-west decrease in the degree of contamination. An overview of the classification is given in Fig. 5. The classification used must be seen as a practically useful division, which in future years will indicate the distinction between the disposal in the sea of class 1 dredged material, the disposal of dredged material of classes 2 and 3 in a large-scale site and of class 4 dredged material in a specially designed site (De Papagaaienbek).

Dredged material is placed in class 4 if it is heavily contaminated with regard to one or more of the parameters used as compared to class 3 material. Class 4

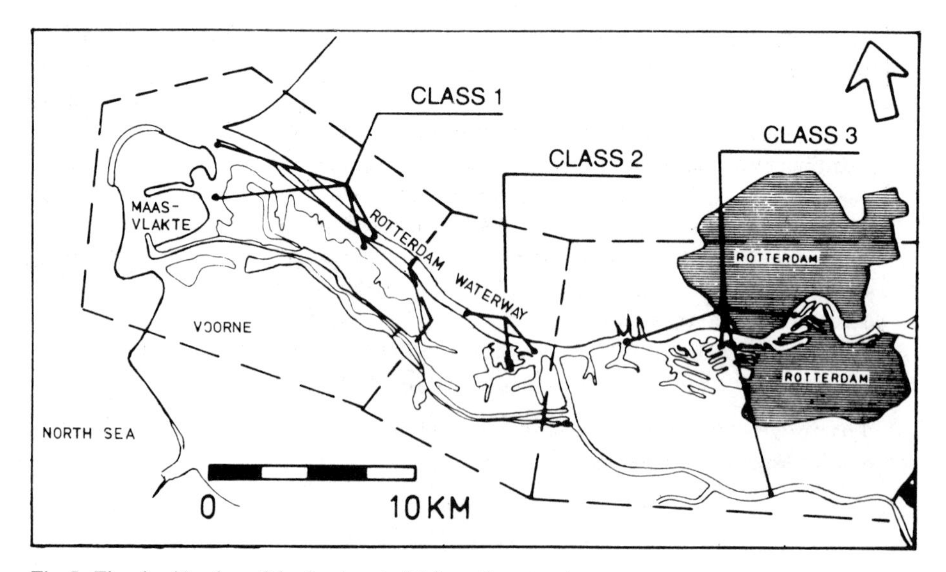

Fig. 5. The classification of dredged material from the estuarine area

dredged material originates as a result of localised activities such as those occurring at shipyards or the discharge points for industrial effluents.

In the policy plan a large number of sites which may be considered for the handling of dredged material are examined against a number of previously formulated aims. The aims are given in Table 4.

On the basis of the examination of the above mentioned aims seven short-term sites (up to 1985), with a capacity which varies from 0.5 to 7 million m^3 were mentioned. Of these one site was finally chosen, the former D.W.L. basins (basins used for the supply of drinking water), with a capacity of 0.5 million m^3.

Table 4. Aims of the policy plan, against which the potential dredged material disposal sites are examined

1. AMOUNT Contribution to adequate possibilities to maintain the depth of harbours and fairways	1.1	Striving towards a sedimentation of suspended silt and sand which favours the management of the harbours and river area
Furtherance of a socially responsible policy with regard to the treatment of the silt and sand that is deposited in rivers and fairways, against the background of the necessity to maintain the required depth of harbours and fairways	1.2	Ensuring that there are adequate possibilities to handle the material dredged from the river and the harbours
	2.1	Striving towards possibilities to limit dispersal of the contaminated dredged material
2. QUALITY Striving towards the most favourable possible relation between social advantages and disadvantages	2.2	Limiting the interference with the ecology and use of the areas where dredged material is handled
	2.3	Contribution to the most favourable possible relation between economic liabilities and assets with regard to the handling of dredged material
	2.4	Striving towards the most favourable possible relation between economic advantages and costs in relation to the handling of dredged material

For the other sites applications were made for the necessary permits, but these were not granted. Even so, in order to obtain one or more sites Article 65 of the WRO was cited. Article 65 makes it possible for the Minister to change the designated use for a site in such a way that the disposal of dredged material on it is still possible. Appeal against this decision is not possible. As a state of impasse had been reached an emergency solution was provided on the still unused industrial terrain on the Maasvlakte. In 1985 a temporary site was created, which will be described later in section 8.2.

In the policy plan it is stated that with regard to the disposal of dredged material in the mid-term (after 1985), it is desirable to restrict disposal in the sea at the Loswal Noord to dredged material from the western harbour area (approx. 13 million m^3 year^{-1}).

The disposal of the remaining dredged material (approx. 10 million m^3) will be provided for by the creation of a large-scale site in the order of size of 100 to 150 million m^3. A preference has been stated for a "closed site" on or in front of the coast of the West Netherlands. Such large-scale sites are not available on land in the province of South Holland. For this reason and in view of the mono-functional character (depot) of a large-scale site, potential sites are sought by the creation of new land in the delta area. For this purpose 11 sites (see Fig. 6) are being generally examined against the previously mentioned aims.

Of the 11 sites investigated those which emerged as the most suitable were: a peninsula discharge site, for which three variations were distinguished (see Fig. 6 and 12) and an island in the sea in front of the mount of the Haringvliet. In the policy plan it was assumed that before the creation of a large-scale site, an environmen-

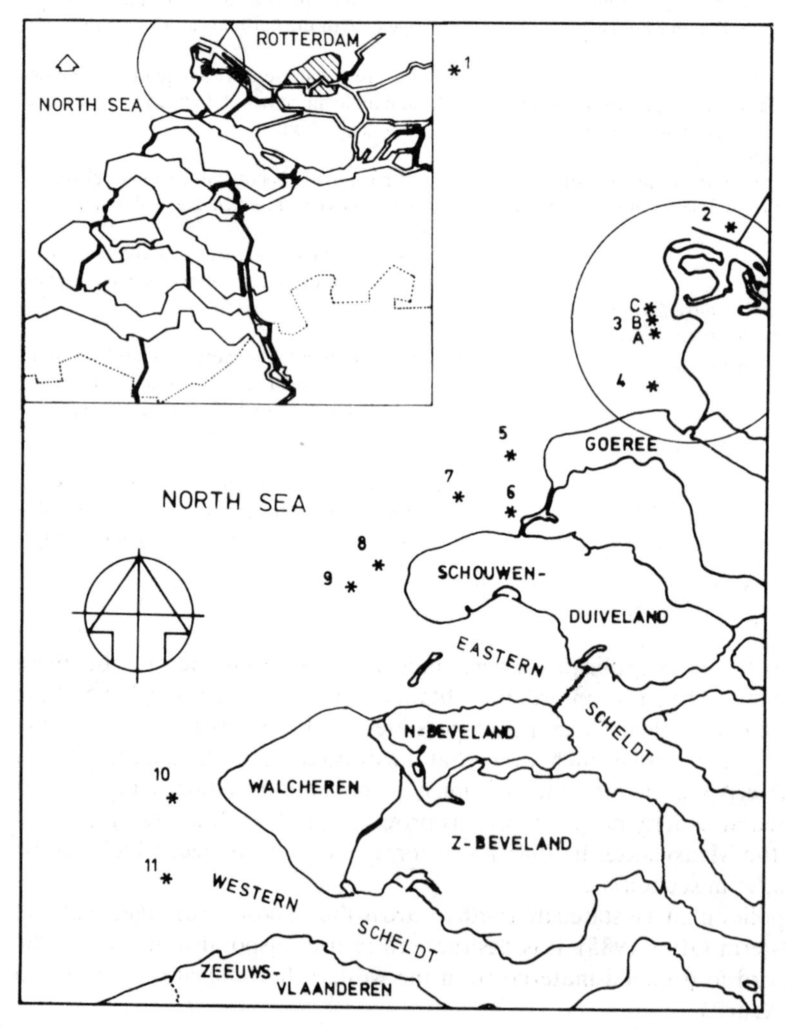

Fig. 6. Potential large-scale dredged material disposal sites (SCDD 1982)

tal impact report would be drawn up, in which at least both alternatives and the variations upon these would be considered. The other potential sites were discarded, as were the sites to the north of the Nieuwe Waterway, because, from the physical point of view, they were difficult to accommodate or because the costs were too high due to the great distance between the dredging and disposal locations.

In accordance with the policy plan, the aforementioned sites are not suitable for the disposal of dredged material which is heavily contaminated (class 4). In the plan, however, an exception is made for the large-scale disposal site, for which the possibility is left open if special measures are taken. The following possible solution to the problem of disposing or handling of this locally heavily contaminated dredged material is offered: storage on the terrain of the industry in question, where space allows and where special measures can be taken to prevent the dispersion of the contaminants. Naturally, attention is drawn to the need to purify effluents at source.

The policy plan concludes with a number of recommendations. The recommendations are:

1. Measures should be taken to limit sedimentation in harbours and fairways;
2. Those responsible should strive towards the progressive purification of efflluents discharged into surface water;
3. There should be an attempt to find constructive uses for dredged material;
4. The gaps in knowledge relating to the mobilisation, dispersion and dose-effect relations of contaminants, which have been indicated in the policy plan, should be filled in;
5. Techniques to minimise the mobilisation and the dispersion of contaminants should be developed;
6. The development of target values for the quality of groundwater, bottom and sea water, if the disposal of dredged material is being considered.
7. There should be research into the possibilities for a coordinated approach to the procedures involved in the granting of permits.

From the above it can be seen that the disposal of dredged material is totally dependent upon its classification. In the following sections this will be given further consideration. The long-term policy and the developments relating to the constructive use and alternative methods of handling the dredged material will also be included.

8 Options

8.1 Disposal at Sea

In the policy plan it is stated that for the mid-term, from about 1985, it is desirable that the disposal of dredged material in the sea, at the Loswal Noord (see Fig. 7), should be restricted to dredged material from the western harbour area (class 1 dredged material). According to present knowledge, this is regarded as being acceptable in relation to the environmental impacts. This idea is based upon the fact that the major part of the sediment is of marine origin. In practice, up to the

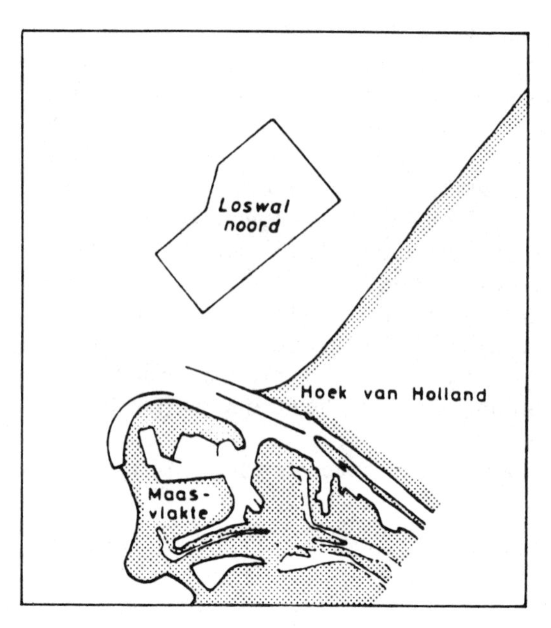

Fig. 7. Location of the dredged material disposal site Loswal Noord

first of January 1985, not only dredged material from the western harbours, but also material from the mouth of the Botlek Harbour (see Fig. 1), were disposed of at sea. Since 1. Jan. 1985, however, the Municipality of Rotterdam has not been given an exemption under the SWA to discharge dredged material from the Botlek (class 2 area), into the sea. That the disposal of class 1 dredged material at sea is considered acceptable is based on present knowledge.

However, in the policy plan it is recommended that further research is necessary and it is also established that the formulation of target values for the quality of seawater and bottom should be developed, in relation to the disposal of dredged material. The policy in relation to the Netherlands part of the North Sea is further explained in the Statement Harmonisation of North Sea Policy (Rijkswaterstaat b 1985) and the (draft) water quality plan (Rijkswaterstaat 1984). These incorporate the knowledge derived from the research so far completed.

Even so it can be established that more research into the dispersion of silt (transport mechanisms, etc.) and the behaviour of the contaminants in the marine environment is required; research that is more realistic, the more so because the quality of the dredged material is changing in such a way that there is a possibility of an anticipated significant improvement in the future.

In the previously mentioned reports no target values for seawater and soil are presented, a development about which there are various opinions and which is disappointing to the Municipality of Rotterdam. With the aid of the means available (appeals to the Council of State) the municipality is pressing for these to be provided in the future. The development of target values for seawater and soil is not as simple as that. For this, consultation, at the very least, with the other North Sea countries, is essential, this being done in the spirit of the conventions of Oslo, London and Paris (Trb 1972, 62; Trb 1973, 172 and Trb 1975, 29).

8.2 Disposal on Land

In 1984 it became clear that of all the sites mentioned in the policy plan for the short term, only the former D.W.L. basins (capacity 0.5 million m^3) had been utilised. It also appeared that the Municipality of Rotterdam would not be granted any further exemptions under the SWA, to put dredged material from the mouth of the Botlek into the sea. Finally, it could be established that the capacity of the disposal sites used up to that time would scarcely be adequate for 1985.

In all haste, a plan was then launched for the realisation of a temporary site (Fig. 8) on the Maasvlakte, with a capacity for approximately 8 million m^3 of dredged material. The site had to bridge the period from 1985 to 1987, when the large-scale site for the disposal of dredged material could be taken into use. The site is on the Maasvlakte, which is intended for large-scale industrial use. For this reason the site will be dismantled after the large-scale site comes into use. This means that the dredged material from the temporary site will be removed to the large-scale site.

The site has a temporary character and in the Public Nuisance Act it is required that the slopes of the dikes (between mean sea level + 3 to 5 m and + 15 m o.d.), within which the dredged material is discharged and the bed of the site should be lined with a 0.25-m-thick clay layer. Investigation into the behaviour and dispersion of contaminants from the dredged material to the wider surroundings (Gemeentewerken Rotterdam 1983), has indicated that contamination of the soil and the groundwater at the site can be prevented for a period of up to 15 years by the application of a clay layer 0.25 m thick.

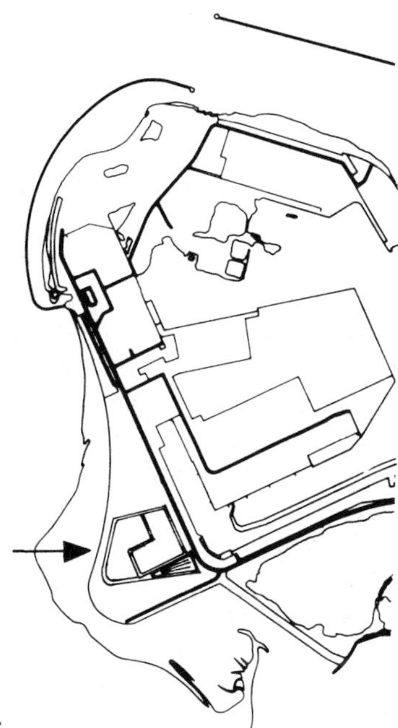

Fig. 8. The location of the temporary site

The effectivity of the clay layer will be followed up by periodic sampling of the groundwater under the site. It is intended that after the completion of the large-scale site the temporary site will be dismantled (the dredged material being transported to the large-scale site) and the site will be prepared for the originally intended function. With this in mind a clay layer of 0.25 m seems adequate.

For the clay cover "ripe" class 1 dredged material from Europoort is employed. The temporary site is located where the former, so-called clay factories were situated. In the clay factories, a layer of class 1 dredged material, with an initial thickness of 1–1.5 m was discharged within bunds 2–3 m high. The dredged material was given a period of 2–3 years in which to ripen, during which time it underwent a series of handling processes. The clay prepared in this manner was intended for use in dykes and the ceramics industry (see Sect. 8.7).

For the temporary site, in addition to a permit under the Public Nuisance Act, an exemption to the SWA is also required in relation to the discharge of the return water (see Sect. 3). In the SWA it is required that the surplus water is discharged into the Mississippi Harbour via a series of basins in which further settling of suspended matter in the return water takes place.

The settling basin (see Fig. 9) is designed on the basis of theoretical principles. The effectiveness of the basin, in relation to the quality of the water to be discharged, is presently under investigation (Public Works Department Rotterdam 1985). At present there is still very little knowledge relating to the quality of the water to be discharged. The settling of material in the surplus water was not unusual in earlier discharge sites, but up to the time of the temporary bridging site (TBS), basins had not been designed to specifically calculated dimensions.

In addition to the TBS a limited number of smaller discharge sites are used for the disposal of dredged material of classes 2 and 3. These include Storm Polder and Jan Gerritse Polder. The storage of dredged material in these sites is largely due to the sailing distance between the dredging location and the fact that for these locations the necessary permits were already obtained some time ago.

8.3 The Disposal of Heavily Contaminated Dredged Material: De Papegaaiebek

On the initiative of the Ministry of Transport and Public Works and the Municipality of Rotterdam, a site on the Maasvlakte, de Papegaaiebek (see Fig. 1), is being prepared for the disposal of heavily contaminated (class 4) dredged material. In the Papegaaiebek 1.5 million m^3 of dredged material can be accommodated.

By means of extensive sampling programmes in the estuarine area the amount and the quality of the class 4 dredged material has been determined (Rijkswaterstaat 1984a). The dimensions of the Papegaaiebek are based on this information.

The site is prepared in such a way that the dredged material is placed 0.5 m above the highest groundwater level (a criterion from the Waste Material Act). In order to protect the quality of the soil and the groundwater, the dredged material is packed in a liner. The effectiveness of the plastic sheet is monitored (by measuring in levelling tubes) and by a horizontal drainage system beneath the liner.

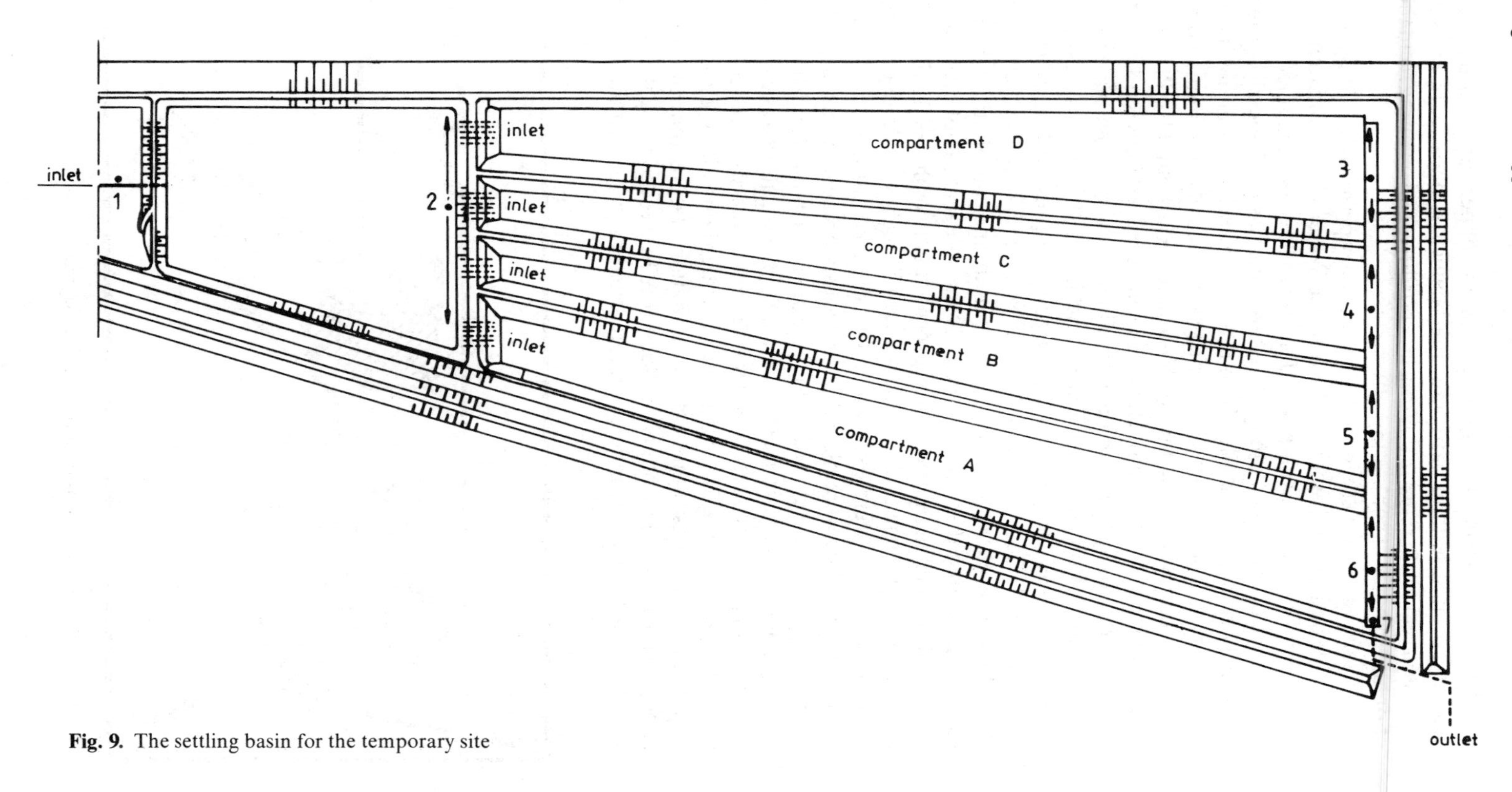

Fig. 9. The settling basin for the temporary site

The surplus water that is released on the upper surface after the disposal is discharged via a settling basin into the Nieuwe Waterway. In the settling basin provision is made to remove floating layers (films of oil resulting from the discharge of dredged material containing oil).

The site came into use in 1986; the necessary permits (including those required by the Waste Material Act and the Surface Water Contamination Act), with the exception of the WCA exemption (Jan. 1986) have been granted.

The use of the class 4 dredged material site has made possible the removal of the heavily contaminated silt which is found in specific localities in the harbours and fairways of the estuary. Up to the present heavily contaminated material has only been removed from the 1st Petroleum Harbour (1st PH; Fig. 10). The sediment in the 1st Petroleum Haven was not dredged between 1979 and 1982. As a result of the discharge of industrial effluents, the dredged material was heavily contaminated by oil and drins (aldrin, endrin, telodrin, etc.). In a single operation the sediment was dredged and discharged into previously excavated pits in the central channel, the mouth and the western branch of the 1st Petroleum Harbour (Kleinbloesen et al. 1983). After intensive study and consultation with all interested parties it appeared that this was the only feasible method of disposal.

The dredging and disposal was executed in two phases. In the first phase the mouth, the central channel and the western branch were dredged to the required depth and the pits were excavated. The dredged silt and the soil from the 1st Petroleum Harbour were placed in previously excavated pits in the Botlek Harbour. The material excavated from the Botlek Harbour was disposed of at sea.

During the first phase, an extensive field investigation was carried out into the dispersion of silt in the surface water as a consequence of dredging and discharge activities (Public Works Department of Rotterdam, 1982). This was specifically requested by, amongst others, the Foundation Nature and Environment, in an appeal case before the Council of State (the highest court in the Netherlands in which an appeal can be made by an interested party against the permit issued by

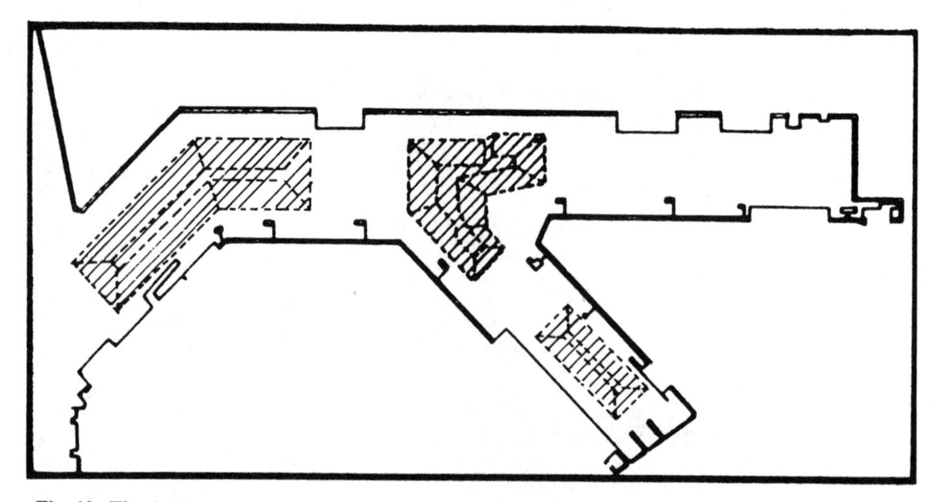

Fig. 10. The 1st Petroleum Harbour, with the disposal pits

the body granting such permits). Amongst others, the Foundation Nature and Environment appealed against the permit issued in this case by the Ministry of Transport and Public Works, Estuarine Directorate. In the first hearing of the appeal it was decided that for the first phase of the work no suspension of work should result from the appeal and that in the second hearing, after the consideration of the results of the desired investigations, the permit and possible suspension should be reconsidered.

Moreover, the Laboratory of Soil Mechanics used the model VERA to determine the dispersion of organic contaminants present in the dredged material to be stored in the pits.

Specific conditions were laid down to cover the methods used to dredge and to discharge the contaminated silt. Thus, it was only permitted to discharge the material with the aid of a diffuser. A schematic diagram of a diffuser is given in Fig. 11. The use of a diffuser leads to a limitation of the turbidity around the discharge point in comparison to the discharge point without special provisions, because the diffuser takes the dredged material to the bottom as a cohesive (density) flow.

It is also necessary to de-gas the dredged material which contains much organic matter. Methanogenic bacteria form methane in the "older" sediments. During the dredging and discharge processes the methane has opportunities to escape (disturbance of the structure) and thus causes an increase in turbidity in the surface water, which is exactly what must be avoided. After completion of the project, within the framework of the MCM, a joint venture between the Municipality of Rotterdam and the Ministry of Transport and Public Works, an extensive study into environmentally responsible dredging and disposal was started (MCM, 1984).

The research into the dispersal of silt during the dredging and discharge indicated that the silt which came into suspension was not dispersed outside the 1st Petroleum Harbour.

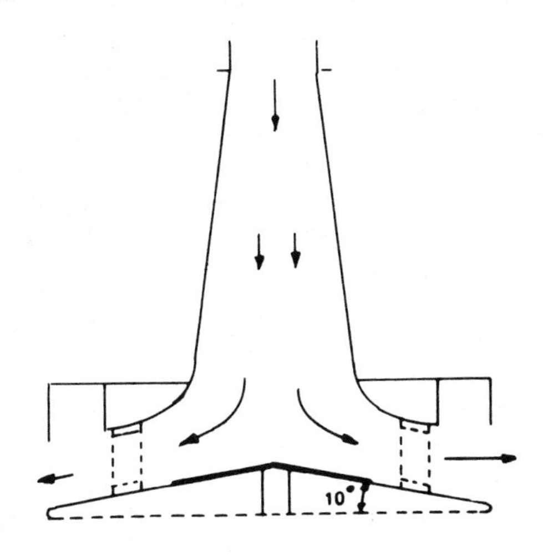

Fig. 11. Schematic illustration of a diffuser

From the research into the dispersion of organic contaminants it appeared that as a consequence of seepage flow, the transport of contaminants in the direction of the surface of the polders lying to the south could occur (Weststrate et al. 1984). In fact, as a result of adsorption in the acquifers (Pleistocene sands) and in particular in the seepage area (approx. 30-m-thick Holocene clay/peat series), it will take more than 10 000 years for the contaminants to reach the surface water. In these processes possible degradation or alteration is not taken into consideration.

In the public hearing of the appeal case before the Council of State, it was also determined that the necesary permits were justly granted. Hereby it must be noted that, on the grounds of the investigations carried out by, or commissioned by the Municipality of Rotterdam, the Foundation Nature and Environment, as most important opponent, reacted positively towards a further continuation of the work.

8.4 Large-Scale Site for the Disposal of Dredged Material from the Estuarine Area

In 1982, within the framework of the experimental environmental impact reports, the Ministerial Council decided to apply the EIR procedure to the "large-scale site for the disposal of dredged material". In the definitive policy plan, mention had already been made of the EIR obligation of the large-scale site. In the plan in question the large-scale site was the recommendation for the middle, long term. In the large scale disposal site can be disposed of 150 million m^3 class 2 and 3 dredged material (annually: 10 million m^3).

The procedure for an Environmental Impact Report is closely defined. The project was presented to the authorised body, the Minister of Tranport and Public Works by the instigators, the Municipality of Rotterdam, the Rijnmond Authority and the Ministry of Transport and Public Works, Estuarine Directorate. On the basis of the information provided by the instigators, including a description of the intended actions, the authorised body, together with a panel of independent experts, laid down guidelines. (A feasibility study by the Rijnmond Public Authority et al. 1982, had preceded the drawing up to the EIR.)

The guidelines indicated the minimum conditions that the EIR must satisfy. The guidelines also determined which alternative sites should be given general consideration.

Discussion of the guidelines was possible. The reactions were largely related to the location of the alternative sites. The locations of the four potential sites investigated are shown in Fig. 12.

During the construction of Europoort and the Maasvlakte, a southern boundary to the port activities to be developed on the Maasvlakte was fixed at government level. This boundary was called the demarcation line. The intention behind the decision embraced the limitation of air and noise pollution by keeping free and suitably laying out a screening strip of sufficient breadth, between the Maasvlakte and the coast of Voorne. In terms of landscape, such a screening strip forms a boundary zone for the harbour and industrial area and taking into account hydrological consideration, forms a sufficiently broad water zone to protect the dune area of Voorne. Possible interference with the dune area of Voorne (young lime-rich, wet dunes) and the sea area directly in front of the coast of Voorne (silt and sand banks exposed at low water and intensively used as feeding grounds by

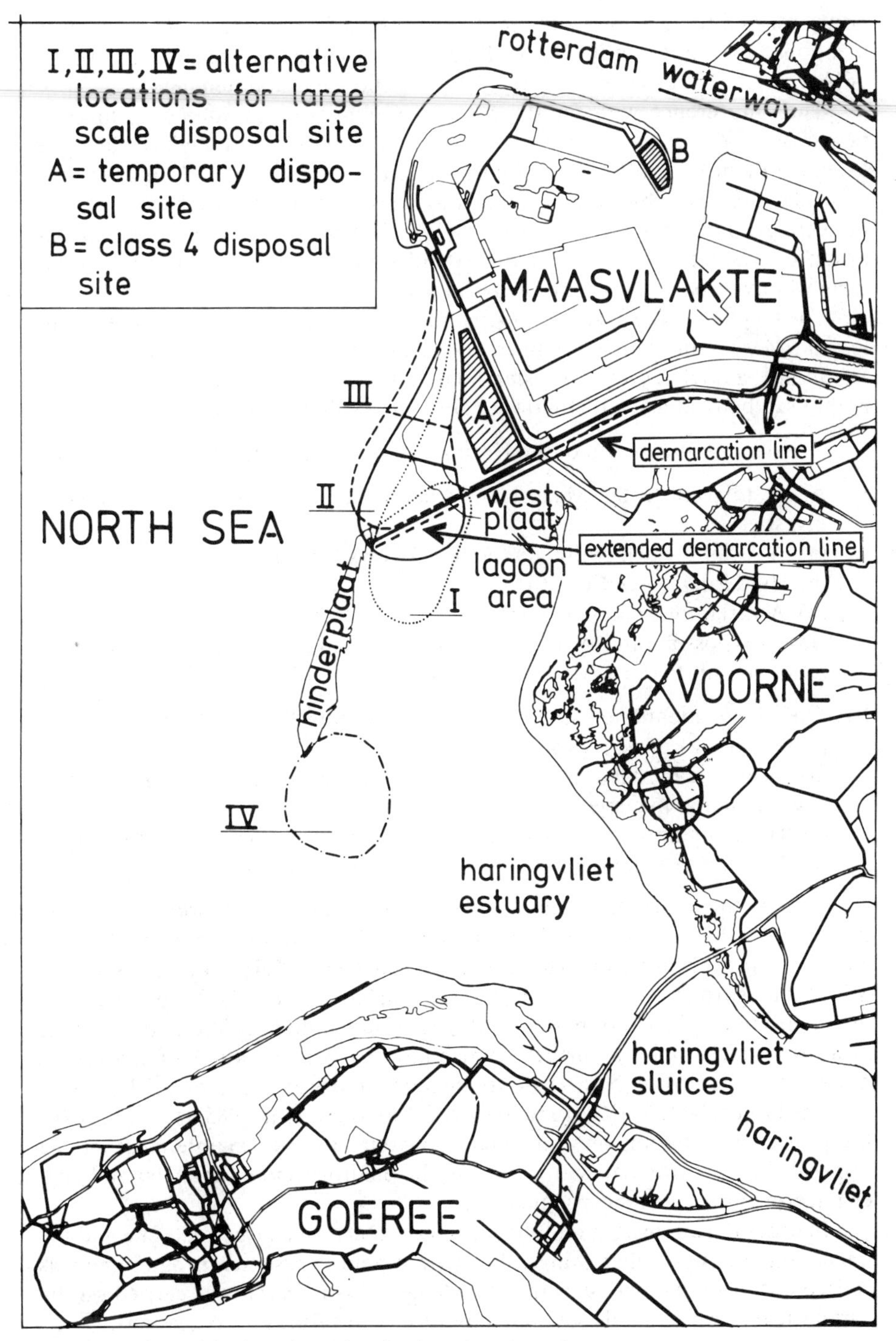

Fig. 12. Locations of the four alternative sites investigated

birds), formed the nucleus of the reactions. It thus followed from the previously mentioned feasibility study (O.I.R., 1982), that the closer the area was to the coast of Voorne, the lower were the costs of construction and maintenance.

The interference with the dune and sea areas was considered by the participants to be greater for two of the three alternatives; although, in fact, the extended demarcation line and the island in the mouth of the Haringvliet also exceeded this. For this reason, there was insistence, especially by the nature-protecting organisations and the parish of West Voorne, on the investigation of one or more alternatives lying above the extended demarcation line and preferably lying wholly or partially on the Maasvlakte.

The fact that in the past the construction of the Maasvlakte had resulted in the loss of a large natural area, the Beer, played an important part in these reactions.

In the guidelines the Minister finally stated that "in the opinion of the authorised body at least the following alternatives should be treated:

I. An island in the mouth of the Haringviet, seawards of the Haringvliet dam;

II. A site adjacent to the Maasvlakte, whereby at least three variants must be investigated:
(a) A variant which entirely or largely follows the existing depth contours of the Hinderplaat (see Fig. 10).
(b) A variant which, on the one hand, follows the existing depth contours as closely as possible and, on the other, affects the Hinderplaat as little as possible;
(c) A variant to the west of the Maasvlakte.

III. The above mentioned alternatives and variants for which the following possibilities were also set down:
(a) Using the dredged material in a way which is responsible with regard to environmental hygiene.
(b) Handling and processing the dredged material before disposing of it.
(c) The best existing methods of protecting the environment.

Finally, the alternatives I and II of the guidelines were worked out in the Project Note/EIR to produce four alternative sites. From alternative III of the guidelines, the variants a and b were considered to be parallel policy lines (see Sect. 8.7). With regard to variant III c, the best existing possibilities for the protection of the environment in combination with the aforementioned alternatives, those measures which make a further contribution to the limitation of the effects described in the Project Nota/EIR are included in the Project Note/EIR.

The Project Note/EIR (Municipality of Rotterdam et al. 1985) is a synthesis of the project information on construction, layout, use and maintenance of the disposal site, in particular with regard to the environmental, financial and civil engineering impacts and consequences.

In order to gain insight into the technical aspects of a large-scale site and the impact on, amongst other things, the environment, a research programme was carried out to supplement existing knowledge. This research was carried out by various teams, which made separate reports on the various investigations.

To formulate the Project Note/EIR a special project organisation was set up (see Fig. 13), consisting of:

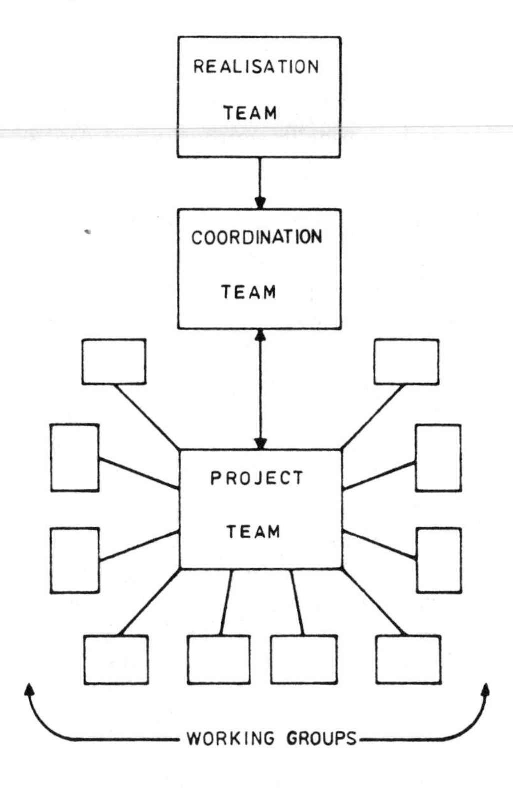

Fig. 13. Schematic diagram to illustrate the project teams

1. Realisation Team Large-Scale Site (RTLS), consisting of managers and mandated national civil servants;
2. A Coordination Team Large-Scale Site (CTLS), consisting of "higher" civil servants, which had the task of leading the project organisation and which was responsible for drawing up the Project Note/EIR;
3. A project Team Large-Scale Site (PTLS), which together drew up the Project Note/EIR. The PTLS carried the responsibility for the research;
4. Ten teams of research workers.

The technical and environmental hygiene aspects were elaborated further by the research teams. Each of these research teams drew up a report which formed the basis of the overall report.

The foundations of the EIR were:

1. Dredged Material from Maintenance Work. The anticipated amount and quality of the dredged material to be disposed of are determined, in particular with regard to the contaminants present during the period of use of the large-scale site.

2. Handling and Constructive Use. The possibilites of handling and constructive use of dredged material of classes 2 and 3 were investigated in order to gain insight into the alternative policies for the period during and after the use of the large-scale site.

3. Tranport of the Dredged Material and of Drainage Water. The methods of transporting dredged material and filling the site were investigated. The possible ways of removing the surplus water and the quality of the water which would be discharged from the site during and after the disposal of the dredged material were also investigated.

4. Geotechnical Investigation. The subjects investigated were the geological structure of the substrata at the alternative sites and in their environs and the various properties of the soil. For this, amongst others, seismic and acoustic measurements were carried out in combination with borings and echo sounding. The geology was important with regard to the necessary insight into both the technical parameters relating to the soil and to describing the transport of contaminants out of the site to the substrata.

The Delft Soil Mechanics Laboratory and the National Geological Service worked on the research and the interpretation of the results.

5. Physical Processes. The research included the changes which occur in the dredged material with regard to the volume of the layers of dredged material. In addition, the amount of water which is drained away via the surface and groundwater flow, the development of the ripening process and the bearing capacity of the top layer, the dredged material and the drainage system of the site were included in the research.

In order to predict the consolidation the numerical model SLIB (van Tol et al. 1985) was developed. With regard to the development of ripening and bearing capacity of the top layer at the site, use is made of the model FRYMO (Rijniersce 1983).

The model was developed for the simulation of the physical ripening process for the ground which becomes dry in the Iijsselmeer polders. These models were calibrated and verified during practical tests and experiments on laboratory scale.

The State Service Ijsselmeer Polders (R.IJ.P.) was amongst those working on this research.

6. Dispersion Processes. The anticipated dispersion of contaminants from the site into the substrata was investigated, i.e. the relation between the concentration, time and place. The groundwater patterns which will form under and in the direct environs of the site are approached both analytically and numerically. Central to this are the density flows resulting from the difference between fresh and saline water.

The concentration/time relation at the site itself was investigated with the help of the model CHARON (de Rooij 1985), as are the processes in the substrata/subsoil, insofar as they are relevant. The dispersion of the contaminants from the site is forecast with the aid of the model VERA (Environmental Consultance Inc. 1982). The parameters used in the making of this model were in part experimentally determined for the purpose of this project as were the simulation calculations which were carried out on laboratory scale.

The period for which the dispersion of the contaminants was studied was 3000 years. The idea behind this was that the greater part of the effects would only become apparent in the long term and that, as it emerged from the consideration, after this period a steady state may have been reached.

The Delft Soil Mechanics Laboratory, the Delft Hydraulics Laboratory and the Delft University of Technology were amongst those working on this research.

7. Coastal Morphology. The changes in wave and currents resulting from the construction of the large-scale site were analysed for the various alternatives. Research was also carried out into the consequences of these for the alternatives (design and maintenance aspects) and on the adjacent coastal area. Thus, a number of mathematical models of the Municipality of Rotterdam and the Ministry of Transport and Public Works and the Delft Hydraulics Laboratory were used.

8. Ecology. Research was carried out on the influence of the construction of the site upon the abiotic and biotic parts of the ecosystem of the estuarine area of the Haringvliet and the dunes of Voorne and Georee, as well as the relation between both parts.

Material collected in the field and relations derived from the literature on this subject were used in the prognosis of the environmental impact. A disadvantage is that long-term research is necessary for most of the prognoses relating to ecological changes. This time was not available to those who drew up the Project Note/EIR. The Ministry of Transport and Public Works-DDMI, Delta Institute, R.I.V.O. and University of Utrecht were amongst those working on this research.

9. Ecotoxicology. The possible influence of contaminated seepage water from the site upon the environmental/hygienic quality of the seabed around the site was investigated as was the effects of the taking up of these contaminants by organisms.

The development of the quality of the soil in the long term is determined with the aid of the chemical model CHARON and by simulation experiments on laboratory scale. The effects upon organisms were simulated experimentally under semi-natural conditions. The Delft Hydraulics Laboratory and the T.N.O. were amongst those working on this research.

10. Landscape, Use and Layout. The research was aimed at the anticipated effects of the site in relation to human experience. Further study was directed towards the way in which the site can be absorbed into the existing environs, while in addition proposals were developed for the most desirable use for and layout of the site, the sand dams and the direct environs. For this purpose, use was made of the usual techniques of landscape planning, such as photo and computer simulations.

In summary, it can be stated that in drawing up the Project Note/EIR, use was made of the best available, existing techniques to forecast the environmental impact for each of the alternative sites. Even so special techniques were developed. This was not necessary within the framework of the EIR. However, those involved in the research were of the opinion that any unnecessary risks resulting from uncertainty should be avoided. It may be concluded that taking into consideration the time that was spent in setting up such an extensive Project Note/EIR, in which so many different aspects had to be elucidated, the project organisation had worked well. The total cost of drawing up the Project Note/EIR, i.e. external costs (excluding costs of the staff of the municipalities and the state) amounted to 6–7 million guilders.

It must also be remembered that in a civil engineering project of the order of 150–250 million guilders, the engineering consultant's costs may be very high (10–15%).

The environmental impacts described in detail in the Project Note/EIR and the subreports are grouped for each aspect which is relevant to the process of decision making.

The aspects are:

1. Geology and geohydrology;
2. Water movement and geomorphology;
3. Quality of the surface water, sediment and soil;
4. Plankton;
5. Bottom fauna in the marine area;
6. Vegetation;
7. Birds;
8. Functional usage.

In the selection of the aspects relevant to the decision-making process, the consideration were as follows:

1. There must be sufficient, general information on the aspects to be studied to permit assessment and evaluation of the predicted consequences;
2. In the decision-making process these aspects must be given specific consideration;
3. The aspects of the ecosystem must be important in relation to the functions of the natural environment.

Finally, the environmental impact for each site is assessed in the divisions water, soil and air and the planned functions.

The project Note/EIR confines itself to a summary of the environmental impacts and costs of each alternative site, without expressing a collective preference of those involved.

Concurrent with the decision of the Ministerial Council that the EIR procedures were applicable, it was decided that for the realisation of a large-scale site an application for a concession (government permit to carry out the works) must be made.

After the appearance of the Project Note/EIR the instigators applied for permits for all the alternatives.

The project Note/EIR and the application for a concession, as well as the necessary application for exemption according to the Waste Material Act etc., were open to public inspection.

Around this time the combined nature-protecting organisations and the Council of Westvoorne presented alternative sites. These sites were wholly or partly located on the Maasvlakte and are shown in Fig. 14.

Most reactions to the public inspection were directed towards the choice of site. The content of the Project Note/EIR, the plan, the elaboration and conclusion were either scarcely discussed or ignored.

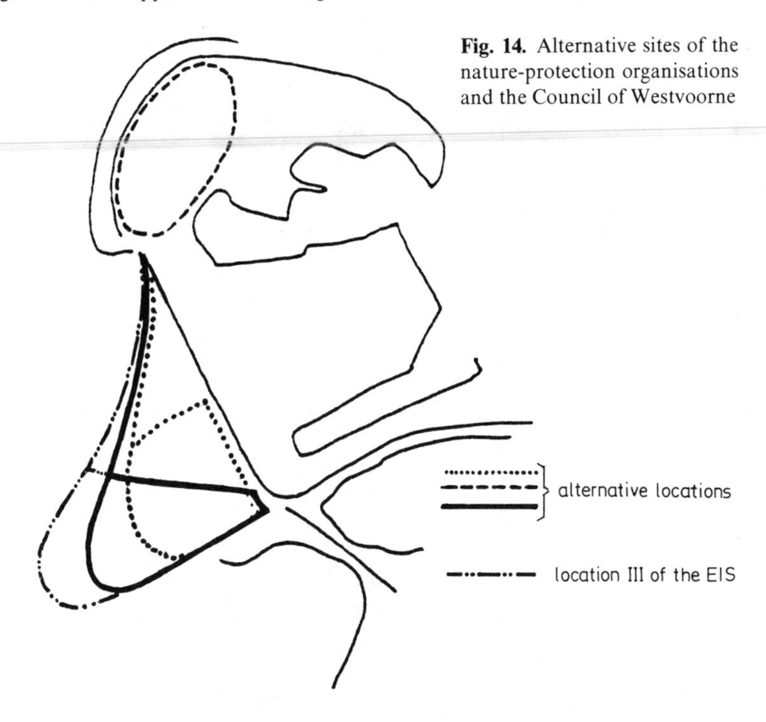

Fig. 14. Alternative sites of the nature-protection organisations and the Council of Westvoorne

This indicates that a sound plan and elaboration thereof, making use of the best techniques, open the way to a reasoned decision, in which the arguments are clear.

Although these sites were not acceptable to the instigators, in particular the Municipality of Rotterdam, the alternative plans played a substantial role in the final choice of the alternative site III. The instigators of the Project Note/EIR may not have expressed any collective preference for one or more sites, but during the processes in the council and advisory committees of the municipality and in the Second Chamber, a preference was stated on the grounds of the information provided. Finally, the authorised body granted a concession for site III, with the restriction that the instigators should optimise the site on the basis of economic and technical aspects. The optimisation study has already been completed (Municipality of Rotterdam et al. 1985) and has led to the reduction in the area of the site (see Fig. 15), and with regard to costs, is relatively cheaper.

No appeal was lodged against the permit granted so that the tenders were drawn up and the contract placed.

8.5 Long-Term Policy for the Disposal of Dredged Material

On the basis of the above, it seems that the disposal of dredged material can be guaranteed for the next 15 to 20 years. The Municipality of Rotterdam has no desire to create a large-scale site after 2002 (the official date by which time the large-scale site is expected to be filled). Apart from the environmental objections to such a disposal site at sea, the cost is far too high. This means that the quality of the harbour

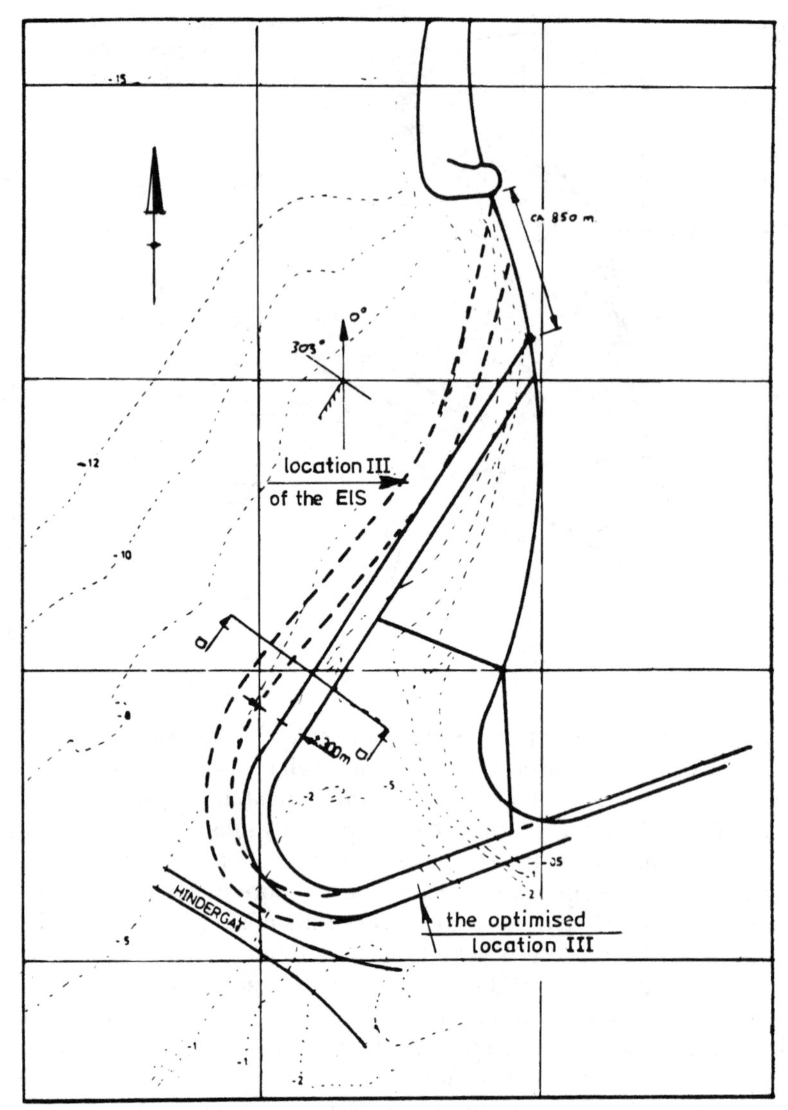

Fig. 15. Optimised site alternative III (for comparison, alternative III of the Project Note/EIR is also shown)

silt must be such that it can be disposed of in the sea or that it can be re-used on a large scale, for example for civil engineering construction work, in the ceramics industry or for agricultural purposes. It is therefore necessary that the chemical pollution of the Rhine should be stopped. International consultation is one of the means of reaching this goal. It is partly due to the consultations of the International Rhine Commission (IRC) and in relation to the European Community, that the quality of the Rhine water has already been improved during recent years.

Still, it is not very likely that the Rotterdam dredging problem will be solved by this international consultation before 2002; the processes of international decision making run too slowly for this. In the opinion of the Municipality of Rotterdam, further measures are needed.

Rotterdam is striving towards a reasoned dialogue with those who discharge effluents. If these efforts come to nothing, the council is considering taking legal action with the aim of recovering the extra costs from those discharging the effluents. It is also the opinion of the Council that the accordance between the FWA and SWA policy leaves something to be desired. The lack of norms is also regretted. Rotterdam will try in all possible ways to obtain national action on these matters in the future.

The consultation, which the Municipality of Rotterdam will hold with the dischargers of effluents, will be supported by an extensive, combined technical and legal investigation.

The aim of the technical research is to identify the dischargers of the various contaminants into the Rhine and to calculate the contribution of each of these to the pollution. In the legal investigation the question is whether the Municipality of Rotterdam will be able to charge the extra costs of disposing of contaminated silt against one or more of those who cause this contamination and if steps to do this can be taken under civil law.

The investigation will be completed in phases and the International Technical Centre for Water Studies (ICWS) in Amsterdam, the Delft Hydraulics Laboratory, the University of Amsterdam will be involved, together, for the legal part, with the Faculty of Law of the Erasmus University Rotterdam, in cooperation with the Rotterdam Advocates Office Kernkamp. The cost of this investigation will amount to 10 million guilders.

8.6 Beneficial Use and Alternative Methods of Treatment

Beneficial use and the application of alternative means of treating dredged material has been the subject of study during recent years. From the inventory studies carried out (MKO 1982; Gemeentewerken Rotterdam 1984) on alternative means of treatment, it appeared that all the techniques applied were selective, in the sense that each is only applicable with regard to a specific group of contaminants. The application of these selective techniques to dredged material would require the concatenation of the techniques, which would involve such high costs that this would be unreasonable for dredged material of classes 2 and 3 (i.e. more expensive than the present methods of treatment).

Hydrocycloning was the only technique which emerged from this study as being potentially feasible (technically and possibly also economically feasible) for the treatment of class 2 and 3 material. After hydrocycloning two streams remain, a relatively coarse fraction (sand) and a relatively dirty, fine fraction. A civil engineering use may be sought for the coarse fraction and the fine fraction can still be disposed of. On the grounds that it scarcely combats the cause, the application of this technique (which in essence involves only a reduction in the volume of the contaminant load), on a large scale, remains open to doubt, the more so since laboratory experiments indicate that higher standards must be set with regard to the quality of the relatively clean fraction than has been assumed up to the present. Moreover, it has been experimentally determined that the consolidation of 1 m^3 of the fine fraction is almost equal to that of 1 m^3 of untreated material (MKO 1986) and this implies that just as much space is required.

With regard to its use for artificial gravel, dredged material is clearly behind competing waste materials such as fly ash and concrete and masonary rubble with regard to both research and in physical composition.

Further research (Rang et al. 1985) has indicated that it is technically possible and, from the standpoint of environmental hygiene, acceptable to make artificial soil, but that the process is too expensive and that only a small demand is anticipated.

With regard to its use as an alternative to clay it appears that for technical reasons, only a limited amount of dredged material (10%) can be added during the production. In addition, for the same technical reasons, the possibility of mixing is limited to the production of yellow and variegated bricks (20% of the total). Here also, the market is limited. A further technical investigation of environmental and market aspects has shown that there are some possibilities, albeit on a limited scale.

The investigation of the environment involved the immobilisation of contaminants in relation to the firing temperature and the emissions during the firing process. It is recommended that the bricks should be fired at a temperature of 1100° C and that use be made of a smoke purifier (SDT 1985).

Although it appears that by mixing more dredged material, bricks can be made which fulfill the requirements for non-load-bearing inner walls; in view of the much lower costs of producing sand-lime bricks, there will be no market for the product.

A completely different means of beneficial use, the use of ripened dredged material for the construction of dikes, has not come into use up to the present, even for ripened class 1 material. In this case it appeared that even when a technically and financially feasible opportunity occurred, it was not socially acceptable. Research has indicated that the use of class 1 material was acceptable from the points of view of finance, civil/technical and environmental hygiene.

9 Conclusions

Since the beginning of the 1970s there has been a growing awareness that dredged material contains contaminants and, due to a lack of cooperation, a shortage of disposal sites has arisen.

Up to this period there was no question of a clearly defined policy, but rather a more or less ad hoc approach.

This came to an end with the appearance of the D-EIR and the policy plan. Both of these were subjected to intense discussion. Technical and environmental studies formed the basis of both studies and as a result of their appearance various studies directed towards one or more of the aspects considered relevant were initiated. This finally resulted in a solution to the disposal problem; a solution which was partly based on a classification of the dredged material.

On the basis of a division into non-contaminated or lightly contaminated, contaminated and heavily contaminated classes, the dredged material is disposed of in the sea, at a large-scale site or at a specially designed site, the Papegaaiebek.

This policy line is administratively ratified. In the case of the large-scale site the administrative decision was preceded by an extensive investigation into a number of alternative locations. The alternatives were chosen in such a way that conflict was largely between economic and environmental interests.

The research was executed soundly, making use of the best techniques. This resulted in a careful description of the costs and effects for each alternative. In the conflict between the instigators of the plan and the various interested parties, the scales were finally tipped in favour of the most expensive, but at the same time the most environmentally friendly option.

For the sake of completeness the past and present costs of the various methods of disposal are shown in Table 5.

Table 5. Past and future costs of the various methods of disposal

	Price per m^3 (Dutch guilders)
Old inland sites	5.00–10.00
Sea	3.00– 7.00
Large-scale site	5.00– 8.00
Papegaaiebek	20.00–40.00

The variation in the prices results from the differences in distance from the site, the measures enforced at the site and the demands made with regard to the dredging work, etc.

The solutions presented here are valid for a period of 15–20 years. A long-term policy will be formulated to follow, the main points of this policy being:

1. Prevention of the creation of heavily contaminated dredged material;
2. Improvement in the quality of the dredged material.

References

Dijkzeul A (1982) De waterkwaliteit van de Rijn in Nederland in de periode 1970–81. RIZA 82–061

EEG-richtlijn (1982) Mededeling van de commissie aan de Raad betreffende gevaarlijke stoffen die dienen te worden opgenomen in lijst I van de richtlijn 76/464/EEG. van de Raad Publ C-176

Environmental Consultance Inc. (INTERA) (1982) Hydrology, contaminant transport model (VERA). Houston, Texas, USA

Gemeentewerken Rotterdam, Rijkswaterstaat (1982) Milieuaspecten onderhoudsbaggerspecie Gehalten aan olie, bestrijdingsmiddelen en zware metalen. Analyseresultaten, monstercampagne 1981, deel B. Code 110.10–R8221

Gemeentewerken Rotterdam (1983) Advies ten behoeve van de inrichting van een klasse IV specieberginglocatie: het verontreinigingsbeheersende effect van een kleilaag op het transport van verontreinigingen. Code 110.20 R8333

Gemeentewerken Rotterdam, Rijkswaterstaat (1984) Granulaire samenstelling baggerspecie en gehalten aan organische en inorganische microverontreinigingen: analyseresultaten monstercampagne 1984. Code 110.11–R8424

Gemeentewerken Rotterdam (1985) Voorstel bemonstering kwaliteit retourwater overbruggingslocatie Maasvlakte. Code 110.70 R8517

Gemeentewerken Rotterdam, Rijkswaterstaat, Openbare Lichaam Rijnmond (1985) Alternatief III, Optimalisering, veiligheid en stabiliteit.

Gemert WJTh van, Laan HM van der, Veen HJ van (1984) Onderzoek reiniging baggerslib. TNO Gravenhage

Groenewegen HJ, Nijssen JPJ (1984) Redevelopment of former disposal sites. Proc Int Conf, Environmental Contamination, London

Grontmij (1983) Milieueffect onderzoek Put van Heenvliet. Verontdieping, Zeist

Heidemij Advies (1983) Verticaal transport van chemische stoffen in schone afdekgrond op gerijpte baggerspecie. 630/3156.1

International Rijn Commissie (IRC) (1982) Jaarverslag van de IRC 1980, Koblenz

KIWA, Gemeentewerken Rotterdam, DWL (1986) Permeatie van kunststof drinkwaterleidingen in verontreinigde baggerspecie

Kleinbloesem WCH, Wijde RW van der (1983) A special way of dredging and disposing of heavily polluted silt in Rotterdam. 10th World Dredging Congr, Singapore

Landbouwhogeschool Wageningen, Gemeentewerken Rotterdam (1986) Ingestie van verontreinigingen door kinderen.

Leeuwen P van, Groenewegen HJ, Kleinbloesem NCH (1983) Policy plan for the disposal of dredged material from the port of Rotterdam. Proc 10th World Dredging Congr, Singapore

Ministerie van Volksgezondheid en Milieuhygiene (1979) Definitief milieueffectrapport berging baggerspecie

MKO (1981) Technische, financiele-en waterkwaliteitsspecten van verontdieping van het Oostvoornse Meer met onderhoudsbaggerspecie. MKO-R80.05

MKO (1982) Verontreinigingsbeheersing baggerspecie. MKO-R82.04

MKO (1984) Milieubewust baggeren en storten. Literatuuroverzicht. MKO-R8315

MKO (1985) Opbouw dichtheid en sterkte in de vertikaal na sedimentatie van slib. MKO-R85

MKO (1986) Verslag consolidatieonderzoek op onderhoudsbaggerspecie uit de bovenloop van een hydrocycloon (laboratorium opstelling). MKO-R-8601

Nijssen JPJ, Wijnen EJE (1985) Redevelopment of polluted dredged material disposal sites: Consequences of pollutant transfer via flora elements. Proc Int Conf Heavy metals in the environment, Athens

Rang MC, Schouten CJ, Ruiter T de, Zoat J (1985) De milieuhygienische en fysische eigenschappen van producten vervaardigd uit baggerspecie. MKO-R8509

Rijkswaterstaat, Gemeentewerken Rotterdam, Openbare Lichaam Rijnmond (1982) Slufterdam Project (Interim Rapport)

Rijkswaterstaat (1984a) Raming klasse 4-specie beneden rivieren gebeid

Rijkswaterstaat (1984b) Waterkwaliteitsplan Noordzee, conceptontwerpplan

Rijkswaterstaat en Waterloopkundig Laboratorium (1985) Waterkwaliteitsplan Noordzee, verslag van onderzoek deel 1 t/m 3

Rijniersce K (1983) Een model voor de simulatie van het fysich rijpingsprocess van gronden in de Ijsselmeerpolders. Staatsuitgeverij, Gravenhage

Rooij NM de (1985) Mathematical simulation of bio-chemical processes in natural waters by the model CHARON. R 1310-10

Stuurgroep Berging Baggerspecie (SCDD) (1982) Verwerking van baggerspecie uit havens en vaarwegen in het benedenrivieren gebied, beleidsplan

Tol AF van, Brassinga HE, Elprama R, Rijt C van (1985) Consolidation of soft fine-grained dredged material (SLIB) Proc 11th Int Congr Soil mechanics and foundation engineering, San Francisco

Vreman K, Driel W van (1985) Beschikbaarheid van zware metalen in rivierslib voor landbouwgewassenen landbouw huisdieren. Proc Symp Onderwaterbodem, Rol en Lot, Rotterdam

Waterloopkundig Laboratorium (1980a) Geochemisch en hydrologisch onderzoek naar het gedrag van zware metalen en pesticiden in de Broekpolder. M1478

Waterloopkundig Laboratorium (1980b) Milieuchemische aspecten van het storten van baggerslib uit de Maasmond, de Botlek en de Waalhaven in het Oostvoornse Meer. M1501/M1549

Waterloopkundig Laboratorium (1980c) Zware metalen en fosfor in afgezet slib in het Rotterdamse havengebied en enkele zoetwater-locaties ten oosten en zuiden daarvan in 1979-1980. M1717

Waterloopkundig Laboratorium (1981) Waterkwaliteitsaspecten tijdens en na het storten van baggerspecie uit de Maasmond, de Botlek- en de Waalhavens in het Oostvoornse Meer M1501.M1549
Waterloopkundig Laboratorium (1983) Modellering waterkwaliteit van de Put van Heenvliet in relatie tot de voorgenomen berging van havenslib. R 142
Waterloopkundig Laboratorium (1985) Grootschalige locatie berging baggerspecie. Modellering van het gedrag van verontreinigingen in de toplaag. R2151
Wegman RCC, Hofstee AWM (1981) Kwalitiets onderzoek baggerspecie 1989/1980. RIVM 638001001
Westsrate FA, Meurs GAM van, Groenewegen HJ, Nijssen JPJ (1986) Developments in the use of numerical contaminant transport modelling in the Netherlands, considering three major cases. Proc Int Conf Geotechnical and geo-hydrological aspects of waste Management, Fort Collins

Environmental Management of New Mining Operations in Developed Countries: The Regional Copper-Nickel Study

INGRID M. RITCHIE[1]

1 Introduction

The Regional Copper-Nickel (Cu-Ni) Study which is described in this chapter is the result of a unique effort by a state government in the United States (US) to guide the development of major mineral resources of Cu, Ni, and Co in an orderly, comprehensive, and environmentally sound manner. In the early 1970s, the state recognized that exploration of the resource, which is located in northeastern Minnesota within the Superior National Forest (SNF) and adjacent to the Boundary Waters Canoe Area (BWCA), could bring both large economic benefits and widespread damage to an environmentally sensitive region of the country.

As a result of these concerns, the Minnesota State Legislature commissioned a regional study in 1974 to provide a comprehensive technical examination of the economic, social, and environmental impacts associated with the potential development of the state's Cu-Ni resources. The regional study, which was completed in 1979 at a cost of $\$4.3 \times 10^6$, produced a final report consisting of five volumes based on 160 technical reports. It is important to note that the Regional Cu-Ni Study, consistent with the direction of the state legislature, does not make policy recommendations, nor does it provide site-specific evaluations.

Rather, it was intended as a basis for conducting site-specific studies as development proposals occurred. Therefore, this study is not the end of mineral policy development in the state, but a beginning which provides a technical basis for establishing public policy relating to management of the state's mineral resources. The evolution of public policy is a continual process shaped by changing economic and socio-political climates, and the success and usefulness of the state's investment in the Regional Study will be determined in the future by the actions of the public, private, and governmental sectors.

This chapter, which is based on the technical documents produced by the study staff, describes the approaches used in developing the regional assessment and highlights some of the major findings. Detailed information can be obtained from the technical reports and summaries which are given in the reference list.

[1]Room 4083, Business/SPEA Building, Indiana University, 801 West Michigan, Indianapolis, IN 46223, USA

2 Historical Background

Although the discovery of Cu and Ni in what is now the Superior National Forest of Minnesota occurred in the late 1800s, it was the discovery of Cu-Ni sulfide-bearing rock near Ely in 1948 that led to extensive prospecting and exploration by state and federal agencies, mining companies, and private individuals. In 1966, the state developed its policy for long-term leasing of minerals and almost immediately offered 5.4×10^4 ha of state-owned land for leasing. Resource exploitation was further encouraged by the Copper-Nickel Tax Act of 1967 which provided a favorable tax policy for mining.

Public concern over development of mineral resources during the late 1960s and early 1970s resulted in a series of studies to identify potential economic and environmental impacts of development. As a result of this concern, the governor of the state appointed an Inter-Agency Copper-Nickel Task Force in January, 1972, to evaluate the impacts that could result from resource exploitation. In response to the Inter-Agency Task Force report, the 1973 legislature recognized that further study was required and the Minnesota Legislative Resources Council (MLRC) released $100,000 for a more extensive study which was to include a plan for a regional monitoring system. In early 1973, the Citizen's Advisory Committee (CAC) to the Minnesota Environmental Quality Council (MEQC) recommended a moratorium on mining development until further studies could be made to ensure that the region's pristine and sensitive ecosystem would be protected.

In 1974, both International Nickel Company (INCO) and American Metals Climax (AMAX) began collecting bulk samples with the intention of developing mining and possibly smelting operations. This action served to coalesce public concern, and in response the MEQC, passed the "Duluth Copper-Nickel Resolution" in October, 1974. The resolution was an important milestone in the development of state policy. It declared Cu-Ni development to be a major state policy issue which should involve the citizens of the state in planning and decision-making. It also created a new Copper-Nickel Inter-Agency Task Force that would function broadly to advise the MEQC on the development of the Cu-Ni industry. Before mining could begin the "Duluth Resolution" required the completion of an adequate regional environmental impact statement (EIS). Furthermore, a site-specific EIS was required for each development proposal.

The Inter-Agency Task Force began planning a 3-year regional EIS in November, 1974. A work plan for the regional EIS was submitted to the MEQC in April, 1975, with a cost estimate of $\$5.3 \times 10^6$. In July, 1975 the legislature released funding to hire an executive director and shortly thereafter, the MEQC, by resolution, assumed complete responsibility for the regional EIS. This action reduced the Inter-Agency Task Force to a nonfunctioning advisory board. The study officially began in 1976, and in August, 1976, the legislature changed the project title from "Regional Copper-Nickel EIS" to the "Regional Copper-Nickel Study" to reflect the broader intent of the project.

3 The Regional Study Organization and Objectives

The project study staff was organized into broad disciplinary areas of planning and natural resources (Fig. 1). In addition, the scientific team of about 90 professionals was supported by operations and a data coordination staff. The first year of the study was used primarily for planning, program implementation, and initiation of the regional characterization studies which continued into the second year. Data analysis and impact analysis studies were conducted during the second and third year of the project.

The work plan was approved by the MEQC and MLRC, included several objectives. First, the existing environmental and socioeconomic conditions of the region prior to any development were characterized to provide baseline data. This required extensive ambient environmental monitoring and data gathering. Second, various hypothetical Cu-Ni development alternatives were modeled to determine the impacts of each on the region's ecosystem. Third, environmental impacts from regional growth without Cu-Ni development were evaluated. Fourth, the impacts of Cu-Ni development on the existing ecosystem and on the projected changes in environmental quality as a result of regional growth without Cu-Ni development were evaluated.

The audience lists for the final study reports and individual technical reports is broad and includes local, regional, state, and federal governmental agencies; mining industries and secondary support industries; the scientific community; environmental interest groups; and the general public and news media. The ultimate policy that the state develops will, of course, be shaped not only by the scientific studies but also by public values, interests, and input into the decision-making process.

4 Study Area Characterization

The first step in developing the regional assessment was to establish the boundaries for each of the areas of study. The focus of the Regional Cu-Ni Study was an area of about 5440 km^2 in northeastern Minnesota (Fig. 2) designated as the Regional Cu-Ni Study Area, or simply, the Study Area. The major Cu-Ni deposits in the region occur along the Duluth Gabbro Contact in a band 5-km-wide and 80-km-long (the Resource Area). This band, for reference purposes, has been divided into a set of seven resource zones (Fig. 3). The slightly larger "development zone" area is expected to contain the bulk of facilities needed to support mining development.

In addition to the Study Area, three other areas were defined for water quality, air quality, and socioeconomic studies. The Air Quality Study Region was defined as the area within 150 km of the center of the development zones (Fig. 2). Special areas of interest in the Air Quality Study Region include the wilderness areas of the BWCA and VNP (Voyageurs National Park), and the proposed site of a coal-fired power plant in Atikokan, Ontario, Canada.

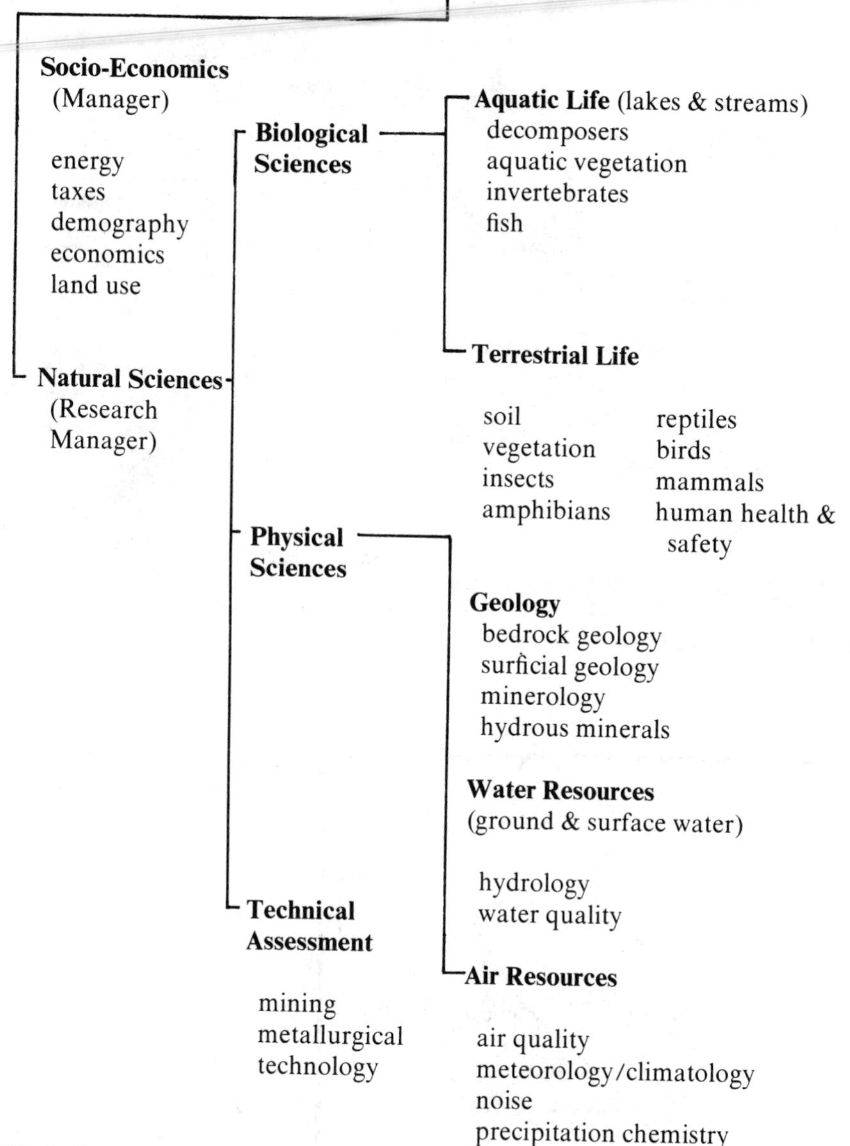

Fig. 1. Phases of project development

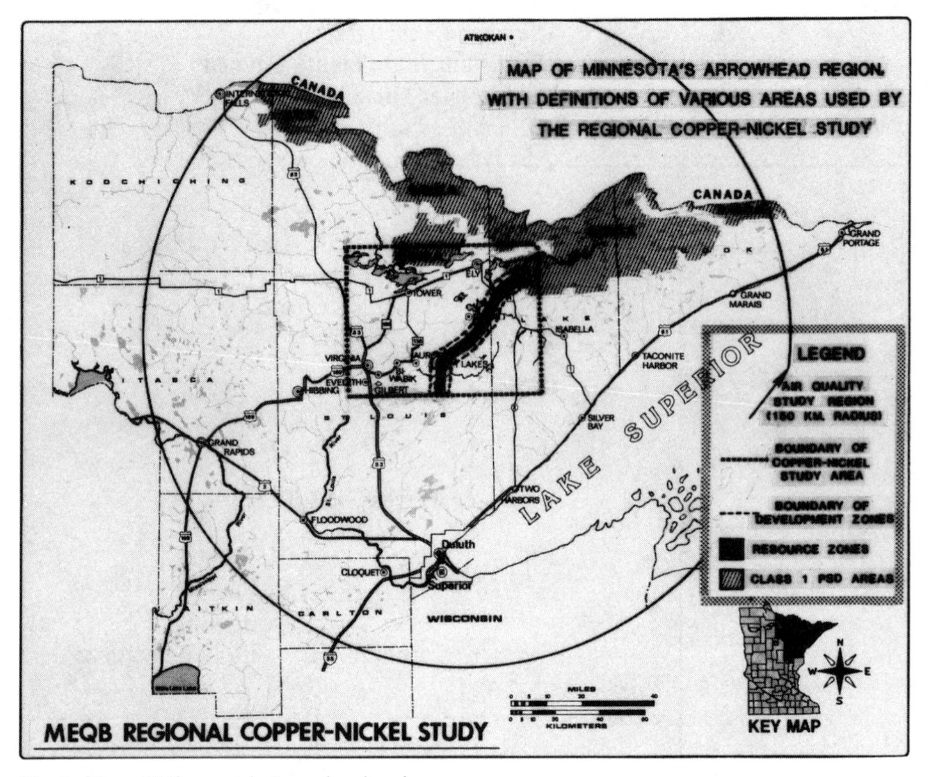

Fig. 2. Map of Minnesota's Arrowhead region

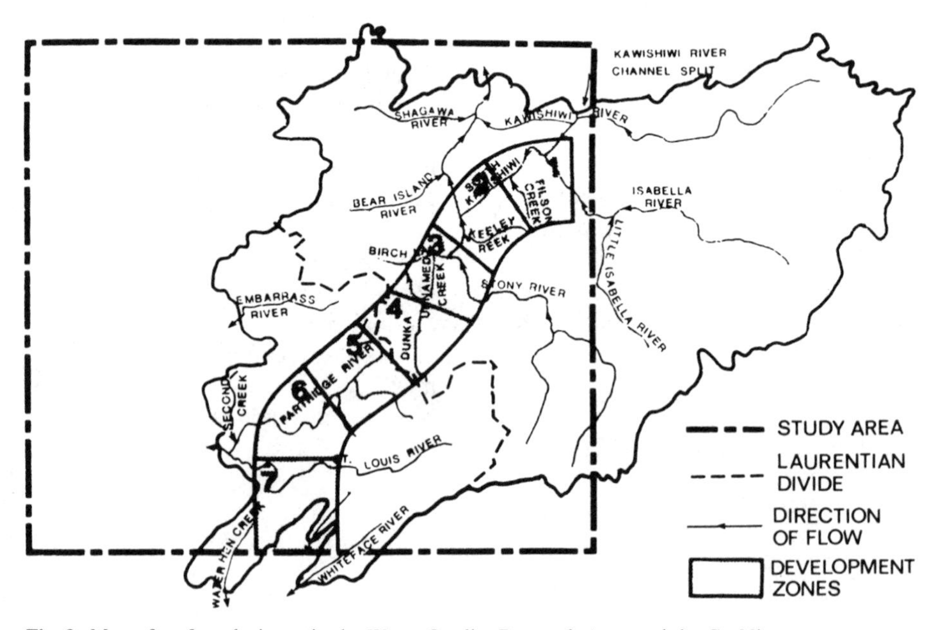

Fig. 3. Map of surface drainage in the Water Quality Research Area and the Cu-Ni resource zones

The Water Quality Research Area includes the complete watersheds of 14 streams (Fig. 3). The Laurentian Divide separates the Rainy River Watershed, which includes a portion of the BWCA, from the St. Louis River Watershed. Waters north of the Divide eventually drain into Hudson Bay and the North Atlantic, while waters south of the Divide drain into Lake Superior and eventually into the Atlantic Ocean via the St. Lawrence River.

Because the socioeconomic impacts of Cu-Ni development would not be limited to the Study Area, some socioeconomic evaluations included the larger seven county Arrowhead Region (Aiken, Carlton, Cook, Itasca, Koochiching, Lake, and St. Louis counties).

4.1 Land Use

Although the Study Area supports multiple land uses including timber management, taconite mining, recreation, and wildlife management, the region has a relatively low population density and is considered to have minimal environmental pollution. The Study Area includes parts of the SNF, the BWCA and is adjacent to VNP, areas that are major national recreation and wilderness resources protected by stringent environmental regulations and zoning laws.

About 52% of all surface lands in the Study Area are under federal, state, or county ownership. Forested areas comprise 77% of the region; aspen-birch stands (57% of all forest lands), spruce-fir stands (30%), and white-red-jack pine stands (10%) make up the forest lands and account for almost half of all the timber harvested in the region. Other major land-use areas are water (8%), mining (3%), bogs/swamps (4%), open pasture (4%), cultivated/agricultural (2%), and residential/commercial (2%).

At the time of the study, it was projected that by the year 2000, taconite development would engulf an additional 809 ha of land, and 90% of this requirement would be taken from the forest lands. However, the region's taconite industry has been severely depressed and it is unlikely that it will recover to projected levels.

4.2 Population and Residential Settlement

The "boom/bust" cycles of population fluctuations in northeastern Minnesota have been well-documented and are typical of a resource extraction-based economy which has been dependent on timber and ferrous ores.

The largest community in the Study Area is Virginia in the southwest corner which has the most concentrated rural/residential settlement. Virginia (population 12,000) currently provides support services to the taconite industry and would be the major service center for the Cu-Ni industry. The middle section of the Study Area from Aurora to Babbitt is primarily a rural/residential area with a number of old farms which are no longer cultivated. The northwest area is forested and unsettled with the exception of seasonal residences around lakes. This part of the Study Area contains some wilderness lands and is a major water wilderness/recreation area. The northeast corner is also a water wilderness/recreation area containing part of the BWCA and widespread rural/residential settlement. Ely (population 5,200) is a center for tourism associated with the BWCA in the northeast section of the Study Area. Rural residential settlement is minimal in the southeast corner which is entirely contained within the boundaries of the SNF.

The baseline population of the Arrowhead Region in 1977 was 377,000 and the population of the smaller Study Area which contains eight communities was 51,200. In 1977, the taconite industry employed about 30% of the Study Area work force (7,300). In general, communities in the Study Area are primarily tied to the taconite industry and to a lesser extent, tourism.

An additional limiting growth factor for the towns of Virginia, Eveleth, Biwabik, and Gilbert is the paucity of land available for expansion due to existing encroachment by mining. These towns are presently surrounded by vast mine pits and overburden dumps produced by the taconite industry. Growth in Hoyt Lakes, Babbitt, Aurora, and Ely is not as limited by land constraints.

During the period from 1950 to 1970, the eight communities in the Study Area have shown a continuing trend toward an aging population. Other communities outside of the Study Area, however, have experienced an influx of younger families in response to expansion of the taconite industry. More elderly people over 65 years of age per 100 persons aged 20 to 64 resided in Ely, Eveleth, and Virginia than in other towns. Ely and Eveleth had the lowest number of people under age 20 per 100 working persons aged 20 to 64; Hoyt Lakes had the highest number. These differences in age distribution can significantly affect the ability of a town to respond to the additional demands for goods and services required by youths (schools, recreational facilities) or the elderly (medical care and nursing facilities, public transportation). A rapid influx of new residents, primarily young adults or families, as a result of Cu-Ni development, could further hamper the ability of the communities to provide adequate housing and public services.

Changes in demographic characteristics were forecast for the Study Area using a computer-based regional economic demographic model of the University of Minnesota Regional Development Simulation Laboratory (SIMLAB). The model includes the effects from regional interindustry linkages and relationships between production by regional industries and their national and regional markets, production per worker, employment, earnings, labor force, and population.

Without Cu-Ni development, the population is expected to increase to 55,000 by the year 1987 and then drop to 43,000 by the year 2000 because of increased worker productivity in the taconite industry. Total regional employment is predicted to decrease from 30,000 in 1987 (27% in the taconite industry) to 24,200 by the year 2000 (15% in taconite). However, the unforeseen depression of the taconite industry has already resulted in decreases in both the regional population and taconite labor force.

4.3 Mineral Resources

The regional geology of the Study Area is characterized by hard, crystalline, Precambrian age bedrock overlain by unconsolidated surficial deposits less than 20,000-years-old. The bedrock geology includes rocks of the Giants Range

Granite, Biwabik Iron Formation, Virginia Formation, and Duluth Complex. The rocks of the Duluth Complex predominate in the Study Area and have been categorized based on mineralogy and texture into the Anorthositic Series and Troctolitic Series.

The Cu-Ni mineralization occurs in a 92-m-thick base zone of the Troctolitic Series in a band that is about 5-km-wide by 80-km-long. About 5 to 10% by weight of the mineralized portion of the Troctolitic Series are sulfides and the most common sulfide-containing minerals are pyrrhotite (a nonore mineral, $Fe_{1-x}S$), pentlandite (35% Ni), chalcopyrite (36% Cu), and cubanite (22% Cu). The sulfide composition ranges from 33 to 40%, and the sulfides occur primarily as disseminated grains within a silica matrix.

Surficial deposits in the Study Area include glacial till and clay, sand and gravel, peat, and interspersed peat and till. Within the development zones glacial till and clay and interspersed peat and till predominate over pockets of sand and gravel. Some of these materials may be used as borrow material in the mining operation for construction of tailings basins and associated dams.

Although Cu and Ni metal are the major metals of economic interest, other metals may be recovered. Titanium in the mineral ilmenite could also be mined. Cobalt which occurs in conjunction with Cu and Ni could be recovered as a by-product of development. In addition. Au, Ag, and other precious metals such as Pt are present in relatively small amounts that could provide additional income to a mining development.

Existing information indicates that the concentration of asbestiform amphibole minerals in the Duluth Complex form is quite low (about 0.1 ppm by weight). But, amphibole minerals in nonasbestiform habits are expected to be much higher, ranging up to 13% by volume with an average of 3% by volume. This raises the possibility that processing the Cu-Ni ore will release fiberlike cleavage fragments of amphibole which might present a hazard to human health.

Estimates of mineral resources were prepared by the Minnesota Department of Natural Resources (MDNR) using core essays from 324 drill holes provided by mining companies over an area of 109.3 km^2. It is recognized that this low density of drill holes can be used only as a general indication of the extent of resources that are present. Tonnage estimates were made for resources meeting one of the two criteria: (1) Type 1 resources have a minimum vertical thickness of 15.2 m and at least 0.5% Cu; (2) Type 2 resources have a minimum vertical thickness of 30.5 m and at least 0.25% Cu in the top 30.5 m of the core (or core less than 30.5 m in length if the bottom of Duluth Complex was reached). Total resource estimates using both open-pit mining (assuming a depth less than 305 m) and underground techniques (assuming a depth greater than 305 m) were developed (Table 1). The average ratio of Cu to Ni was found to be 3.33 to 1.

The total recoverable Cu resources are estimated to be 28×10^6 t and two-thirds is present as an underground resource. Over 55% of the underground resource is in zone 2 with the next largest amount, 19%, in zone 4. The open-pit resource is more uniformly distributed with the largest amount, 28%, in zone 1. The total Ni resource is estimated to be 9×10^6 t.

Overall, recoveries by open-pit mining with subsequent processing, smelting, and refining are estimated to be 86% for Cu and 68% for Ni. These recoveries decrease for underground mining to 66% for Cu and 52% for Ni. Therefore, the

Table 1. Cu ore tonnage, grade[a], and metal content estimates by resource zone

Resource zone	Open-pit resource (10^6t)		Underground Resource (10^6 t)	Total resource (10^6 t)	Contained Copper metal (10^6 t)	Total Copper metal (%)
	0.25–0.50% Cu Near surface	≥0.50% Cu Above 1,000 ft.	≥0.50% Cu Below 1,000 ft.			
1	110	370	370	850	5	18.0
2	–	340	1,600	2,000	13	46.8
2 and 3	15	110	–	130	0.8	2.9
3	250	19	76	340	1	3.6
4	180	49	550	780	5	18.0
5	38	73	230	340	2	7.2
6	60	52	49	160	0.9	3.2
7	–	11	–	11	0.07	0.3
Total	650	1,000	2,900	4,600	28	100

[a]It is assumed that resources of 0.25–0.50% Cu average 0.34% Cu, and ≥0.50% Cu average 0.66% Cu. The numbers have been rounded.

overall resource estimate after losses is 20×10^6 t of recoverable Cu and 5×10^6 t of recoverable Ni. Recoverable Co is estimated to be 80,000 to 90,000 t. At the average 1977 market prices of \$1.50 kg^{-1} for Cu and \$5.07 kg^{-1} for Ni, the recoverable resource represents a gross value of over \$$50 \times 10^9$.

If a single development operation produced 10^5 t yr^{-1} Cu it would take 200 yr to recover the estimated 20×10^6 t Cu. It is possible that larger or multiple operations would exhaust the resource more rapidly. For example, two operations each producing 10^5 t yr^{-1} could operate for 100 yr; four operations for 50 yr. These estimates indicate that the resource does have the potential to support a regional industry with a life expectancy of many decades. The potential importance of the resources is illustrated by estimates that Minnesota's Cu deposits comprise 25% of world reserves. Nearly all of US Ni reserves and about 13% of world Ni reserves are estimated to be in this region. At the present time, Co is not produced in the US, and the Co contained in Minnesota's Cu-Ni deposits represents the largest domestic resource of this metal.

4.4 Air Quality

Characterization of regional air quality was the first step in the regional air analysis. A regional air-monitoring network was established based on an existing network operated by the Minnesota Pollution Control Agency (MPCA). Two years of ambient air quality data were collected at 11 sites representative of rural (three sites), community (six sites), and industrial areas (two sites). The network consisted of 11 high-volume particulate samplers, 9 SO_2 and 9 NO_2 bubblers, 2 continuous coulometric SO_2 analyzers, 15 canopy throughfall samplers, and 4 bulk deposition samplers. Sampling methodology and quality assurance audits followed Environmental Protection Agency (EPA) designated methods. Metals and nutrient analyses from the bulk and throughfall samples were performed by the Minnesota Department of Health (MDH).

A regional emissions inventory of existing and proposed industrial sources was used to project future air quality without Cu-Ni development. The regional emissions inventory included 46 point sources (within 150 km radius of the orebody), each emitting more than 100 t yr^{-1} SO_2 or particulate matter for the base period 1975–76 and the projected year 1985. The year 1985 was selected because it was the earliest anticipated date that a smelter complex might be operational, and it was also the latest year for which emissions projections could be based on expansion plans for industrialization in the region. An emissions inventory was also developed for 1977 to comply with requirements for the Prevention of Significant Deterioration (PSD) analysis.

The increase in SO_2 emissions of 130% (from 8.48×10^4 t to 1.97×10^5 t) from 1975–76 to 1985 is due primarily to growth in the power generation industry and the conversion of the taconite industry from natural gas to coal. Emissions from power generation are expected to increase from 4.12×10^4 t to 1.41×10^5 t; taconite emissions are expected to increase from 3.33×10^4 t to 4.49×10^4 t. Particulates during the same period are expected to decrease by about 40% (from 9.25×10^4 t to 5.76×10^4 t) largely due to improved control efforts by the taconite industry.

The modeling of existing and projected regional sources was based on the mesoscale-modified Gaussian dispersion model (MGM) which provides average atmospheric concentrations and ground deposition patterns. The MGM, which was developed by the study staff, was verified in the range of 5 to 250 km using 1.5 yr of monthly deposition data, 1 yr of 24-h ambient air quality data, and two standard dispersion models. The MGM deposition rates were found to be within a factor of two of the measured rates; calculated concentrations were within a factor of two to ten depending on the averaging time; and comparison to existing dispersion models gave results to within a factor of two.

The air-monitoring and modeling data were analyzed within a regulatory framework that includes national air quality standards (NAAQS) which are designed to protect public health (primary standards) and welfare (secondary standards). In addition to these standards, the region is also subject to the amendments of the Clean Air Act of 1970. These amendments mandated the establishment of a baseline reference level of pollutants present in 1977 due to major point sources emitting more than 100 t yr^{-1} SO_2 or particulates, and established incremental increases above the baseline levels that could be allowed for industrial development. The entire US portion of the air quality Study Area (Fig. 2) is designated as a Class II PSD area with the important exceptions of the BWCA and the VNP which are designated as Class I areas and are subject to more stringent regulation (Table 2).

The monitoring program revealed that the present air quality in the region is generally representative of remote or rural areas. Annual SO_2 levels were less than 10 $\mu g\,m^{-3}$ at the remote location and less than 13 $\mu g\,m^{-3}$ in community and industrial areas. It does not appear that SO_2 NAAQS will be exceeded for years if Cu-Ni development does not occur. However, there is a possibility that the 24-h Class I PSD increment for SO_2 will be exceeded in portions of the BWCA. This possibility

Table 2. United States ambient air quality regulations

Pollutant	Wording of standard	Ambient air quality standards	
		Primary standard	Secondary standard
Suspended particulate matter	Maximum annual geometric mean	75 $\mu g\ m^{-3}$	60 $\mu g\ m^{-3}$
	Maximum 24-h concentration not to be exceeded more than once a year	260 $\mu g\ m^{-3}$	150 $\mu g\ m^{-3}$
Sulfur oxides	Maximum annual geometric mean	80 $\mu g\ m^{-3}$ (0.03 ppm)	
	Maximum 24-h concentration not to be exceeded more than once a year	365 $\mu g\ m^{-3}$ (0.14 ppm)	
	Maximum 3-h concentration not to be exceeded more than once a year		1300 $\mu g\ m^{-3}$ (0.50 ppm)

Pollutant	Maximum allowable PSD increments ($\mu g\ m^{-3}$)	
	Class I areas	Class II areas
Particulate matter:		
Annual geometric mean	5	19
24-h maximum	10	37
Sulfur dioxide:		
Annual arithmetic mean	2	20
24-h maximum	5	91
3-h maximum	25	512

may prevent any new sources, such as a Cu-Ni smelter or taconite expansion facility, from locating in northeastern Minnesota unless the PSD restrictions are waived by a variance.

Annual geometric mean concentrations of total suspended particulates (TSP) averaged about 10 $\mu g\ m^{-3}$. TSP concentrations in the region ranged from 1 $\mu g\ m^{-3}$ to 367 $\mu g\ m^{-3}$) on a 24-h basis. Elevated 24-h TSP concentrations (18 readings above 150μgm^{-3} and 3 readings above 260 $\mu g\ m^{-3}$) were generally associated with population centers and taconite mining operations. During the taconite mining strike of August-December 1977, the regional TSP concentrations decreased an average of

60% over the preceding spring and summer, and the decreases at individual sites were at least 45%.

Because the primary 24-h NAAQS have been exceeded, portions of the Study Area have been designated as nonattainment areas for TSP. This action could also restrict new industrial development if TSP concentrations continue to remain elevated in future years.

Industrial growth in the region without Cu-Ni development is not expected to result in SO_2 concentrations that would exceed the annual or 24-h NAAQS nor will annual TSP NAAQS be exceeded. However, modeling results predict that point sources could exceed the 24-h TSP NAAQS (Table 3).

Table 3. Predicted SO_2 and particulate concentrations resulting from regional growth and Cu-Ni development

	SO_2 ($\mu g\ m^{-3}$)		Particulate ($\mu g\ m^{-3}$)	
	Annual	24 h	Annual	24 h
Regional growth				
Region average	2.3	–	0.1	–
High	5.6	78	0.7	160
2nd high	4.9	77	0.6	150
Class I PSD	0.81	16	0.02	3.0
Class II PSD	3.4	55	0.3	50
Cu-Ni development				
Base case	5.7	120	< 1.0	40
Option 1	2.1	54	< 1.0	20
Option 2	0.8	20	< 1.0	3

Although ambient NAAQS may be exceeded as a result of regional growth, the PSD increments pose the greatest limitations to siting a smelter in the region. Considering a factor of two accuracy in the modeling, the maximum 24-h SO_2 PSD increments are expected to be exceeded in both the Class I and Class II areas as a result of regional growth, and the PSD increment for particulates is predicted to be exceeded in the Class II area but not in the Class I area (Table 3).

The elemental composition of atmospheric particulate matter was typical of remote, midcontinental areas (Table 4). Sulfur, Fe, and Pb which can be derived from both natural and anthropogenic sources were selected to differentiate between urban and nonurban sites. The average concentrations of Pb and Fe were higher, respectively, in urban/industrial sites than in rural sites. The major sources of enhanced Fe concentrations are probably derived from the taconite industry, while enhanced Pb is primarily the result of local combustion of leaded gasoline. The average S concentration of particulates was not significantly different between urban and nonurban sites. Since the SO_2 concentrations in the region are low, it is likely that most of the S is present as SO_4 and the result of transport from distant sources.

Table 4. Elemental composition of atmospheric particulates in northeastern Minnesota, arithmetic means (ng m^{-3})

Elements	Rural	Urban/industrial	Region	
			Mean	SD
Al	100 - 260	75 - 500	200	140
Cl	20	10 - 70	30	20
V	ND - 1.0	ND - 1.0	< 0.5	< 0.5
Cr	0.1 - 10	1.0 - 6.0	4.0	4.0
Mn	6.0 - 20	10 - 20	10	7.0
Fe	280 - 750	990 - 1500	830	480
Co	0.3 - 1.0	2.0 - 3.0	1.0	1.0
Zn	10 - 80	10 - 20	20	20
As	2.0 - 5.0	4.0 - 8.0	4.0	2.0
Br	3.0 - 7.5	5.0 - 30	8.5	9.0
Sb	ND - 2.0	ND - 2.0	0.5	1.0
Pb	20 - 30	10 - 140	40	40
Cu	1.0 - 5.0	1.0 - 6.0	2.5	2.0
S	470 - 750	360 - 700	590	230
Ni	ND - 1.0	0.3 - 4.0	1.0	1.0
Cd	ND - 2.0	ND - 4.0	1.0	1.5
P	30 - 50	20 - 60	40	10
Si	450 - 930	530 - 1500	760	400
K	100 - 160	80 - 190	130	40
Ca	170 - 300	80 - 350	230	90
Ti	10 - 30	10 - 40	20	10
Sn	0.5 - 2.0	ND - 4.5	1.5	1.5
Ga	ND - 0.4	ND - 1.0	0.2	0.4
Ge	ND - 0.2	ND - 0.1	< 0.04	0.07
Se	0.1 - 0.7	0.2 - 0.5	0.4	0.2
Rb	0.1 - 1.7	0.2 - 1.0	0.6	0.5
Sr	0.3 - 2.1	0.6 - 4.9	1.8	1.8
Ba	8 - 30	30 - 50	30	10
W	ND - 0.4	0.2 - 0.6	0.4	0.2
Hg	ND - 0.4	ND - 0.3	0.1	0.1

The atmospheric deposition of trace elements to the region was estimated by comparing bulk, throughfall, and calculated dry inputs obtained from data collected during 1977 (Table 5). In general, the trace metal deposition rates were also typical of remote midcontinental areas. Although pH values are reported in Table 5, measurement of pH in bulk deposition is not meaningful with respect to evaluation of acid-forming species because of neutralization processes, reevaporation, microbial growth and oxidation/reduction reactions that can occur during the collection period.

Dry deposition rates of elemental particulates (Table 6) were calculated using the relationship: $D = V_g \times C$, where D is the dry deposition rate, V_g is the deposition velocity, and C is the air concentration. The data in Table 6 represent minimal values since the values of V_g used in the calculations were based on a deposition surface of filter paper rather than a forest canopy.

Table 5. Bulk and canopy throughfall deposition rates in northeastern Minnesota, arithmetic means ($kg\ ha^{-1}\ yr^{-1}$)

Elements	Bulk deposition			Canopy throughfall	
	Range	Mean	SD	Mean	SD
Al	0.250 - 0.800	0.410	0.260		
As	0.004 - 0.010	0.010	0.003		
Ca	8.40 - 13.0	11.0	2.00		
Cd	0.002 - 0.010	0.004	0.001	0.0007	0.0007
Cl	11.0 - 17.0	13.0	2.80		
Cu	0.008 - 0.020	0.010	0.004	0.006	0.004
Fe	0.270 - 0.690	0.440	0.210	0.33	0.13
F	0.500 - 1.20	0.850	0.290		
K	1.50 - 2.00	1.70	0.220		
Mg	4.50 - 56.0	21.0	24.0		
Ni	0.005 - 0.010	0.005	0.0001	0.007	0.004
Pb	0.050 - 0.20	0.100	0.070	0.021	0.005
Zn	0.040 - 0.090	0.060	0.020	0.050	0.020
Alk ($CaCO_3$)	13.0 - 28.0	18.0	6.50		
TOC	21.0 - 35.0	28.0	5.90	150	110
P-total	0.0001 - 0.001	0.0004	0.0004		
NO_2-NO_3	0.003 - 0.004	0.004	0.0006		
TDS	54.0 - 150	83.0	43.0		
S as SO_4	18.0 - 31.0	21.0	6.40	21.0	10.0
pH	4.00 - 4.80	4.60	0.20		
Sp cond. 25°C	23.0 - 39.0	29.0	7.30		
Suspended solids	50.0 - 81.0	62.0	15.0		
Na	1.50 - 2.50	2.10	0.400		

Table 6. Average dry deposition rates in northeastern Minnesota ($g\ ha^{-1}\ yr^{-1}$)

Elements	Rural	Urban/industrial	Region	
			Mean	SD
Al	410 - 1080	306 - 2030	810	550
Cl	1.58 - 75.7	44.2 - 230	82.0	56.8
V	ND - 1.58	ND - 0.946	<63.0	<63.0
Cr	0.315 - 18.9	1.26 - 10.1	5.68	5.68
Mn	10.7 - 29.0	20.2 - 47.3	25.9	13.6
Fe	2250 - 6130	8120 - 12220	6760	3960
Co	0.631 - 2.52	3.15 - 5.36	2.52	1.58
Zn	18.0 - 143	17.3 - 31.2	37.5	40.1
As	1.26 - 3.15	2.52 - 4.73	2.52	1.26
Br	3.78 - 9.15	5.99 - 40.4	10.7	11.4
Sb	ND - 0.631	ND - 0.631	0.315	0.315
Pb	12.6 - 28.7	12.6 - 129	34.1	36.0
Cu	2.52 - 15.1	4.10 - 18.6	8.20	5.68
S as SO_4	442 - 700	337 - 659	555	221
Ni	ND - 5.36	2.21 - 15.8	4.42	4.73
Cd	ND - 2.52	0.315 - 6.30	0.946	1.58

The data in Tables 5 and 6 suggest that dry deposition of Al and Fe was greater than bulk deposition, and that the wet deposition of S as SO_4 was about 35 times that of dry deposition. These data very likely reflect the fact that the bulk sampler does not provide representative data for some species and/or uncertainty in deposition velocities used for those species.

The effects of growth in regional point sources without Cu-Ni development on ambient SO_4 concentrations and deposition rates were also considered. The modeling results for the baseline case predicted a regional average SO_4 concentration of 5.4 ng m^{-3}. This value, when compared to a measured average of 1.8 μg m^{-3}, suggests that major point sources in the region are not primary contributors to ambient SO_4 levels in the Study Area. Regional growth is predicted to increase ambient annual SO_4 concentrations by a factor of about 15, and maximum 24-h concentrations of 1.2 μg m^{-3} are predicted. At the present time, there are no NAAQS for SO_4.

Sulfate deposition rates in the region due to local sources are predicted to more than double from 2.2 kg ha^{-1} yr^{-1} in 1977 to 4.5 kg ha^{-1} yr^{-1} in 1985. If it is assumed that background SO_4 would increase accordingly, then the total regional SO_4 deposition could be about 30 kg ha^{-1} yr^{-1} by 1985. If the modeling assumptions about wet and dry deposition are correct, it appears that dry deposition in the region is primarily due to local sources. However, about 85% of the total (wet plus dry) SO_4 deposition is not accounted for by regional sources.

Mineral fibers are a potentially serious environmental health hazard for the occupational and nonoccupational population in northeastern Minnesota. Asbestos is a mineralogical term that includes the asbestiform varieties of various silicated minerals including chrysotile (a member of the serpentine group) and the asbestiform varieties of actinolite-tremolite, anthophyllite, cummingtonite-grunerite, and riebeckite (members of the amphibole group). Chrysotile always occurs in the asbestiform habit, while amphiboles usually occur in a nonasbestiform habit. Asbestiform minerals occur as fibers, but cleavage fragments, such as those produced from crushing and processing nonasbestiform minerals, do not satisfy the definition of fibers and should be considered "fiberlike." However, when asbestiform and nonasbestiform minerals are subject to crushing and processing in typical mining operations, the resulting fragments have minor differences in morphology and physical properties that are very difficult to distinguish under a transmission electron microscope (TEM). Asbestos fibers and nonasbestiform cleavage fragments have different characteristics in terms of tensile strength, flexibility, durability, and surface properties. The extent to which these differences are related to the harmful properties of asbestos is uncertain.

Fibers were measured at seven sites in the region throughout 1977, using 24-h membrane samplers and a cascade impactor sampler. Fibers were classified as amphibole, chrysotile, nonamphibole or noncrysotile, and ambiguous.

Overall, fiber concentrations varied greatly over the days of sampling at each site. Fiber levels at rural sites averaged 26,500 fibers m^{-3}, and fibers at urban/industrial sites averaged 62,400 fibers m^{-3}. Of these fiber counts, chrysotile fibers comprised about 18% of the rural samples and 34% of the urban/industrial samples, but analysis of blank filters by the MDH suggested that millipore filters contained significant levels of chrysotile fibers and nucleopore filters sometimes contained amphibole fibers. Therefore, it was suspected that most, if not all, of the chrysotile found in the air samples could be attributed to contamination of the filters.

Amphibole fibers averaged 2,860 fibers m^{-3} at rural sites and 13,300 fibers m^{-3} at industrial sites. Most of the fibers in the samples were nonamphibole or nonchrysotile (12,400 fibers m^{-3} at rural sites and 17,300 fibers m^{-3} at urban/industrial sites).

From a public health perspective, particles in the respirable range of 1 to 2 μm are of the most concern. Based on a single sampling site, about 60% of the amphibole fibers and about half of all fibers observed were in the respirable range. Overall, amphibole fiber levels appeared to be related to wind direction and to come from a source in the eastern portion of the Study Area. Other types of fiber categories showed no correlation with wind direction.

4.5 Water Quality

Currently, major point-source discharges in the Study Area are either municipal discharges from sewage treatment plants or industrial discharges related to the taconite industry. There are currently about ten industrial dischargers and five major sewage treatment plants in the research area. Important area-source discharges include runoff which is exposed to sulfide minerals.

Annual average flow for 12 streams studied by the United States Geological Survey (USGS) ranged from 0.65 to 29 cm s^{-1}. High flow generally occurs after precipitation and following the spring snowmelt. Average low flow for 7 days is 0.06 to 5.3 cm s^{-1} compared to an average high flow of 2.5 to 135 cm s^{-1}.

Groundwater yield is generally low, limited by the low permeability of the area bedrock and often shallow overlaying glacial deposits. Yields generally average less than 19 l min^{-1}. Three relatively small areas have high volume aquifers yielding up to 3788 l min^{-1}.

Current industrial use of surface water is primarily for electric power generation. Mine-pit dewatering is the greatest groundwater use. Taconite mining operations occupy about 0.3% of the total drainage area. At current levels, water use does not cause significant impacts on the region's water resources, although withdrawal from some streams must be reduced during periods of low flow.

Hydrologically the Water Quality Research Area is a headwater region; that is, surface flow originates within the region. Nine watersheds containing a total drainage area of 3,489 km^2 are contained north of the Laurentian Divide and five watersheds containing a total drainage area of 1,249 km^2 are contained south of the Divide. Soils north of the Divide are shallow (0–3 m) and bedrock outcropping is common; soils are poor in nutrients, high in trace metals, and acidic. The soils south of the Divide are deeper (3–30 m), but still low in nutrients.

Surface water is abundant in the Study Area. Average annual runoff in the region is about 10 in. The Study Area includes 360 lakes larger than 4 ha in addition to 14 small rivers and streams. Nearly 75% of the Water Quality Research Area, and an even larger proportion of the surface water (91%) is north of the Laurentian Divide. Lakes north of the Divide are more numerous and larger, and the volume of stream flow is greater because a larger area is being drained. Because some of these waters are inside the BWCA, not all of the water north of the Divide is directly available for use.

The surface water quality of the Study Area was sampled from March, 1976 through September, 1977. Thirty-two stream sites on 13 different river systems and 35 lake substations on 26 lakes were established. Criteria for lake and stream site selection were based on location, size, shoreline characteristics, depth, and other factors. Sampling was conducted on a monthly, bimonthly, quarterly, or event basis depending on the importance of the parameter and site in question. The primary intent was to establish sites that would provide a representative cross-sampling of all the factors which presumably affect water quality in the Study Area.

Groundwater samples were collected for chemical analyses quarterly during 1976–1977 for 12 observation wells finished in glaciofluvial sand and gravel, 11 wells finished in the Rainy Lobe till, and 2 wells finished in peat material. An additional sampling occurred during a drought period in October, 1976.

Water samples were collected and analyzed according to acceptable methods of the EPA, MDH, and the USGS. Table 7 summarizes the parameters that were analyzed. The quality of the region's water resources is generally very good except for several streams with watersheds affected by extensive taconite mining activities. Streams draining largely undisturbed watersheds can be described as containing soft water, having low alkalinity, low total dissolved solids, low nutrients, high color, very low trace metal concentrations, and low fecal coliform counts. Streams draining disturbed watersheds would be considered to contain moderately hard to hard waters with elevated dissolved solids, nutrients, and trace metal concentrations compared to undisturbed watersheds. Color and fecal coliform concentrations are not significantly different in the two watershed classifications. Most water quality parameters tend to be much less variable in undisturbed streams than in disturbed streams.

The quality of the lakes which were studied is variable though similar to the quality of the undisturbed streams. However, the lake values may be less meaningful for determining baseline concentrations than values in streams because of the limited number of samples.

In general, concentrations of most chemical constituents are higher in the groundwater than in streams and lakes of the area. Groundwater from wells close to the Duluth Gabbro contact were found to have higher levels of trace metals and SO_4 than wells located at a distance from the contact.

Trace metal concentrations in surface waters are a major concern related to Cu-Ni development. At background stream stations, Cu, Ni, and Zn levels are generally very low, with median concentrations of Cu and Zn in the range of 1–2 $\mu g\ l^{-1}$ and Ni around 1 $\mu g\ l^{-1}$. Other trace metals of biological importance (As, Cd, Co, Hg, and Pb) have median concentrations significantly below 1 $\mu g\ l^{-1}$. There is little variability in the levels of As, Co, Cd, Hg, Ti, Se, and Ag across almost all surface waters monitored. Variability of metal levels does not appear to be related to watershed size. As expected, Fe, Mn, Cu, Ni, Zn, Pb, F, and Cr concentrations in streams are significantly higher in disturbed watersheds than in undisturbed areas.

The dynamics of metals in lakes are somewhat different from those in streams because of the large surface area of bottom sediments with varying oxidation-

Table 7. Water Quality of Lakes and Streams in Norhteastern Minnesota

	Streams			Lakes		
	Range	Median	N	Range	Median	N
Ca (mg l^{-1})	1.8 - 80.0	7.4	333	1.9 - 46.0	7.2	129
Cl (mg l^{-1})	0.1 - 88.0	2.0	462	0.1 - 9.3	1.6	94
Color (PCU)	4 - 500	90	463	1 - 400	80	141
Mg (mg l^{-1})	1.0 - 40.0	3.8	333	0.6 - 12.2	3.1	129
K (mg l^{-1})	0.2 - 8.4	0.6	310	0.2 - 2.2	0.6	84
Silica (mg l^{-1})	1 - 34	7.0	465	1 - 19	4.8	135
Na (mg l^{-1})	0.2 - 45.0	1.8	304	0.8 - 18.0	1.8	77
Sp Cond[a]	12 -1198	65	463	24 - 389	65	141
Hardness (mg l^{-1})	5.3 - 310.0	31	206	9.0 - 142.0	28.9	92
Turbidity (JTU)	0.5 - 64.0	2.1	463	0.4 - 7.0	2.0	135
TOC (mg l^{-1})	0.4 - 45.0	15.0	80	4.6 - 38.0	14.0	142
Total P (μg l^{-1})	0.8 -2100	20.1	260	0.8 - 220.0	22.9	140
NO_2+NO_3(mg l^{-1})	0.01 - 13.0	0.1	260	0.01- 1.50	0.02	120
Total N (mg l^{-1})	0.01 - 14.1	0.91	250	0.12- 9.9	0.63	78
Alk (mg l^{-1})[b]	1 - 190	23	457	1 - 73	19	141
HCO_3 (mg l^{-1})	6 - 151	25	257	9 - 88	20.5	38
SO_4(mg l^{-1})	1 - 630	7.4	434	1 - 140	7.8	138
pH	4.7 - 8.4	6.91	458	5.7 - 8.8	7.1	141
Al (μg l^{-1})	1.6 - 760	90	270	4 - 610	77	60
As (μg l^{-1})	0.2 - 5.0	0.8	185	0.4 - 2.1	0.6	41
Cd (μg l^{-1})	0.01 - 0.45	0.03	443	0.01- 0.77	0.03	114
Co (μg l^{-1})	0.1 - 11.0	0.4	410	0.1 - 2.2	0.4	72
Cu (μg l^{-1})	0.2 - 12.0	1.5	447	0.2 - 10.0	1.5	129
Total Fe (μg l^{-1})	99 -5500	610	457	16 - 2300	350	138
Pb (μg l^{-1})	0.1 - 6.4	0.5	443	0.1 - 1.9	0.4	114
Mn (μg l^{-1})	5 -1200	50	309	1 - 5400	29	135
Hg (μg l^{-1})	0.01 - 0.60	0.08	183	0.08- 0.64	0.28	96
Ni (μg l^{-1})	0.2 - 210	1.2	447	0.4 - 6.0	1.0	129
Zn (μg l^{-1})	0.1 - 30.0	2.2	446	0.2 - 35.5	1.8	113

[a]Micro-mhos/cm @ 25°C.
[b]As $CaCO_3$.

reduction potentials. Lakes can act as sinks for metals so that the chemistry of outflowing waters is different from that of inflowing waters. The concentration of metals can also vary within the lake itself. The most elevated metals in the Study Area lakes were Fe, Al, and Mn. Copper, Ni, and Zn have median levels between 1–2 μg l^{-1}; As, Co, and Pb have median levels of below 1 μg l^{-1}, respectively. Cadmium levels were an order of magnitude lower than those for As, Co, and Pb. The greatest variabilities in concentrations were exhibited by Mn, Zn, Cd, and Al; As was the least variable in concentration.

Maximum recommended water quality levels (Table 8) were exceeded for Cd, color, Cu, Fe, Pb, Mn, Hg, Ni, N, pH, and Zn in one or more of the streams monitored. In most cases, these elevated levels occurred at locations adjacent to mining activities.

Almost all of the streams which were monitored exceeded the EPA water quality criteria for Hg (0.05 μg l^{-1}). The median concentration of Hg for all streams

Table 8. US EPA water quality criteria[a]

Parameter	Criteria[b]
Color (PCU)	75
pH	5–9; 6.5–9.0[c]
As	50
Cd	10; 0.4–4[c]
Cl	250,000
Cu	1000
Cr	50
Total Fe	700; 1000[c]
Hg	2; 0.05[c]
Mn	50
N	10,000
Ni	0.01 times 96-h LC_{50}[d]
Pb	50
Zn	5000

[a] μ l^{-1} unless otherwise noted.
[b] Domestic consumption.
[c] Fisheries.
[d] Aquatic life.

monitored was 0.08 $\mu g\ l^{-1}$ with a range of 0.01–0.6 $\mu g\ l^{-1}$. Standards for Hg are based on US Food and Drug Administration guidelines for edible fish. Certain freshwater species concentrate Hg by a factor in excess of 10,000, and high Hg levels have been found in fish from some of the area's lakes and streams. The source of Hg in the region is unknown.

Because acid precipitation is a potential problem, the quality of precipitation in the Study Area was monitored at several sites in the Study Area on an event basis. Eighty-four percent of the samples (32) had a pH less than 5.7 which means that most of the precipitation measured can be considered acidic. Twenty-four percent of the samples had a pH of 3.6–4.4. The geometric mean pH of samples collected in the Study Area was 4.6. These values are comparable to, or even less than values measured in areas of the world where ecological damage has already occurred.

If patterns of increasingly acidic precipitation continue, it is likely that many of the poorly buffered small streams will have noticeable decreases in aquatic populations during and following spring melt. Stream systems are very sensitive because the flush of water from the spring snowmelt can represent a majority of the water than the stream carries through the whole year. Recovery from these episodes can be expected to be fairly rapid until the sources of the organisms are themselves affected. Recovery would be very slow once the source areas are affected.

Because acidic precipitation and SO_4 deposition are primarily related to air pollution sources outside the region and are projected to increase significantly over the next 10 to 20 years, acidification may represent a serious threat to the aquatic ecosystems of northeastern Minnesota, even if Cu-Ni development does not occur. Long-term changes in the aquatic communities may already be underway based on the general decrease in the pH of precipitation.

One crucial parameter related to acidification that was monitored is the water's buffering capacity. Calcite Saturation Indices (CSI) were calculated for all study lakes and 30 lakes in the BWCA to measure buffering capacity. Lakes with a CSI less than 3.0 are well-buffered; lakes with an index between 3.0 and 5.0 are poorly buffered with the possibility that acidification may already be occurring; and an index over 5.0 denotes lakes with little or no buffering ability and a strong possibility that severe acidification has already occurred.

The poorly buffered lakes in the region are with few exceptions headwater lakes. This may be explained by the fact that buffering is a function not only of atmospheric processes, but also of watershed geology. The chemistry of headwater lakes often reflects that of precipitation. Watershed contributions to headwater lake chemistry assume secondary importance. The ability of the lakes to assimilate $H+$ ions generally increases from headwater to downstream lakes. Headwater areas of the region (which includes half of the BWCA lakes studied) are generally not well buffered and have limited capacities to assimilate existing acid loadings. Of the 26 lakes in the Study Area, 19 had CSIs that were less than 3.0; 7 of the lakes had CSIs between 3.0 and 4.1. Of the 30 lakes in the BWCA, 15 of the lakes had CSIs from 1.5 to 3.0 and 15 had CSIs from 3.1 to 4.1.

The presence of Cu-Ni sulfide mineralization has directly affected water quality in two areas. In one case, human disturbance of the mineralization has accelerated the chemical/physical weathering of this material as Filson Creek flows naturally over exposed mineralized gabbro. Within the Filson Creek watershed, total concentrations of Cu and Ni during the sampling year generally increased from headwater locations (less than 1 $\mu g\ l^{-1}$ Ni) to the creek's point of discharge into a river (3 to 5 $\mu g\ l^{-1}$ Ni). The smaller Cu and Ni concentrations at the headwater locations reflect the smaller percentage of sulfide-bearing material in the till and the greater distance from the mineralized contact zone.

At another location on this same creek, exploration activities prior to 1977 resulted in a small volume surface discharge at the foot of a bulk sample site. Elevated levels of metals (1×10^4 to 1.3×10^4 $\mu g\ l^{-1}$ Ni, 3.6×10^2 to 1×10^3 $\mu g\ l^{-1}$ Cu, and 1.9×10^2 to 5.3×10^3 $\mu g\ l^{-1}$ Zn) discharged into a small tributary of Filson Creek and raised the Ni and Cu concentrations by about 9 $\mu g\ l^{-1}$ and 5 $\mu g\ l^{-1}$, respectively. This change in trace metal concentrations, however, was not sufficient to result in measurable, biological changes in the creek.

In the second case, a small watershed contained several waste piles of mineralized gabbro from a nearby taconite mining operation. Seeps from the waste piles flowed into a stream which had elevated Ni levels (85 $\mu g\ l^{-1}$ compared to an undisturbed stream 1 $\mu g\ l^{-1}$). Field studies in this watershed have demonstrated that extensive disturbance of the mineralized gabbro without corrective can result in significant water quality degradation.

Total mineral fiber counts for stream samples range from 3.8×10^5 to 7.92×10^6 fibers l^{-1} with a median of 3.31×10^6 fibers l^{-1}. These levels are comparable to literature reports for other streams, as well as to levels in beer, wine, and other bottled drinks. Large variations in fiber levels were observed in both disturbed and undisturbed streams. Since it is not known whether the Reserve Mining Company was discharging during the sampling periods, the large variations may be due to natural variability.

5 Cu-Ni Development Alternatives

Assessment of the environmental, social, and economic impacts of Cu-Ni development requires a full understanding of all phases of development from mining to shipment of refined products. In order to understand the potential impacts, a series of hypothetical models were developed.

The developed mining models were intended to be representative of mining alternatives; they were not intended to predict or recommend choices for development. Mining models were developed within a framework that covered the possible range of sizes for Minnesota Cu-Ni development. The models included all basic types of mining (open pit, underground, and combination of both), and typical resource inputs and material outputs for the various scales of development.

Three basic mining models integrated to the production of 10^5 t of Cu and Ni metal from the smelter/refinery operations were developed. These included an underground mining model with an annual capacity of 12.35×10^6 t of ore, an open-pit mining model with a capacity of 20×10^6 t of ore, and a combination of a 5.35×10^6 t underground and a 11.33×10^6 t open-pit mine, totaling 16.68×10^6 t ore y^{-1}. These large-scale operations are necessary to compete economically because Minnesota's resource is low-grade.

Because of the costs of transporting the mined ore, it is likely that processing facilities will be located no more than a few miles from the mine site; Fig. 4 provides a general flowsheet for a Cu-Ni development. The final size of milled particles will be less than 0.2 mm in diameter. Water is added during and after grinding and the slurry (65% solids) then proceeds to flotation. Depending on the flotation process that is selected by a mining company, a single bulk Cu-Ni concentrate or two concentrates may be produced. The Cu-Ni concentrate from a bulk flotation process will probably contain from 10 to 22% Cu and 2 to 3% Ni. If two concentrates are formed, a Cu concentrate could range from 11 to 24% Cu with a negligible amount of Ni, and a Ni-Cu concentrate would probably contain from 4 to 7% Cu and 4 to 7% Ni. About 95 to 97% of the ore would be disposed of as tailings.

Regardless of the type of concentrate produced, it is likely that pyrometallurgical technology (continuous, flash, or conventional smelting) rather than hydrometallurgical technology will be used prior to refining. For simplicity in the remaining discussion only operations generating a single bulk concentrate (6.35×10^5 t yr^{-1} assaying 13.825% Cu, 2.647% Ni, 30.001% Fe, and 25.870% S) are modeled as input to a smelter using flash-smelting technology with an H_2SO_4 plant to clean SO_2 gas from the smelter gases and produce a product for sale.

Cu-Ni matte from the flash furnace is further treated in Pierce-Smith converters by a modified process involving selective fluxing, blowing, slagging, and skimming to produce blistered Cu and a slag which contains the bulk of the recoverable Ni and Cu. The blistered Cu is then fire-refined and prepared for casting in anode furnaces. Anodes are shipped to the Cu refinery for final purification by electrorefining procedures. Slag from the flash furnace will be cleaned along with Cu converter slags in two electric furnaces to produce a Ni-Cu matte and a discardable slag. Another converting step, followed by leaching and

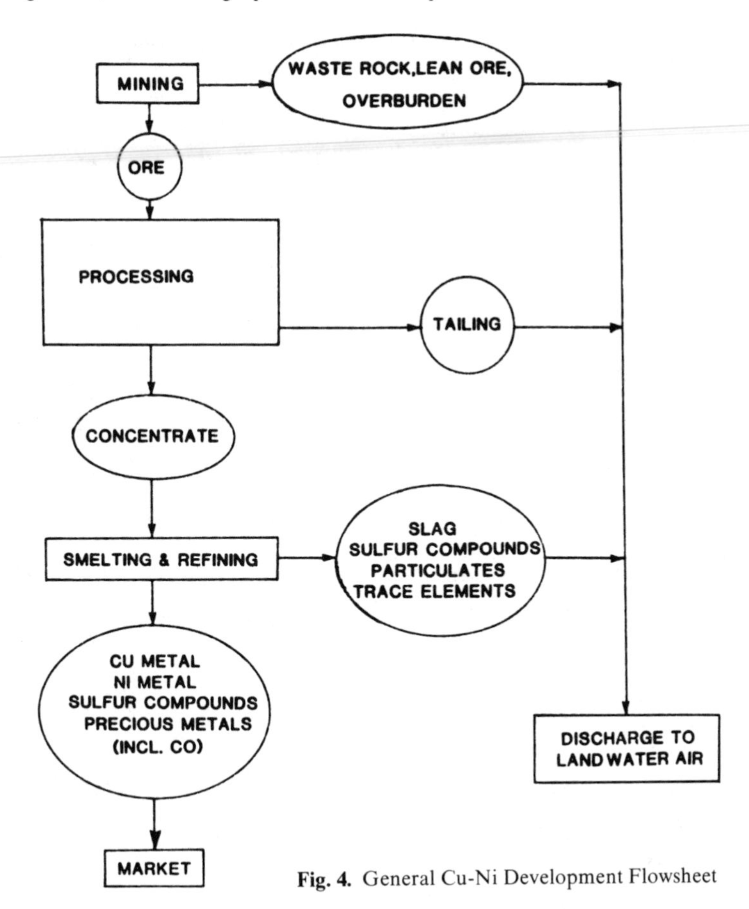

Fig. 4. General Cu-Ni Development Flowsheet

electrowinning techniques at the refinery, will recover Ni, Cu, and Co from the Ni-Cu matte.

Although the smelter/refinery facilities could be located elsewhere in Minnesota, the US, or another country, both the capital and operating costs would be increased. For the purposes of identifying total impacts, it was assumed that all facilities would be located at a common site. The development models (Table 9) are based on an overall life of 30 years for each facility.

For each of the mining models, about half of the total land requirements would be devoted to tailings disposal. The underground model requires about 10% less capital investment because of the higher-grade ore and smaller processing plant facilities. Labor requirements during mining consume a larger share of the operating costs and range from 33% of the total from the open-pit model to 53% of the total in the underground model resulting in a higher total operating cost for the underground model of 3 to 7%. The smelter and refinery facilities dominate the capital costs (42 to 50% of the total).

Table 9. Summary of Minnesota Cu-Ni development model requirements over the total life for each model.

Totally integrated operation comparison parameter	Mining operation type and size, 10^6 tonnes ore		
	12.35 Underground	16.68 Combination	20.00 Open Pit
Total land use, hectare	2,023	3,310	4,047
Total materials produced			
Overburden, 10^6 m^3	0	0.54	0.71
Waste rock-lean ore, 10^6 tonnes	28	380	650
Ore, 10^6 tonnes	280	410	500
Concentrate, 10^6 tonnes	15	15	16
Tailings, 10^6 tonnes	270	390	480
Metal (85%Cu + 15%Ni) 10^6 tonnes	2.3	2.4	2.5
Slag, 10^6 tonnes	14	15	15
Sulfuric acid, 10^6 tonnes	10	11	11
Total capital cost, \$ 10^6 [b]	660	760	760
Total cumulative operating cost, \$$10^6$ [b]	3,200	3,100	3,000
Total cumulative energy requirement, 10^{12} BTU	330	370	400
Water requirement			
Process,[c] 10^9 l yr^{-1}	125	136	144
Make-up, 10^9 l yr^{-1}	3.8	3.0	3.0
Potable, 10^9 l yr^{-1}	4.9	6.8	8.0
Full production operating workforce	2,480	2,220	1,990
Mine	1,560	1,220	960
Mill	300	380	410
Smelter/refinery	620	620	620
Peak construction workforce	2,520	2,760	2,820
Mine	280	360	320
Mill	990	1,150	1,250
Smelter/refinery	1,250	1,250	1,250

[a]Based on an effective operating life of 23.0, 24.4 and 25.0 years for the 12.35, 16.68, and 20.00 × 10^6 t yr^{-1} operations, respectively. All values have been rounded.
[b]Includes a smelter/refinery with acid plant and secondary hooding control of SO_2 (\$1977).
[c]Process includes total water needs for processing, smelting, and refining.
[d]Excludes water in ore and mine discharge water.

Smelting and refining operations also dominate the total energy requirements and range from 57 to 64% of the total. Again, the reduced size of the processing facility needed to handle the higher grade ore of the underground mine results in a total energy requirement that is about 17% lower than for the open-pit model. Electricity will supply about 150 MW or 75% of the energy needs. About one-half of the electrical demand is created by smelting and refining, and the remainder is used by the mining and milling operations.

From an environmental perspective, the underground mining development would result in lower air, water, and land discharges, lower energy demand, and greater reclamation potential. Economically, the underground development has greater employment opportunities and lower capital costs, but it also has a greater total production cost, greater occupational hazards, and lower resource recovery. Overall, the most environmental and economically cost-effective development model would be the combined, open-pit/underground development.

5.1 Air Emissions

The air quality impact analysis was based on an integrated unit (i.e., mining, milling, processing, smelting, refining) that would produce 10^5 t yr^{-1} of Cu plus Ni metal. Although a development project would result in emissions of CO, NO_x, and HC, these emissions are negligible when compared to SO_2 and particulate emissions. Therefore, these pollutants were not modeled and are not discussed further.

Several hypothetical smelter models (all utilizing flash furnace smelting technology) with varying degrees of pollution control were used to evaluate potential impacts of particulates and SO_2. The Federal New Source Performance Standards (NSPS) provided the framework for quantifying emissions from the hypothetical models. To meet these standards, a new smelter must have the equivalent of double-contact, acid-plant control (99.5% removal by volume) for all process gases from roasters, smelting furnaces, and converters. Three pollution control configurations were selected to provide a range of SO_2 and particulate emissions for modeling (Table 10).

In addition to the SO_2 emission rates noted for the smelter models during periods of normal operation, a model was generated to represent emissions during an upset period. To simulate such conditions, it is assumed that some sort of failure forces the bypassing of the acid plant and any subsequent S-removal equipment. The resulting stack emissions contain all of the SO_2 in the strong gas stream, as well as 90% of the SO_2 in the weak gas stream for an SO_2 emission rate of 10,326 g s^{-1}.

Fugitive SO_2 and particulate emissions from the model smelters are estimated to be about 0.2% of the total particulate input based on mass balance analysis of concentrate from the Duluth Gabbro Complex. Fugitive emissions are constant for all three models, but they become relatively more important as the degree of control increases, ranging from 8% of the Base Case emissions to about 70% for the Option 2 Case with Best Available Control Technology (BACT). Figure 5 illustrates the S balance for the Option 1 Case. Figure 6 illustrates the stack particulate emissions balance for the Options 1 and 2 smelter models. Tables 11 and 12 summarize cost estimates for removing SO_2 and particulates from smelter process gases.

Contributions to fugitive particulate emissions from mining activities were also evaluated using an open-pit mine development (2800 t yr^{-1}) and a smelter meeting NSPS. Based on this development, the hauling activities are expected to contribute nearly 50% of the particulate emissions, crushing and grinding operations about 10%, and the smelter about 40%. Emissions factors for milling and

Table 10. Summary of particulate and SO_2 emissions ($t\,yr^{-1}$) for a smelter producing $10^5\,t\,yr^{-1}$ of Cu plus Ni metal[a]

Smelter model	Particulate emissions			SO_2 emissions		
	Stack	Fugitive	Total	Stack	Fugitive	Total
Base case	2385	1500	3885	385[b]	31	389
Option 1	358	1500	1858	143[c]	31	174
Option 2	358	1500	1858	32[d]		

Particulate composition[e]

Constituent	Distribution (wt. %)	Constituent	Distribution (wt. %)
Cu	13.8	Fe (sulfides)	0.30
Ni	2.6	Sb	0.0002
S	25.9	Cl	0.011
As	0.004	F	0.0004
Cd	0.004	SiO_2	15.5
Co	0.13	Al_2O_3	3.4
Be	0.00006	MgO	2.6
Pb	0.006	CaO	1.6
Hg	0.00001	Other[f]	4.3
Zn	0.11	Total	100

[a] All models assume normal operating conditions.
[b] Acid plant control of strong SO_2 gas to 650 ppm SO_2 plus redirection of weak SO_2 with secondary hooding.
[c] Same as b plus scrubbing of collected weak SO_2 gas to 650 ppm SO_2.
[d] Acid plant control of strong SO_2 gas to 300 ppm SO_2 plus scrubbing of acid plant tail gas and collected weak SO_2 gas to 143 ppm SO_2; represents best available control technology.
[e] The model assumes that the particulates have the composition of the mill concentrate. Normal operating conditions were assumed (Coleman, 1978).
[f] Includes oxides of Na, K, Ti, P, Mn, Cr, and Fe.

mining activities (waste rock piles, tailing basins, hall roads, open pits) were developed based on data for taconite mining.

Fiber studies conducted on potential ore materials indicate that although the occurrence of true asbestiform minerals is quite rare, cleavage fragments are generated during processing which meet current state definitions of mineral fibers observed under TEM analysis. For modeling purposes, it was simply assumed that all smelter particulate emissions and all tailings basin and industrial emissions contain 10^9–10^{10} fibers g^{-1} based on analysis of bulk materials in bench-scale tests.

5.2 Water Balance

The overall Cu-Ni development operations were separated into three parts to develop a water balance. Subsystem A consists of the land areas which are part of the mine/mill operation, but it excludes the tailing basin. Movement of water

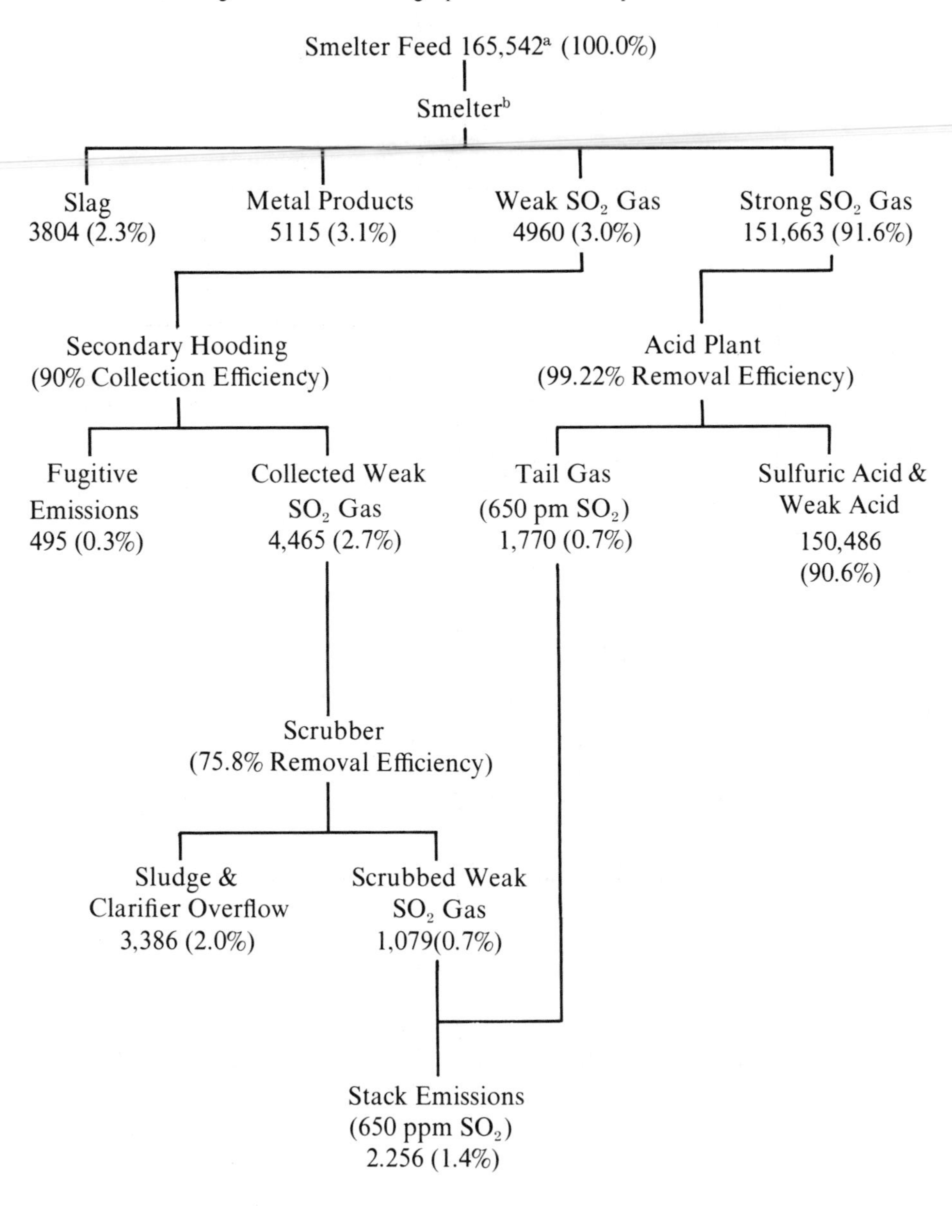

Fig. 5. Sulfur balance for an Opinion 1 smelter with a flash furnace
[a]All emissions are in t yr^{-1} S.
[b] Acid plant control of strong SO_2 gas to 650 ppm SO_2; secondary hooding collection of weak SO_2 gas followed by scrubbing to 650 ppm SO_2; normal operating conditions are assumed.

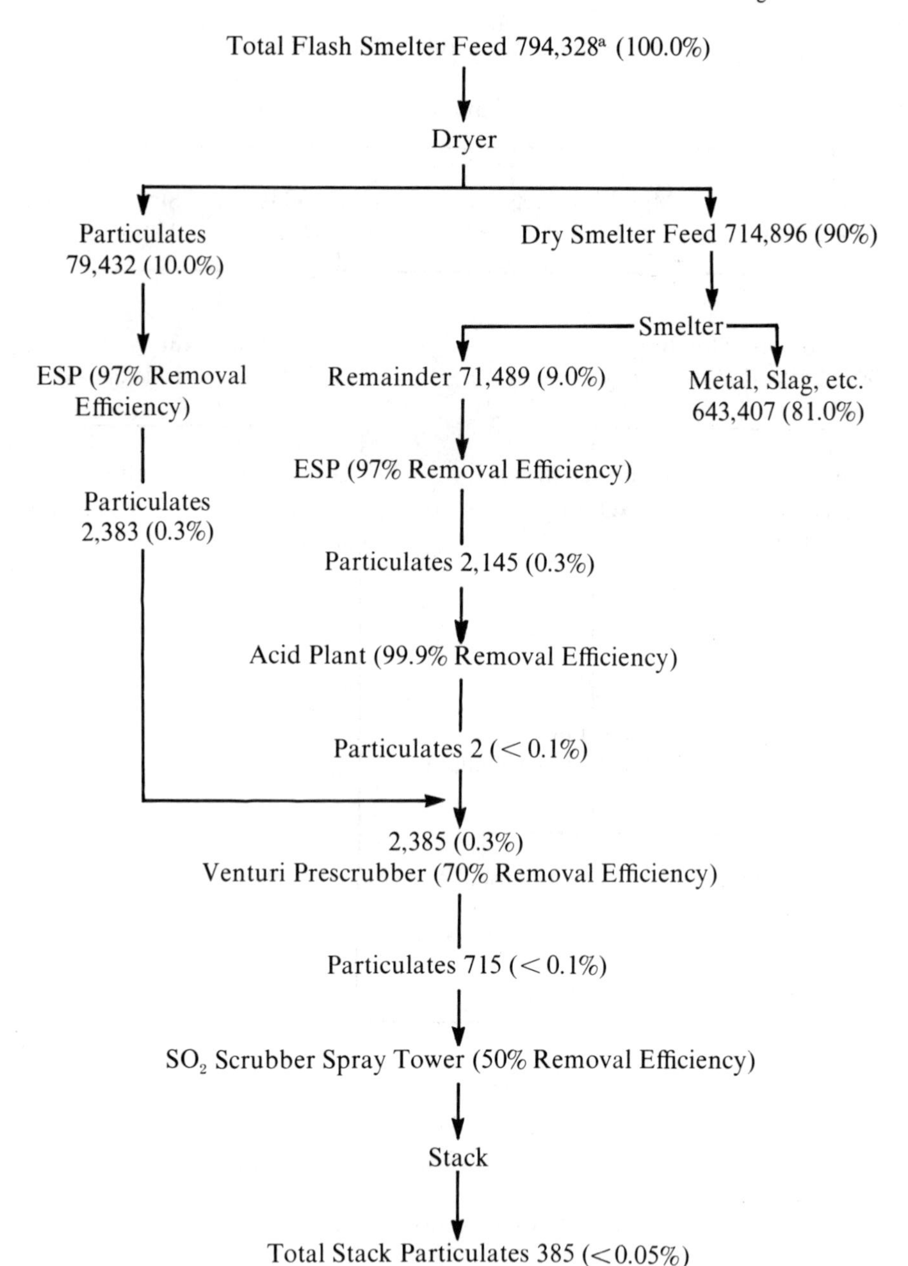

Fig. 6. Particulate balance for Option 1 and 2 smelters

[a]All emissions are in t yr^{-1}.

Table 11. Estimated costs[a] for removing SO_2 from smelter gases

	Option 1 smelter	Option 2 smelter
System description	Scrubber treats only weak gas streams Both acid plant and scrubber gases exit at 650 ppm Acid plant removal = 99.22% Scrubber removal = 75.8%	Scrubber treats both wet gas stream and acid plant tail gas Both acid plant and scrubber work at SO_2 removal efficiencies representative of best available technology Acid plant removal = 99.64% Scrubber removal = 90%
Total capital costs, $\$10^6$	4.6	6.0
Total annual operating costs, $\$10^6$	2.0	2.6
Resulting annual stack emissions, S, t yr^{-1}	2,256	501

[a] 1977 dollars.

Table 12. Estimated capital costs[a] for removing particulates from smelter gases

	Gas flow	Capital cost ($\$10^3$)[b] at various efficiency levels (%)			
Gas stream	(10^3 m^3 min^{-1})	97	99	99.5	99.9
Flash furnace model					
Dryer	1.70	417.6	541.8	630.0	828.0
Roaster	–	–	–	–	–
Smelting furnace	1.33	327.1	424.4	493.5	648.6
Copper converters	1.39	342.4	444.3	516.6	679.0
Nickel converters	0.30	73.8	95.7	111.3	146.3
Weak SO_2 gas[c]	5.05	1241.0	1610.0	1872.2	2460.5

[a] 1977 dollars.
[b] At \$323 m^{-2} of collecting surface (installed).
[c] If particulate removal is not performed in SO_2 scrubber.

through the waste rock piles, open pit, and other elements in this subsystem is controlled by natural processes. Annual water balances were calculated using areas equal to the final size of the pit, rock piles, and other components at the end of the mining operation, recognizing, of course, that all of these components start at zero size and grow to the indicated sizes over the life of the operation. The most significant source of poor-quality runoff from the mine site will be the stockpiles. The hydrologic behavior of the stockpiles was modeled based on 3 years of study at a taconite mine and at a Cu-Ni test site.

Subsystem B includes the mill and associated tailing basin and reservoirs. Water requirements for the mill total 2.46×10^3 l for each tonne of crude ore processed and much of this water is continually recycled. Make-up water to replace that lost in the system can come from precipitation onto the tailing basin, appropriation from lakes, streams, groundwater, or the use of runoff from subsystem A. Losses from the mill system are primarily spills and evaporative losses around buildings, and range from 2.6 to 4.2% of the total water required. Water is used to transport concentrate to the smelter (538 l t^{-1} concentrate) and to transport tailing materials to the basin (2.46×10^3 l t^{-1} tailing).

Potable water requirements for the entire complex average 34 l min^{-1} for each 10^6 t yr^{-1} capacity. Potable water is required for sanitary facilities and for some specialized cooling requirements in the smelter/refinery complex. Sanitary water will be treated before discharge.

The minimum amount of water that must be stored in the tailing basin for proper mill operation is a 5-day supply, or roughly 6 ha of pond for each 10^3 t day^{-1} of tailing discharge. Based on standard tailing basin practices, it is estimated that 80% of the basin will be covered by water for dust control. This would supply many times the minimum pond area required for proper mill operation. Losses include evaporative losses, surface outflow, and seepage.

Subsystem C consists of the smelter/refinery complex; water requirements include those for contact and noncontact waters. Contact water in the complex is used for scrubbing of gases going to the acid plant; direct cooling of anodes, cathodes, and other products after casting; granulation of molten slag by water immersion; and electrolytic solution, scrape anode rinsing, and other uses. Estimates of appropriation requirements for the smelter/refinery complex range from a low of about 2.65×10^3 l min^{-1} to a high of about 1.70×10^4 l min^{-1}. About 1.89×10^3 l min^{-1} are actually consumed, primarily through evaporation during contact cooling and as a raw material consumed in the production of H_2SO_4. In addition, the water which enters the smelter/refinery complex with the concentrate is lost to evaporation and the dryer before the concentrate is smelted. These losses are independent of recycling. The minimum water requirements assuming recycling of water indicate the complex would need to appropriate 2.63×10^3 l min^{-1} of which 1.87×10^3 l min^{-1} is lost in the plant. The remainder must be discharged or treated before reuse in order to prevent the buildup of undesirable constituents in recycle water systems.

Noncontact water is required for cooling of furnace walls and doors and other refractory equipment; condensing of steam from waste heat and turbines; cooling of bearings; and, cooling of the acid plant. If water is used on a once-through basis for cooling, a continuous appropriation of about 1.69×10^6 l min^{-1} would be required. This would result in an evaporative loss of 2.39×10^3 l min^{-1}. Blowdown requirements for the cooling system could range from 227 l min^{-1} if the make-up water is of high quality to 7.91×10^3 l min^{-1} if the make-up water has high levels of dissolved solids.

Overall, the total contact and noncontact water system in a 10^5 t yr^{-1} smelter/refinery complex assuming maximum recycling would require a total appropriation of 1.29×10^4 l min^{-1} of which 4.26×10^3 l min^{-1} is consumed primarily by evaporation. The remaining 8.67×10^3 l min^{-1} is not consumed and could be treated and discharged. If the smelter/refinery complex is located close to the mill, this discharge could be used to meet part of the mill make-up requirement.

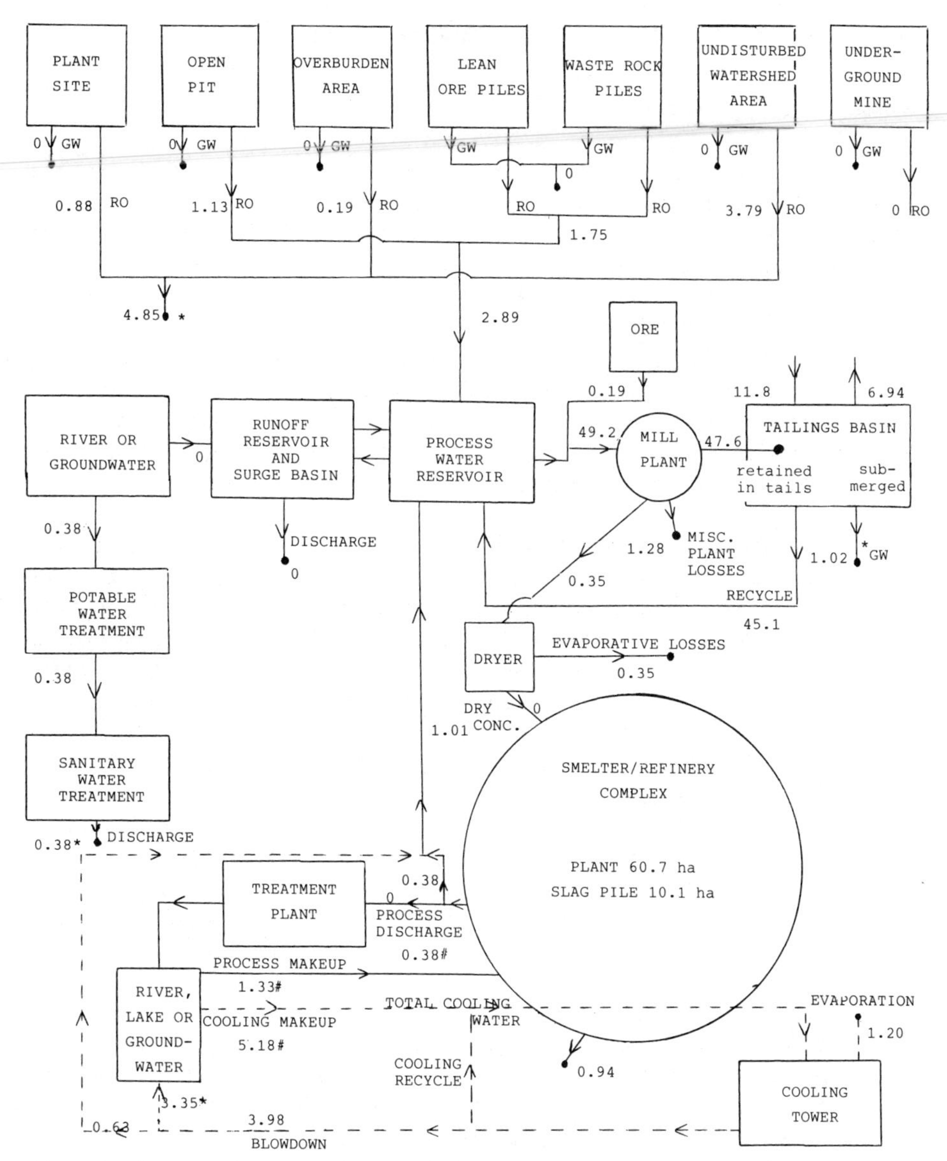

Fig. 7. A daily water balance for an integrated mine/mill and smelter/refinery complex. Key to figure: Water quantities in 10^9 l. * = discharges to environment. # = appropriations from outside the system. 20×10^6 t open pit mine; 100,000 t smelter/refinery. Average precipitation and evaporation. Tailings basin on semipermeable base

The water balance for a total integrated mine, mill, and smelter/refinery system (Fig. 7) which is based on a 20×10^6 t yr^{-1} open-pit mine model would require a net water appropriation of 6.50×10^9 l yr^{-1} and release a total discharge of 9.22×10^9 l yr^{-1}. Of the net discharges, 4.85×10^9 l yr^{-1} come from subsystem A, 1.02×10^9 l yr^{-1} from subsystem B (primarily tailing seepage); and 3.35×10^9 l yr^{-1} from subsystem C (noncontact water).

6 Cu-Ni Development Impact Analysis

6.1 Economic and Social Impacts

Because of uncertainties involved in predicting conditions 10 to 20 years in the future, growth studies were designed to evaluate impacts that could result under certain conditions of Cu-Ni development, but growth studies were not intended to forecast precisely what would happen. These growth studies are an important component of the regional study because advance planning can avoid some problems of mining boom towns.

Actual employment opportunities in the region will depend on the type of development, the size of the development, and the extent of development integration. Peak construction work force requirements would be around 2,500 workers for a fully integrated development, while the peak operating work force requirements would vary between 2,000 and 2,500 workers depending on the type and size of mine.

The sequencing of construction and operation of the facilities could vary the work force requirements. For example, the construction and operation simultaneously of three fully integrated operations requires a peak construction work force of almost 10,000 jobs. At the end of construction, the work force requirements would decline to about 6,000 and then gradually increase to about 6,800. These abrupt population changes could have a significant impact on land use, demand for housing, energy, and highways, as well as the full range of government services.

Demands on services can be mitigated by phasing the sequence of development. The transitional effects will be minimized by sequencing development of small operations as opposed to developing large operations all at once. Sequential development throughout the Cu-Ni resource area would provide for the greatest growth-related benefits and the fewest growth-related detrimental impacts. The location of a smelter/refinery complex outside of the Study Area but still in the state of Minnesota could take advantage of already established industrial infrastructure and a trained labor force, while reducing the rate of growth in the Study Area.

Population growth in the Study Area is expected to come from new and existing construction workers and service workers that will immigrate to fill jobs in retail, wholesale, manufacturing, and service sectors associated with Cu-Ni development. Again, a wide range of potential population growth impacts could occur depending on the timing and scale of development.

A maximum development case based on four mine/mills and four smelter/refineries all starting production within an 8-year period shows a doubling of the Study Area's 1976 population. Although the resource zone contains enough Cu-Ni ore to support this degree of development, it appears unlikely to occur. It does, however, demonstrate the potential changes that could result from Cu-Ni development.

A single, fully integrated mine/mill/smelter/refinery model suggests a population increase over 1976 levels of 20,000 by 1990 to a total area population of 67,000. The population increase that would be due specifically to Cu-Ni amounts to 12,000; the remaining 8,000 of the 20,000 population increase would be the result of other regional growth.

The fiscal impacts of development could be very significant. It is expected that between 60 to 70% of all Cu-Ni immigrants are expected to live in and around existing urban areas. There are eight cities and seven school districts in the Study Area. The estimated average operating cost per year ($1977) is $212 per person for the cities and $1,477 per pupil unit in the school districts. The range of per capita operating costs for the cities in 1977 was $156–241. The average operating cost per pupil unit for the state in 1977 was $1,059 as compared to $1,477 for the Study Area.

In order to determine the fiscal impact of Cu-Ni development, it was assumed that the facilities of both the communities and the school districts will be at full capacity prior to Cu-Ni development, and that capital expenditures will be required to accommodate an increased demand for public services. The analysis shows that each of the communities will have greater annual expenditures than income derived from taxes on development. These deficiencies range from $\$5 \times 10^3$ yr^{-1} to $\$1.3 \times 10^5$ yr^{-1} for the smelter/refinery complex located in zone 4, and from $\$1.2 \times 10^4$ yr^{-1} to $\$7.2 \times 10^5$ yr^{-1} for a combination open-pit/underground mine and mill located in zone 4. It appears that revenues generated by the mining operation will not balance the additional public expenditures required to meet the demands of an expanded population. Although this assumption may not be totally accurate, it serves to illustrate the direct relationship between demand-induced public expenditures, on the one hand, and additional public revenues generated by the source of the demand, on the other. This analysis indicates the need for careful evaluation of the present Cu-Ni tax generation and distribution policies.

Cu-Ni development could generate significant tax and royalty revenues for the state of Minnesota. Royalty payments could be a major income source for the state depending on the state's share of minerals that are mined. About 15% of the Cu-Ni mineral rights in the resource zones are administered by the state of Minnesota. The state's share (10.6%) of the total Cu-Ni gross value based on 1977 market prices would be $\$5.3 \times 10^9$. If an average royalty rate of 6% is used for state lands, the potential royalty payments to the state could amount to over $\$3.0 \times 10^9$. In addition, the tax on royalties which would be paid to the 32% of mineral ownership in the private sector could amount $\$9.6 \times 10^9$.

Under present tax laws, a smelter is taxed differently than a mine and mill and results in large, direct payments to local government. Total direct and indirect tax revenue to school districts and city and county governments of the Study Area from the mine/mill models that were analyzed range from $\$63 \times 10^6$ to $\$108 \times 10^6$ over a life of an operation; 80% of this total would be in the form of state aids. A large smelter/refinery complex could generate from \$100 to $\$200 \times 10^6$ in local revenues over its life, 15 to 30% in the form of state aids.

Minnesota's present Cu-Ni tax laws produce revenues comparable to other major mineral-producing states. State tax policy can significantly affect the profitability of marginal Cu-Ni operations, especially if taxes result in higher front-end costs. In general, however, state taxes are expected to have less impact on profitability than the market price of metals and the grade of the ore mined.

6.2 Air Impacts

6.2.1 SO_2

The model projections of pollutant concentrations (Table 3) for the smelter cases alone and when added to regional 1985's sources are less than the SO_2 primary annual and 24-h standards. When the smelter sources are combined with regional sources, maximum 24-h SO_2 concentrations of 143 $\mu g\ m^{-3}$, 78 $\mu g\ m^{-3}$, and 44 $\mu g\ m^{-3}$ are predicted for the Base Case, Option 1 and Option 2 smelters, respectively, at 5 km from the sources. Corresponding second high values are 125 $\mu g\ m^{-3}$, 71 $\mu g\ m^{-3}$, and 43 $\mu g\ m^{-3}$.

Fugitive emissions, which are the same for all the smelter cases, result in high SO_2 concentrations close to the smelter (within 1 to 2 km) that decrease rapidly with distance. The stack emissions by contrast have little effect close to the smelter.

PSD modeling analyses for Cu-Ni development indicate that siting limitations based on annual averages may exist. The Class I annual increment for SO_2 was exceeded at 5 to 10 km from the source using the Base Case and Option 1 smelters. The most severe restrictions, however, to siting a smelter in the region are the 24-h SO_2 PSD increments. None of the smelter models could be built within 20 km of a Class I area. Only the smelter with the BACT (acid plant and scrubber) appears to meet the increment between 20 to 40 km of the Class I area. The smelter meeting NSPS could be built at distances greater than 40 km from the Class I area and the Base Case model would require at least 70 to 80 km distance from Class I areas. A smelter located in the Class II area could be built with Option 1 or NSPS control levels.

The sensitivity of northeastern Minnesota's environment to acidic precipitation raises serious issues regarding siting decisions for a Cu-Ni smelter in the region. The emissions of SO_2 from a smelter, while significant, are not expected to be a primary regional factor in the area's acidic precipitation problem by itself; but emissions will aggravate the existing problem.

During normal operating conditions, no vegetation damage is expected from SO_2 released by the smelter. However, the air-dispersion models used by the study indicate the possibility for 3-h average SO_2 concentrations of 1,000 to 2,000 $\mu g\ m^{-3}$ within a distance of 5 km and concentrations of 500 to 1,000 $\mu g\ m^{-3}$ at distances up to 8 km if stack emissions should increase due to pollution control equipment failure for a period of 3 h. These levels would be potentially injurious to all major forest species except white spruce.

Animals are not expected to experience any direct impacts from potential air pollution, although they are subject to indirect impacts if air pollution results in significant changes in vegetation and soils.

6.2.2 Particulates

Potential particulate emissions from Cu-Ni development include point sources from the mill and smelter and nonpoint sources from haul roads, blasting, and blow-off from waste rock piles and tailing basins. The results of modeling particulate emissions from smelter facilities (Table 3) are similar to those for SO_2. Annual ambient increases (less than 1 $\mu g\ m^{-3}$) due to a smelter are negligible relative to both ambient and PSD standards, but the Class I 24-h particulate increment of 10 $\mu g\ m^{-3}$ could be exceeded. Although some restrictions to smelter siting may exist because of particulate emissions, the SO_2 emissions would still dictate siting based on existing regulations.

The particulate concentration increases that were projected adjacent to a large open pit and open-pit mine and mill operation ($20 \times 10^6\ t\ yr^{-1}$) indicate that annual levels equal to the presently measured background concentrations (10 $\mu g\ m^{-3}$) could occur within 1 to 3 km of the site. Additionally, based on the 24-h standard, it appears that an open-pit mine could not be located within 10 to 15 km of the Class I area.

It is also possible that the Class II 24-h PSD increments may be exceeded in close proximity to the mine and mill areas (within 1 km); however, it is likely that the mine/mill development property would extend to this distance. The use of dust control methods, particularly on haul roads, would result in corresponding decreases in ambient air concentrations.

Fugitive dust presents less of a problem in the case of an underground operation, although Class I standards may still be exceeded within several km of the operation as a result of emissions from the tailing basin and mill site areas. Again, reasonable dust control measures should extend the area of compliance.

Three other major air quality issues include the potential impact from trace metals, SO_4, and mineral fiber emissions. Table 13 presents a summary of predicted regional concentration and deposition rates for selected chemical species; the highest concentrations were predicted for a site which is 5 km east of an Option 1 smelter (meets NSPS) location. When the predicted data are compared to measured data (Tables 4 and 5), it appears that a smelter meeting NSPS could result in Cu, Ni, and SO_4 concentrations that are an order of magnitude higher than existing concentrations; As, Cd, Pb, and Hg projections are at least two orders of magnitude lower than existing concentrations.

The "worst case" deposition projections for SO_4, Pb, and Hg are an order of magnitude lower than existing levels; As and Cd are an order of magnitude higher.

Table 13. Predicted[a] regional dry deposition rates ($g\ ha^{-1}\ yr^{-1}$) and ambient air concentrations ($ng\ m^{-3}$)

	Range		Regional average	
	Concentration	Deposition	Concentration	Deposition
Cu	1.5 -130	22 -760	13	80
Ni	0.37 - 28	3.7 -117	2.9	13
As	0.0008- 0.054	0.002 - 0.050	0.006	0.006
Cd	0.001 - 0.034	0.0009- 0.024	0.004	0.003
Pd	0.0009- 0.063	0.002 - 0.065	0.007	0.008
Hg	0.0009- 0.014	0 - 0.015	0.002	< 0.001
SO_4	1.5 -120[b]	0.10 - 3.60[b]	13[c]	0.052[c]

[a] Based on the Option 1 smelter (meets NSPS); average of 33 sites.
[b] $\mu g\ m^{-3}$.
[c] $kg\ ha^{-1}\ yr^{-1}$.

Projections for Cu and Ni are more than three orders of magnitude higher than existing levels. It is also predicted that concentrations of Cr, Co, Mn, Cd, and Zn may double existing levels.

The human health impacts of trace metals emissions are not known. The deposition of trace metals is probably the most severe potential impact of air pollution on terrestrial ecosystems that can be expected from a Cu-Ni smelter. The eventual buildup in trace metals in the soil over the lifetime of the mine could result in long-range effects. Some of these effects include slowed decomposition of leaves on the forest floor resulting in deep leaf layers that are poor seedbeds for species which require mineral soil for establishment such as red pine, jack pine, and other conifer species. Decreased rates of nutrient recycling may affect forest growth and productivity. Areas managed for conifer production by the US forest service may be particularly vulnerable, since leaf litter layers are often removed before planting to uncover mineral soils. Forest stands with no leaf litter layer will be most susceptible to trace metals deposition.

Metals deposition onto the soil may also have significant affects on germination and growth of seedlings of native species of vegetation. Experiments conducted by the regional study demonstrated reduced growth rates for germinating seeds of several indigenous scrub and tree species when they were exposed to high levels of trace metals in their growth medium. If metal loadings reach levels where they decrease growth or germination, natural patterns of succession may be altered.

The extent of all these effects depends, of course, on smelter stack and fugitive emissions. The areal extent of deposition could be reduced to within 2.5 km using the Option 1 or Option 2 smelters. Gas collection systems can be used to minimize fugitive emissions, and devices such as fabric filters and electrostatic precipitators can remove over 99% of the particles in a gas stream. The potential environmental consequences of increased pollutant deposition onto waters in the region from a smelter meeting NSPS are discussed in Section 6.3.2.

As currently defined by the MDH, mineral fibers will probably be present in the products of mineral processing. Modeling studies suggest that ambient fiber concentrations in the air around a Cu-Ni development could double within 0.8 km of a tailings basin and may increase by a factor of ten within 8 km of a smelter. Although the health implications of these fiber releases are poorly understood at present, there is a possibility that they pose a significant hazard.

6.3 Water Impacts

6.3.1 Water Supply

The adequacy and availability of water for Cu-Ni development are different in the seven development zones. The development models used in the analysis of water supply are the 20×10^6 t yr^{-1} open-pit model with a tailing basin on a semi-permeable base and the 10^5 t yr^{-1} capacity smelter/refinery; both facilities have water for maximum recycling. Water adequacy was assessed for cases with and without storage; for the average annual flow or the average low flow; and wet and dry years.

The water source was considered to be adequate if the specified flow exceeded the appropriation needed.

The largest water requirement is for the smelter/refinery, and only three zones have an adequate water supply for this complex (zones 1, 2, and 3) if no storage is used. The mine/mill without storage would also be restricted to the same three zones.

From a standpoint of water supply, zones 2 and 3 would be the best areas in which to locate a Cu-Ni operation. These zones have adequate water sources even without storage and a lake which could be used as a natural storage reservoir. Zones 4, 5, and 6 appear to be poorly suited to support Cu-Ni development; zone 4 does not have natural storage available, and zones 5 and 6 are likely to have conflicts with existing municipal and industrial users.

6.3.2 Water Quality

The potential water quality impacts of Cu-Ni development were a major environmental concern. At the time of the study, some streams in the Study Area were affected by taconite mining. Impacts of Cu-Ni development on water quality are expected to be greater than those experienced from taconite development. Water quality source models for discharges that might result from a Cu-Ni industry were developed to evaluate potential environmental impacts. These models consisted of four development components (mine, stockpile, tailings basin, and smelter/refinery), and they were based on actual field and laboratory studies conducted during the study.

It should be noted that the modeling data do not allow precise statements on the quality of water produced from Cu-Ni sources or on the effectiveness of controls for specific effluent parameters; rather, the modeling results provide a range of discharge concentrations.

The method that the Cu-Ni study developed to evaluate the impact of trace metal discharges was based on the combined toxic effect of the trace metals: Cu, Ni, Co, and Zn. The concentrations of the trace metals in $\mu g\ l^{-1}$ are assumed to be additive and normalized to "copper equivalent units" (CEU), by the formula: $1 * (Cu) + 0.1 * (Ni) + 1 * (Co) + 0.1 * (Zn) = CEU$.

The potential impact for a given CEU concentration was then evaluated based on ranges established for aquatic biological impacts. A CEU of 0–5 has no effect; 5–30 CEU has a low probability of measurable impact; 30–100 CEU could result in possible chronic effects; 100–600 CEU could result in potential acute effects depending upon the total organic carbon and hardness present in receiving waters; and a CEU of more than 600 would result in definite acute effects. A CEU of 10 was designated as the level that was environmentally acceptable; this level is equivalent to that of the four metals individually.

Trace metal and sulfate discharges for the four development elements are summarized in Table 14. The mine water discharges were derived from actual discharges from existing pits. Case 1 is based on data from a Cu-Ni exploration site and a taconite pit. Case 2 which is a "worst case" is based upon a bulk Cu-Ni sampling site which is a small (30.5 × 15.2 × 3 m) abandoned pit that has filled with rain and surface runoff. The actual quality of open-pit mine water is expected to be between the two cases.

The filling time for the open-pit mine model (20×10^6 t yr^{-1}) is 300 yr assuming a seepage rate into the pit of 0.02 m s^{-1}. The post-operational water quality will be a function of many factors including water sources, pit size, composition of pit walls and floor, and total surface area available for leaching. Even if a neutral pH is maintained in the pit, it is anticipated that Ni concentrations may exceed toxic levels. If the pit water is acidic, then other metals will mobilize and could pose potential threats to water quality.

The water quality models for lean-ore stockpiles and waste rock piles were also based upon actual concentrations measured at gabbro stockpiles at a taconite pit (Table 14). The models reflect an assumption that acid mine drainage problems will not occur because of the natural buffering capacity of the waste materials which were tested in laboratory studies. Case 1 is based on data from a waste rock pile, and Case 2 is based on leachate from a lean-ore stockpile.

It should be noted that these data are based on relatively limited observations and may not represent actual worst case concentrations. During the time the data were collected, metal concentrations were observed to increase dramatically at the field taconite sampling site. For example, during the period of July 1976-August 1977, Ni values ranged from 580–2420 μg l^{-1} compared to Ni concentrations of

Table 14. Water quality source emissions estimates for Cu-Ni development[a,b]

	Tailing basin discharge (μg l^{-1})[c]		Wasted rock pile and lean-ore stockpile discharge (μg l^{-1})	
	Case 1	Case 2	Case 1	Case 2
Cu	38	1500	53	1710
Ni	50	4700	2420	39800
Co	10	220	21	2400
Zn	9	190	40	2400
CEU[b]	54	2200	320	8300
SO_4 (10^5)	3.6	3.8	12.6	36.2
Flow rate (m^3 s^{-1})	0.13	0.13	0.03	0.03
	Mine water discharge		Smelter/refinery discharge	
	Case 1	Case 2		
Cu	4	21,000	16,600	
Ni	60	25,000	39,800	
Co	3	620	2,400	
Zn	58	220	450,000	
CEU	19	24,100	68,000	
SO_4 (10^5)	0.11	4.38	128	
Flow rate (m^3 s^{-1})			0.26	

[a]Assumes the 20×10^6 t yr^{-1} open-pit mine model; 1×10^5 t yr^{-1} smelter/refinery model.
[b]Case 1 and Case 2 represent the "best" "worst" cases, respectively.
[c]Average year of precipitation.

4600–7100 $\mu g\ l^{-1}$ during July-September 1978. In addition, pH decreased from 7.2 in 1976 to 6.7 in 1978. The reason for these changes is not known. However, these data are potentially significant.

In addition to the parameters listed in Table 14, the following ranges for other parameters are estimated from the stockpile discharges (ranges are given for Case 1 to Case 2): Fe, 208 to 7200 $\mu g\ l^{-1}$; Ca, 200 to 346 $mg\ l^{-1}$; Mg, 123 to 268 $mg\ l^{-1}$; Mn, 2.85 to 11.2 $mg\ l^{-1}$; Cl, 41.3 to 56.7 $mg\ l^{-1}$; Alk as $CaCO_3$, 137 to 79.5 $mg\ l^{-1}$.

During the postoperation phase, the lean-ore piles could be processed through the mill. If not, both the lean-ore piles and the waste rock piles may be leachate sources for many years.

Discharges to the tailing basin (Table 14) include inputs from precipitation into the basin, water from the mill, and collected runoff from subsystem A (Sect. 5.2). The tailing basin models assume that all runoff and stockpile leachate is collected and routed to the basin, and that a closed system is used in which water is recycled from the tailing basin to the mill complex. The worst case (Case 2) assumes that no removal reactions occur and pollutant concentrations were based on the lean-ore stockpile discharges. Case 1 assumes that the tailing basin/mill system will dominate water quality, and it is based upon water quality data from a pilot mill plant operated by the University of Minnesota Mineral Resources Research Center.

During the postoperational phase, the least impact on water quality would result if the tailing basin continued as a neutralizing influence on runoff from the site and if lean-ore piles were processed and removed.

Because field data were not available from which to estimate water quality from a smelter/refinery system, the water quality model was developed by scaling data from domestic Cu operations to produce the hypothetical smelter/refinery model with a $10^5\ t\ yr^{-1}$ capacity. It should be noted that because of the lack of data for an Ni refinery, the model was based on data for electrolytic Cu refineries.

The most likely source of water input to the smelter would be water from streams in the area. Major sources of discharge water include anode casting water, slag granulation water, acid plant blowdown, Cu refinery water, and Ni refinery water. Used potable water which is a small part of the total water produced will be treated and is assumed to be uncontaminated. These different output streams were combined to generate a single smelter/refinery effluent. Using a mass balance approach, final concentrations were computed as a weighted average based on the volume of output water from each source (Table 14).

In addition to the values listed in Table 14, the following smelter discharges are predicted for other parameters: TDS (total dissolved solids), 79,700 $mg\ l^{-1}$; As, 3.0 $mg\ l^{-1}$; Cd, 2.3 $mg\ l^{-1}$; Fe, 16.6 $mg\ l^{-1}$; Hg, 0.017 $mg\ l^{-1}$; Pb, 5.2 $mg\ l^{-1}$.

In general, all of the model values for As, Hg, Pb, SO_4 may overestimate concentrations that would result from a Minnesota development because they occur in higher proportions in domestic smelter feeds on which the model is based.

The discharges estimated in the water quality models were used to evaluate environmental impacts on the Study Area's water resources, especially streams. Cu, Ni, Co, and Zn were selected for impact analysis because they are the major trace metal constituents of the mineralized gabbro, and because they could have the greatest potential impact on aquatic life in the Study Area. Other trace metals such as As, Pb, Cd, and Ag are not expected to have significant environmental impacts because of their low concentrations.

Because smelter process waters will be treated before discharge, the lean-ore stockpile runoff and tailing basin discharges from Table 14 are used to discuss the impacts of water discharges in the Study Area.

A direct discharge from a lean-ore stockpile could result in CEU concentrations as high as 8300 $\mu g\ l^{-1}$. A discharge of this quality without any dilution would be acutely toxic to all aquatic life. At a dilution factor of three to four, some species of fish would still survive. Long-term exposure, however, would decimate fish reproduction and few invertebrates would be available as a food source. When the dilution factor approaches 10 to 20, some species of invertebrates could be expected to survive short-term exposure and more would remain alive. Dilution by factors of 200 to 300 would be necessary to reduce the chronic impacts of this discharge to levels that could not be measured.

Discharges from the tailing basin are expected to pose fewer problems than the lean-ore stockpile runoff; but dilution would still be required to reduce concentrations to acceptable levels.

The ability of the region's water resources to handle the discharges was evaluated by calculating the sizes of dilution watersheds that would be needed to reduce the model stockpile and tailing basin discharges to the composite CEU criterion of 10 $\mu g\ l^{-1}$. It should be noted that although SO_4 is of concern because of surface water acidity, SO_4 discharges are small compared to trace metal discharges.

During the operational phase, the worst case tailing basin discharges (Case 2, Table 14) of 2200 $\mu g\ l^{-1}$ CEU indicates that a dilution area of 1.04×10^4 km^2 would be required if the discharge occurred at a constant rate. This area can be reduced to 3.23×10^3 km^2 if the rate of discharge is varied proportionately with stream flow. A comparison of these requirements to available dilution watersheds shows that even if dilution proportional to flow is used, only one of the Study Area's watersheds has sufficient dilutional area available to dilute worst case discharges to the 10 $\mu g\ l^{-1}$ CEU standard. That watershed, Kawishiwi River, is located north of the Laurentian Divide. None of the watersheds south of the Divide could provide sufficient dilution for the worst case discharge. A number of watersheds north and south of the Divide could provide the 64.8 km^2 dilution area required for Case 1 with proportional flow. Case 1 with constant discharge would require 492 km^2 and only a few watersheds could meet this requirement.

During the postoperational phase, the dilution area required to dilute waste rock pile discharges from Case 1 ranges from 101 km^2 with discharge proportional to flow to 689 km^2 with constant discharge. If, however, the discharge resembled the lean-ore stockpile of Case 2, a discharge area of 8.96×10^3 km^2 would be required for constant discharge and 2.66×10^3 km^2 for discharge proportional to flow. Removal of lean-ore piles before mine closure would result in discharge water quality resembling Case 1, and the two largest watersheds in the Study Area could accommodate this discharge.

The impact of discharge waters on lakes was also calculated using an approach similar to streams. The calculations were based on an assumption of mass balance and complete mixing, recognizing that these assumptions are probably less reliable for the lake analysis than for stream analysis.

During the operational phase, waste rock and lean-ore stockpile discharges into lakes would require a minimum lake watershed area of 62 km^2 for Case 1 to 3.8×10^3 km^2 for Case 2. During the postoperational phase, the minimum dilutional watershed areas for the tailing basin discharges range from 98 km^2 for Case 1 to 3.1×10^3 km^2 for Case 2. None of the lakes in the region could accommodate the worst case discharge, but many lakes north and south of the Divide could accommodate the operational and postoperational Case 1 discharges.

The worst situation would occur if the discharge resembled lean-ore stockpile Case 2. None of the Study Area's streams would have a sufficient dilutional watershed to dilute this discharge to the desired 10 $\mu g\ l^{-1}$ CEU. The largest stream would reduce the concentration to 34 $\mu g\ l^{-1}$ CEU.

Although no quantitative analysis was performed on possible on groundwater impacts, some qualitative conclusions can be drawn. The potential groundwater impacts are greatest in the high permeability Duncan-Embarrass sand plain and buried sand in the Aurora area south of the Stoney River. Groundwater in these areas can travel long distances, whereas in most parts of the Study Area flow is restricted by bedrock topography and peat formations.

Groundwater use in the Study Area is limited to individual homes with the exception of Babbitt and Aurora, where groundwater is used as a municipal supply. These water sources may become contaminated if they are connected to the mine site by an aquifer of sufficient permeability. Sand and gravel outwash is particularly conducive to providing this unfavorable flow condition. Groundwater impacts can be minimized by preventing seepage of mining discharges by siting and construction techniques.

Overall, mitigation techniques will be required in the operational and postoperational phases of development. These include various water management techniques and removal of trace metals. Trace metal concentrations can be lowered by using a closed system approach and discharge proportional to stream flow. Control of water flow through waste rock piles is another important technique which may include revegetation or the placement of peat or tailings with the pile to remove metals from the leachate. Careful location of mining discharges within a watershed is another important mitigative tool. A sufficiently large upstream watershed would ensure adequate dilutional flow. It would also protect against the effect of an accidental release if routine treatment or containment procedures fail.

Aqueous mine wastes can also be treated to remove trace metals using lime precipitation; lime precipitation and effluent polishing; xanthate precipitation; cementation; and activated carbon absorption. These mitigation techniques could result in a CEU of about 50 $\mu g\ l^{-1}$ which could be easily handled by many of the lakes and streams in the Study Area.

The location of Cu-Ni developments will be limited in the Study Area by the proximity of the development to the waters of the BWCA and location with respect to the Laurentian Divide. Development south of the Divide (zones 5, 6, and 7) would have a low probability of impact because the streams flow away from the BWCA (Fig. 3). Development within a small watershed in zone 1 which drains directly into the BWCA would have the highest potential for adverse impacts. The remainder of zone 1, plus zones 2, 3, and part of 4, are classified as having medium impact potential because discharges, although considerably diluted, still flow into the BWCA and have the potential to damage this sensitive area.

7 Summary

The most important findings of the regional Cu-Ni study were presented in the previous sections. At the time of the study, no mining company had proposed specific mining development plans. Therefore, the study developed hypothetical mine/mill/smelter/refinery models in order to put various environmental impacts into perspective. Actual development will probably be different from any of the models that were used in the assessment. However, it is important to emphasize that the impact assessment tools and baseline environmental data have now been developed that will be used to evaluate future site-specific development proposals. Major trade-offs to be considered include: (1) mining or delayed mining; (2) open-pit mining or underground mining; (3) smelter in the Study Area, a remote smelter, or no smelter; and (4) development north or south of the Laurentian Divide.

Currently, federal and state laws prohibit Cu-Ni development within the BWCA, but significant and direct impacts could still occur even if development is located outside the BWCA. Based on air quality and water quality modeling predictions, a no impact requirement for the BWCA probably precludes siting a smelter in the Study Area (even if best available air pollution control systems are used). Although a smelter could be located elsewhere, a no impact requirement would still probably preclude any Cu-Ni mining in zone 1 and open-pit mining in zone 2.

Acknowledgement. The author gratefully acknowledges the assistance of Ms. Jere Hopkins and Ms. Peggy Harvey in the preparation of this manuscript.

Reference List

The Regional Copper-Nickel Study is documented in the five volume reports listed below which were based on 160 technical reports. A complete listing of the technical reports and/or library loans of documents can be obtained from the Environmental Conservation Library of Minnesota at the Minneapolis Public Library, 300 Nicollet Mall, Minneapolis, MN 55401-1992.

Volume 1 - Executive Summary
Volume 2 - Technical Assessment
- Chapter 1 Exploration
- Chapter 2 Mineral Extraction
- Chapter 3 Mineral Processing
- Chapter 4 Smelting Refining
- Chapter 5 Integrated Development Models

Volume 3 - Physical Environment
- Chapter 1 Geology and Mineralogy
- Chapter 2 Mineral Resources Potential
- Chapter 3 Air Resources
- Chapter 4 Water Resources
- Chapter 5 Noise

Volume 4 - Biological Environment
- Chapter 1 Aquatic Biology
- Chapter 2 Terrestrial Biology

Decision-Making Framework for Management of Dredged Material Disposal

CHARLES RICHARD LEE[1] and RICHARD KNOX PEDDICORD[2]

1 Introduction

1.1 Background

Navigable waterways of the United States have played a vital role in the nation's economic growth through the years. The US Army Corps of Engineers (CE), in fulfilling its mission to maintain, improve, and extend these waterways, is responsible for the dredging and disposal of large volumes of sediment each year. Dredging is a process by which sediments are removed from the bottom of streams, rivers, lakes, and coastal waters; transported via ship, barge, or pipeline; and discharged to land or water. Annual quantities of dredged material average about 290 million m^3 in maintenance dredging operations and about 78 million m^3 in new, work-dredging operations with the total annual cost now exceeding $250 million.

Over 90% of the total volume of material dredged is considered acceptable for disposal at a wide range of disposal alternatives. However, the presence of contamination in some locations has generated concern that dredged material disposal may adversely affect water quality and aquatic or terrestrial organisms. Since many of the waterways are located in industrial and urban areas, some sediments may be highly contaminated with wastes from these sources. In addition, sediments may be contaminated with chemicals from agricultural practices.

The chemistry of contaminants in sediments, and thus their mobility and potential to adversely impact the environment, is controlled primarily by the physicochemical conditions under which the sediment exists. Fine-grained sediments that are saturated with water typically are anoxic, reduced, and near neutral in pH. These conditions exist in typical open-water aquatic dredged material disposal sites, and may exist in other disposal options such as marsh creation and disposal in shallow water along shorelines. In this chapter, the term "aquatic disposal" is used in a general sense to refer to all disposal conditions in which fine-grained material remains water-saturated, anoxic, reduced, and near neutral in pH. In contrast, when a fine-grained sediment is taken out of the water

[1]Chief Contaminant Mobility and Regulatory Criteria Group, Environmental Laboratory, US Army Corps of Engineers, Waterways Experiment Station, P.O. Box 631, Vicksburg, MS 39180, USA
[2]Battelle, Ocean Sciences, 397 Washington St., Duxbury, MA 02332, USA

and allowed to dry, it becomes oxic and the pH may drop considerably. In this chapter all disposal conditions in which a fine-grained sediment has these characteristics are referred to generally as "upland disposal," even though such conditions can occur on the surface of dredged material islands, the above-tide portions of fills, etc. Nearshore confined disposal sites could have a combination of anoxic, reduced conditions below tide elevation and oxic conditions in the dredged material placed above tidal elevation.

Potential concerns associated with aquatic disposal include contaminants released into the water during and following disposal and the subsequent toxicity and/or bioaccumulation of contaminants by aquatic organisms. Consequences of bioaccumulation may include a wide range of effects from organism toxicity to sublethal genetic abnormalities, food-web biomagnification, and possibly eventual consumption by man. Potential concerns associated with upland disposal include water-quality impacts from effluent discharged during disposal, surface runoff, and leachate following disposal, and uptake of contaminants by plants and animals inhabiting the area following disposal operations, with contaminants possibly reaching man by direct or indirect routes. Each of these potential problems can be minimized by one or more management practices.

Since the nature and magnitude of contamination in dredged material may vary greatly on a project-to-project basis, the appropriate method of disposal may involve any of several available disposal alternatives. Further, control measures to manage specific problems associated with the presence or mobility of contaminants may be required as a part of any given disposal alternative. An overall management strategy for disposal of dredged material has been developed by Francingues et al. (1985). Peddicord et al. (1986) developed a framework for decision making to select the environmentally preferable disposal alternative.

The lead responsibility for the development of specific ecological criteria and guideline procedures regulating the discharge of dredged and fill material at the national level was legislatively assigned to the US Environmental Protection Agency (EPA) in consultation or conjunction with the CE.

The enactment of Public Laws 92–532 (the Marine Protection, Research and Sanctuaries Act of 1972) and 92–500 (the Federal Water Pollution Control Act Amendments of 1972), concerned with the discharge of dredged and fill material, required the CE to participate in developing guidelines and criteria for regulating dredged and fill material disposal. The focal point of research for these procedures is the CE Dredged Material Research Program (DMRP), which was completed in 1978; the ongoing CE Dredging Operations Technical Support (DOTS) Program; the Long-Term Effects of Dredging Operations (LEDO) Program; and the ongoing CE/EPA Field Verification Program (FVP). These research programs have provided much of the technical bases for this chapter.

The framework's major focus is on the question of how should dredged material be tested, the results interpreted to evaluate the degree of potential contaminant impact and the disposal conditions in which the dredged material would have minimal adverse impact on the overall environment. The framework indicates which type of disposal should be considered for a given dredged material and when restrictions on disposal are warranted. The framework is fully comprehensive as to the present state of the art in technical knowledge, but does not address economics/cost feasibility of the recommended criteria or public acceptance/sociopolitical factors. In addition, testing required to address design of a disposal site or selection of necessary control or treatment has not been incorporated into the framework at this time, but will be in the near future.

2 Evaluation and Management of Dredged Material Disposal

2.1 Management Strategy

The following discussion is cited directly from Francingues et al. (1985) and serves as a focus point for this chapter. The selection of a disposal management strategy must consider the nature of the sediment to be dredged, potential environmental impacts of the disposal of the dredged material, nature and degree of contamination, dredging equipment, project size, site-specific conditions, technical feasibility, economics, and other socioeconomic factors. This discussion presents an approach to consider the nature and degree of contamination, potential environmental impacts, and related technical factors. The approach, shown in the flowchart in Fig. 1, consists of the following:

1. Initial evaluation to assess contamination potential.
2. Selecting a potential disposal alternative.
3. Identifying potential problems associated with that alternative.
4. Testing to evaluate the problems.
5. Assessing need for disposal restrictions.
6. Selecting an implementation strategy.
7. Identifying available control options.
8. Examining design considerations to evaluate technical and economic feasibility.
9. Choosing appropriate control measures and technologies.

2.1.1 Initial Evaluation

The initial screening for contamination is the initial evaluation outlined in the proposed testing requirements for Section 404 of the Clean Water Act (EPA 1980). The evaluation is designed to determine whether there is reason to believe the sediment contains any contaminants "in forms and amounts that are likely to degrade the aquatic environment, including potential availability to organisms in toxic amounts." This evaluation also allows identification of specific contaminants of concern in the particular sediment in question, so that testing and analyses may be focused on the most pertinent contaminants. The initial evaluation section is quoted as follows from EPA (1980), Section 230.61, p. 85362:

230.61 Initial evaluation of dredged or filled material

(a) An initial evaluation shall be conducted and documented to determine if there is reason to believe that any dredged or fill material to be discharged into waters of

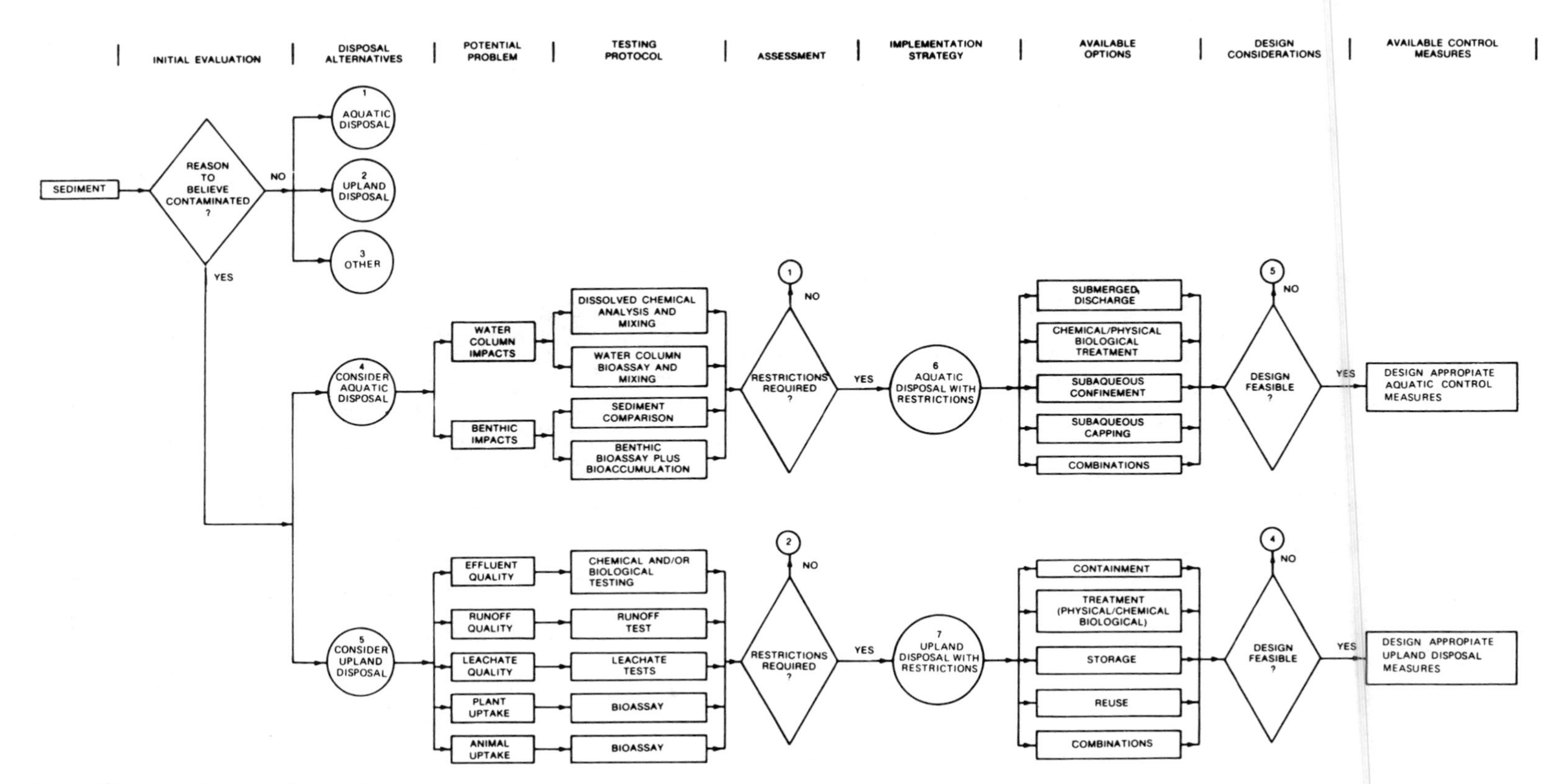

Fig. 1. Management strategy flowchart

the United States contains any contaminant above background level. This initial evaluation will be used in assigning the proposed discharge to a category for testing. This evaluation should be accomplished with existing data file with or readily available to the permitting authority; Regional Administrator, EPA; and other public and private sources, as appropriate.

Factors which may be considered for the extraction site and, if appropriate, the disposal site, include, but are not limited to, the following:

1. Potential routes of introduction of specific contaminants. These may be identified by examining maps, aerial photographs, and other graphic materials that show watercourses, surface relief, proximity to tidal movement, private and public roads, location of buildings, agricultural land, municipal and industrial sewage and storm outfalls, etc., or by making field inspections.
2. Previous tests on the material at the extraction site or on samples from other similar projects in the vicinity, when there are similarities of sources and types of contaminants, water circulation and stratification, accumulation of sediments, general sediment characteristics, and potential impact on the aquatic environment, as long as no known changes have occurred to render the comparisons inappropriate.
3. The probability of past substantial introduction of contaminants from land runoff (e.g., pesticides).
4. Spills of toxic substances or substances designated as hazardous under Section 311 of the Clear Water Act (see 40 CFR Part 116).
5. Substantial introduction of pollutants from industries.
6. Source and previous use of materials proposed for discharged as fill.
7. Substantial natural deposits of minerals and other natural substances.

(b) Before the permitting authority concludes that there is reason to believe that contaminants are present in the discharge above background levels, he should consider all relevant, available information which might indicate its presence. However, if there is no information indicating the likelihood of contamination, the permitting authority may conclude that contaminants are not present above background levels. Examples of documents and records in which data on contaminants may be obtained are:

1. Report of Pollution Caused Fish Kills (U.S. EPA)
2. Selected Chemical Spill Listing (U.S. EPA)
3. Pollution Incident Reporting System (U.S. CS)
4. Surface Impoundment Assessment (U.S. EPA)
5. Identification of In-Place Pollutants and Priorities Removal (U.S. EPA)
6. Revised Status Report Hazardous Waste Sites (U.S. EPA)
7. Waste Management Facilities in the United States – 1977 (U.S. EPA)

8. Corps of Engineers studies of sediment pollution
9. Sediment tests for previously permitted activities (U.S. CE/District Engineers)
10. Pesticide Spill Reporting System (U.S. EPA)
11. STORET (U.S. EPA)
12. Past 404(b)(1) evaluations
13. USGS water and sediment data on major tributaries
14. Pertinent and applicable research reports
15. NPDES permit records

Contaminant concentrations in the sediment to be dredged can be compared to those concentrations of a reference and/or background sediment to assist in evaluating a sufficient cause for concern. The determination of a critical level of contamination above the reference and/or background should be made on a site-by-site basis and will depend on the administrative goal established for the site such as maintaining nondegradation, achieving cleaner conditions, or returning to background conditions. Under some circumstances, contamination factors of 1.5 above reference have been proposed as an acceptable approach. The acceptability of elevation factors must be established through deliberations with appropriate concerned parties and will be a Local Authority Decision (LAD).

If there is available information indicating contaminants are not present above background levels, restrictions are not required. In this case any disposal alternative may be selected, though the possibility of other environmental impacts such as effects of salinity, substrate alteration, and low dissolved oxygen concentrations must be considered in the final selection. Three disposal operations alternatives are shown in the flowchart (Fig. 1) for uncontaminated or so-called clean sediments: [1][3] aquatic, [2] upland, and [3] others, which include marsh or wetland development and other beneficial uses. The final selection is based on environmental considerations, available dredging alternatives, site-specific conditions, technical feasibility, economics, and other socioeconomic considerations.

If there is reason to believe that contaminants are present, the sediment must be evaluated in relation to the conditions that would be present at the disposal site to examine the potential for environmental impacts. Either aquatic [4] or upland disposal [5] could be initially considered and appropriately evaluated, or both alternatives could be evaluated concurrently. The selection of the disposal alternative to be considered is dependent on the potential problems posed by contaminants, available dredging equipment, site-specific conditions, technical feasibility, economics, and socioeconomic considerations. The evaluation of aquatic or upland disposal of contaminated sediment may not necessarily require that additional tests be conducted. As EPA (1980) Section 230.60 points out,

[3] Numbers in brackets refer to the respective disposal alternative as numbered in Fig. 1.

"Where the results of prior evaluations, chemical and biological tests, scientific research, and experience can provide information helpful in making a determination, these should be used. Such prior results may make new testing unnecessary."

2.1.2 Consideration of Disposal [4]

Consideration of aquatic disposal [4] for a contaminated sediment requires evaluation of the potential impacts on the water column and the benthic environment. Other special disposal problems such as effect on health of disposal operations personnel would be a rare occurrence but should also be considered. Water-column impacts can be evaluated by chemical analysis of dissolved contaminants for which water-quality criteria exist. Bioassays are used when water-quality criteria exist or when there is concern about possible interactive effects of multiple contaminants. The effects of mixing and dilution should be considered during assessment of the test results.

Potential benthic impacts of deposited sediment are first evaluated by comparing both contaminant concentrations and toxicity of the sediments in the dredging and disposal sites. If contaminant concentrations and toxicity in the dredging-site sediment are lower than the concentrations in the disposal site sediment, it can be concluded that disposal will not have further unacceptable adverse impacts on the benthic environment. If contaminant concentrations or toxicity are greater in the dredging-site sediment, a bioaccumulation test should be performed. If the initial evaluation for contaminants and initial sediment characterization indicates a potential for special dredging problems (e.g., noxious emissions), appropriate tests must be performed.

If the impacts are acceptable, the dredged material can be disposed in aquatic sites without restrictions [1]. If unacceptable, options for aquatic disposal with restrictions [6] must be evaluated.

2.1.3 Aquatic Disposal with Restrictions [6]

Four options are available for implementing aquatic disposal with restrictions [6]. These options include bottom discharge; treating the material by physical, chemical, or biological methods; confining the dredged material subaqueously; and capping the dredged material subaqueously. Each option may be used separately or in combination with other options. The design considerations for these options must be examined to evaluate the technical feasibility of the disposal alternative based on effectiveness, availability, compatibility, cost, and scheduling. If the design is feasible, the appropriate aquatic control measures and technologies can be chosen and implemented. If the design is not feasible, upland disposal [5] should then be considered.

2.1.4 Consideration of Upland Disposal [5]

Consideration of upland disposal [5] for a contaminated sediment requires evaluation of the following potential problems: effluent quality, surface runoff quality, leachate production and quality, and contaminant uptake by plants and animals. Impacts of effluent, runoff, and leachate quality can be evaluated by chemical analysis of contaminants released in modified elutriate, runoff, and leachate tests, respectively. If the contaminant levels exceed applicable criteria after considering mixing and dilution effects, bioassays are performed to determine the potential toxicity. Plant and animal uptake can be evaluated by appropriate bioassay and bioaccumulation tests. If initial evaluation and sediment characterization indicates a potential for special dredging or disposal problems (e.g., noxious emissions), appropriate tests must be performed. If the impacts are acceptable, the dredged material can be disposed in upland areas without restrictions [2]. If unacceptable, options for upland disposal with restrictions [7] must be evaluated.

2.1.5 Upland Disposal Restrictions [7]

Four basic options are available for implementing upland disposal: biological treatment, reuse, storage, and rehandling. Combinations of the options exist for this strategy. The selection of the appropriate option is dependent mainly on the nature and level of contamination, site-specific conditions, economics, and socioeconomic considerations. The design considerations for these options must be examined to evaluate the technical feasibility of the disposal alternative based on effectiveness, availability, compatibility, cost, and scheduling. If the design is feasible, the appropriate upland disposal control measures and technologies can be chosen and implemented. If the design is not feasible, aquatic disposal [4] should be considered.

2.2 Description of Test Procedures

2.2.1 Aquatic Disposal

2.2.1.1 Physicochemical Conditions

When sediments are dredged from a waterway and placed in stable deposits in a low energy aquatic environment, very little change occurs in the physicochemical nature of the dredged material. In other words, when a reduced anaerobic sediment with a pH value near neutral is disturbed, removed, and placed in a similar aquatic environment, it will remain anaerobic with a pH near neutral. Consequently, contaminant mobility at the aquatic disposal site will be very similar to that occurring at the original dredging site in the waterway. There will be a minor tendency for limited oxidation to occur as the dredged material is mixed with oxygenated water during the dredging operation. However, the oxygen demand of the reduced sediment is usually so great that any oxygen added via the dredging water will be consumed immediately and will not have any important effect on the physicochemical nature of the sediment. The sediment will therefore remain reduced and maintain a near-neutral pH similar to that originally found at the dredging site.

When highly contaminated dredged material is placed in an aquatic environment, there is a conceptual potential for impacts due to release of contaminants into the water column during disposal, although this potential has rarely been realized in practice. In addition, there is potential for physical effects on benthic organisms and for long-term toxicity and/or bioaccumulation of contaminants from the dredged material. These biological effects are best determined at present by site-specific bioassays. Other special disposal concerns such as potential impacts on health of operating crews would be a rare occurrence, but should be evaluated when considered appropriate.

2.2.1.2 Aquatic Bioassay and Bioaccumulation

It must be recognized that aquatic bioassays of dredged material cannot be considered precise predictors of environmental effects in the field. They must be regarded as providing qualitative estimations of these effects, making interpretation of the potential for environmentally adverse effects in the field somewhat subjective. This interpretative uncertainty increases when a parameter, whose ecological meaning is uncertain, is used as the bioassay end point. In view of the interpretative difficulties, most of the animal bioassays in this chapter specify death, or occasionally the ecologically important parameters of development or reproduction, as the response to be measured. The term "toxicity" is defined in American Public Health Association, APHA (1980) as an "adverse effect to a test organism caused by pollutants" and is used in this chapter in a more restricted sense to refer to ecologically important bioassay end points such as those directly related to survival, development, and reproduction.

The environmental interpretation of bioaccumulation data is even more difficult than for bioassays, because in many cases it is impossible to quantify either the ecological consequences of a given tissue concentration of a constituent that is bioaccumulated or even the consequences of that body burden to the animal whose tissues contain it. Almost without exception there is little technical basis for establishing, for example, the tissue concentration of zinc in an organism that would be detrimental to that individual, not to mention the uncertainty of estimating the effect of that organism's body burden on a predator. Research is under way at WES, the EPA Environmental Research Laboratory at Narragansett, and other laboratories in the United States and abroad to determine the relationship, if any, between body burden of contaminants and important biological functions. Dillon (1984) provides an initial step in this process, but the data base is still inadequate to allow evaluation of the potential ecological consequences of a particular body burden of a specific contaminant(s). Therefore, at present, bioaccumulation data can be interpreted only by comparison to levels in organisms exposed to reference sediment, and to levels determined to be safe for human consumption. Such levels have been established by the US Food and Drug Administration (FDA) and Australian National Health and Medical Research Council for some contaminants in seafood, as presented in Table 1. There are no such levels for aquatic organisms not commonly eaten in these countries. However, there is a potential for contaminants in nonfood organisms to reach some seafood organisms through predation. Although trophic transfer of contaminants from an aquatic predator is known to occur, food-web biomagnification of contaminants to higher concentrations in the predator than in the prey has been established in aquatic systems for only a few contaminants, including polychlorinated biphenyls (PCB), DDT, and mercury and possibly selenium, zinc, kepone, mirex, benzo(a)pyrene, and naphthalenes (Biddinger and Gloss 1984, Kay 1984). The above considerations lead to the recommendation that levels in predatory organisms considered safe for human consumption should be applied to aquatic species that are seldom directly consumed by man, in order to protect against possible human impacts. The interpretative guidance assumes that any statistically significant bioaccumulation relative to animals not in dredged material, but living in reference material of similar sedimentological character, is potentially undesirable. The evaluation of experimental results using this approach requires the user to recognize the fact that a statistically significant difference cannot be presumed to predict the occurrence of an important impact in the field.

Interpretive guidance for environmental tests of dredged material was the subject of a working group convened by the WES on 15–17 May 1984. The participants were all recognized scientific experts in a wide variety of relevant disciplines who also have experience in the practical application of environmental science to regulatory decision making. They included Dr. R. Chaney, US Department of Agriculture – Agriculture Research Service; Dr. J. Anderson, Battelle Northwest Laboratories; Dr. W. Adams, Monsanto Co.; Mr. N. Rubenstein, EPA; Dr. J. O'Connor, New York University; Dr. W. Peltier, EPA; Dr. W. Pequegnat, Consultant, College Station, Texas; Dr. J. Rogers, North Texas State University; Dr. J. Skelly, Pennsylvania State University; Mr. K. Phillips, CE, Seattle District; and Mr. J. Krull, Washington State Department of Ecology. After 3 days of discussion, consensus was reached on the following two major points related to regulatory interpretation of properly conducted aquatic bioassay and bioaccumulation testing of dredged material:

1. There is a cause for concern about unacceptable adverse toxicity impacts in the field when laboratory tests result in greater than 50% toxicity attributable to the dredged material.
2. Bioaccumulation data can be interpreted in relation to human health, but evaluation of ecological impacts of bioaccumulation is much less certain at present. Tentative assessment of the potential for such impacts must consider concentrations in tissues of reference animals, and other effects of the sediment, such as degree of toxicity.

Table 1. Action levels and maximum concentrations for contaminants in aquatic organisms for human consumption

Chemical	Food	Action level[a] mg kg^{-1} (wet w. edible portions)	Maximum concentration[b] mg kg^{-1}(wet w. edible portions
Aldrin	Fish and shellfish	0.3	
Antimony	All nonspecified foods (including seafood)		1.5
Arsenic	Fish, crustacea, molluscs		1.0
Cadmium	Fish		0.2
	Molluscs		1.0
Chlordane	Fish	0.3	
Copper	Molluscs		70.0
	All nonspecified foods (including seafood)		10.0
DDT, DDE, TDE	Fish	5.0	
Dieldrin	Fish and shellfish	0.3	
Endrin	Fish and shellfish	0.3	
Heptachlor, heptachlor epoxide	Fish and shellfish	0.3	
Hexachlorocyclohexane (benzene hexachloride)	Frog eggs		0.5
Kepone	Fish and shellfish	0.3	
	Crabmeat	0.4	
Lead	Molluscs		2.5
	All nonspecified foods (including seafood)		1.5
Mercury	Fish, crustacea, molluscs		0.5
Methylmercury	Fish, shellfish, other aquatic animals	1.0	
Mirex	Fish	0.1	
PCB (total)	Fish and shellfish	2.0[c]	
Selenium	All nonspecified foods (including seafood)		1.0
Tin	Fish		50.0
Toxaphene	Fish	5.0	
Zinc	Oysters		1,000.0
	All nonspecified foods (including seafood)		150

[a]Unitad States Food and Drug Administration (FDA) action levels for poisonous or deleterious substances in human food.

[b]Australian National Health and Medical Research Council Standards for Metals in Food, May 1980. Action level is for these chemicals individually or in combination. However, in adding concentrations, do not count any concentrations below the following levels:

Chemical	Minimum level (mg kg^{-1})
DDT, DDE, TDE	0.2
Heptachlor, heptachlor epoxide	0.3

[c]This is not an action level, but a tolerance limit established through the rule-making process.

2.2.1.3 Water Column

The standard elutriate (EPA/CE 1977) is appropriate for evaluating the potential for dredged material disposal to impact the water column. Since this test includes contaminants in both the interstitial water and the loosely bound (easily exchangeable) fraction in the sediment, it approximates the fraction of chemical constituents that is potentially available for release to the water column when sediments are dredged and disposed through the water column. The standard elutriate is prepared by mixing the sediment and dredging-site water in a volumetric sediment to water ratio of 1:4. Mixed with agitation and vigorous aeration for 30 min, it is then allowed to settle for 1 h. The supernatant is then centrifuged and/or filtered to remove particulates prior to chemical analysis. This procedure is followed because the water-quality criteria apply only to dissolved contaminants, and chemical analyses of an unfiltered water sample cannot identify the bioavailable fraction of sediment-sorbed contaminants. A detailed description of the procedure, including sample preparation, is provided in EPA/CE (1977).

Chemical Evaluation. Water-column impacts of dredged material may be evaluated either in this paragraph or as specified in the next paragraph, depending on the situation. Where Section 2.1.1 identifies concern about the presence of specific contaminants that may be released in soluble form, the standard elutriate may be analyzed chemically and the results evaluated by comparison to water-quality criteria for those contaminants after allowance for mixing at the disposal site. This provides an indirect evaluation of potential biological impacts of the dissolved contaminants since the water-quality criteria were derived from bioassays of solutions of the various contaminants. Chemical analyses of the standard elutriate are quantitatively interpretable in terms of potential impact only for those contaminants for which specific water-quality criteria have been established.

Biological Evaluation. If the water-quality criteria approach is not taken, the potential for water-column impacts must be evaluated by bioassays, with consideration given to mixing. An aquatic bioassay should also be used to determine the potential interactions among multiple contaminants. In this way, elutriate bioassays can aid in evaluating the importance of dissolved chemical constituents released from the sediment during disposal operations. The standard elutriate is prepared just as for chemical use, but the filtrate is used as a bioassay test solution rather than for chemical analysis. A series of experimental treatments and controls are established using graded dilutions of the elutriate. The test organisms are added to the test chambers and exposed under standard conditions for a prescribed period of time. The surviving organisms are examined at appropriate intervals to determine if the test solution is producing an effect. Any bioassay protocol designed for use with solutions can be used by substituting the standard elutriate for the original solution. A useful general protocol is presented in EPA/CE (1977).

Mixing. All data from chemical analyses and bioassays of the standard elutriate must be interpreted in light of mixing. This is necessary since biological effects (which are the basis for water-quality criteria) are a function of biologically available contaminant concentration and exposure time of the organism. In the field, both concentration and time of exposure to a particular concentration change continuously. Since both factors will influence the degree of biological impact, it is necessary to incorporate the mixing expected at the disposal site in the interpretation of both chemical and biological data.

Precise prediction of the shape and areal configuration of the plume, within which the required dilution will be achieved, is a very difficult problem involving hydrodynamic and sediment-transport considerations. Although developmental work is continuing on sophisticated numerical models that will provide this capability, all are expensive because of intensive data input requirements and there is no appropriate verified model that can be suggested for routine use at this time. Consequently, a simplified approach for calculating the projected surface area of the mixing zone is described by the Environmental Effects Laboratory (1976). The approach is based on assuming particular geometrical shapes for the disposal plume, depending upon the mode of discharge and the disposal-site environment. In practice, it is not necessary to calculate the mixing zone for every contaminant in the discharge, but only the one requiring the greatest dilution. All others will be encompassed within its mixing zone.

Use of the simplified approach will indicate the maximum portion (volume) of the total aquatic environment and the surface area projection that would be considered necessary for the proposed discharge activities, because it assumes that the dredged material discharge will be completely mixed at the disposal site and that chemical constituents measured in the standard elutriate will behave conservatively following disposal. Included in the discussion by the Environmental Effects Laboratory (1976) are methods for estimating the mixing zone for scow, hopper, and continuous pipeline discharges, as well as for several hydrodynamic conditions in the receiving water.

At this time, there is no fully satisfactory simple and rapid technique that can be used to determine the size and configuration of the acceptability of the mixing zone required to accommodate a discharge into an aquatic system. However, there are several important concepts that should be considered in determining the acceptability of a mixing zone. The size of a designated mixing zone should be limited, but each mixing zone should be tailored to a particular receiving water body and no attempt should be made to apply a single size limitation in any water body. In other words, a decision should be based on a case-by-case evaluation at each proposed disposal site and the beneficial use(s) to be protected. In addition to the considerations listed below, a relatively larger mixing zone can be tolerated for intermittent discharges (compared to continuous discharges) without having an important adverse impact on the receiving waters. Concern over acceptability of the calculated mixing zone increases in proportion to:

1. Size.
2. Configuration.
3. Proportion of volume of receiving water body occupied.
4. Proportion of cross-sectional area of receiving water occupied.
5. Time required to achieve desired dilution for each discrete discharge event.
6. Frequency of discharges during the dredging and disposal operation.
7. Duration of the dredging and disposal operation.

8. Proximity to municipal water intakes.
9. Proximity to sources of recharge for drinking water intakes.
10. Proximity to areas of high, human water-contact activities at the time of major use.
11. Proximity to shellfish beds with commercial or recreational importance.
12. Proximity to unique or concentrated fish or shellfish spawning areas at the time of major use.
13. Proximity to major sport or commercial fishery areas at the time of major use.
14. Proximity to unique or concentrated fish or shellfish nursery areas at the time of major use.
15. Proximity to major fish or shellfish migration routes at the time of major use.
16. Proximity to other major disposal sites or discharges at the time of their use.

Several authors have defined mixing zones in terms of biological effects. However, the mixing zone calculated by the method described should not be equated with a zone of adverse biological impact. The basis for the recommended approach is the fact that the effects of a discharge are a function of exposure concentration and exposure time. Although appropriate and applicable water-quality criteria or bioassay results are used to define the volume of water in which acceptable concentrations may be equaled or exceeded, the duration of mixing zone conditions cannot be easily quantified at this time. Therefore, the method should only be used to estimate the volume and surface area at a disposal site within which discharge concentrations will exceed a particular value during the actual discharge.

Benthic. It is generally felt that if a dredged material is going to have an environmental impact, the greater potential for impact lies with the deposited sediment at the disposal site. This is because it is not mixed and dispersed as rapidly or as greatly as the dissolved material; most contaminants remain associated with the particulates; and bottom-dwelling animals live and feed in and on the deposited material for extended periods. Therefore, the major evaluative efforts should be placed on the deposited material. No chemical procedures exist that will determine the environmental activity of the chemicals found in the material.

Scientific studies conclusively indicate that most subaqueous disposal of dredged material in low-energy aquatic environments where stable mounding will occur, will generally minimize changes in mobility of most contaminants (Brannon 1978; Gambrell et al. 1978; Neff et al. 1978; Wright 1977). The potential for accumulation of a contaminant in the tissues of an organism (bioaccumulation) may be affected by exposure concentration and factors such as duration of exposure, salinity, water hardness, temperature, chemical form of the contaminant, sediment characteristics such as organic carbon content, and the particular organism under study. The relative importance of these factors varies. Elevated concentrations of contaminants in the ambient medium, or associated sediments, are not always indicative of high levels of contaminants in tissues of benthic invertebrates of biological effects.

Potential benthic impacts are best evaluated by a combined consideration of total or bulk chemical analyses of the sediment to identify contaminants present and toxicity test(s) to determine their bioavailability. If results of these tests do not provide sufficient information for decision making as discussed later in this chapter, a bioaccumulation test should be performed to determine the potential for contaminants to accumulate in the tissues of animals exposed to the dredged material.

Benthic or deposited sediment bioassays are derived from more traditional techniques for testing contaminants in solution. While there are many variations, those most useful for this chapter all involve exposure of aquatic test organisms to deposits of whole sediment for a specified period, followed by quantification of the responses. For reasons of regulatory interpretation and implementation, the response of choice here is mortality (and occasionally development or reproduction). A technique widely used and suitable for a wide variety of aquatic macroorganisms is given in EPA/CE (1977). This technique should be utilized to test effects on a finfish, a crustacean, a mollusc, and an annelid acceptable to local interests as sufficiently sensitive and adequately representative of the local aquatic environment. Many other exposure designs, species, and life stages can also provide useful information and may be utilized in addition to, or instead of, those described in EPA/CE (1977). All widely recognized sediment bioassay techniques of regulatory utility involve toxic effects of exposure of a few days to a few weeks. Tissues of surviving organisms which exceed about 1 g in weight could be analyzed for contaminants at the end of the exposure to indicate the potential for bioaccumulation from the sediments. The contaminants to be analyzed should be those for which there is a sufficient cause for concern. In order to best interpret bioaccumulation data, it is necessary to know concentrations in tissues at steady state rather than only at some intermediate point on the uptake curve. This can be achieved by extending the exposure period until steady state is reached, although this can raise serious questions about the representativeness of uptake after extended time in the laboratory unless elaborate precautions are taken. Another alternative is to calculate steady-state tissue concentration based on sequential data collected over a few days and a first-order uptake-depuration kinetics model. This has been shown to give acceptable estimations of steady state based on a few days exposure by Branson et al. (1975) and McFarland et al. (1984). A third approach, probably the best under the circumstances where it is possible, is the use of field data as discussed in EPA/CE (1977). There is presently no generally accepted quantitative means of assessing potential long-term changes in sediment effects due to possible breakdown of some organic compounds into compounds of greater or lesser bioavailability and effect.

2.2.2 Upland Disposal

2.2.2.1 Physicochemical Conditions

When dredged material is placed in an upland environment in which it does not remain water-saturated, drastic physicochemical changes occur. As soon as the dredged material is placed in a confinement area and allowed to be exposed to the atmosphere, oxidation processes begin. The influent slurry water initially is dark in color and reduced with little oxygen as it is discharged into the confinement area from a hydraulic dredge. Mechanically dredged sediments such as with a clamshell will have sediment pore water that will initially be dark in color and reduced. As the slurry water passes across the confined disposal site and approaches the discharge weir, the water becomes oxygenated and will usually become light gray or yellowish brown. The color change indicates further oxidation of iron complexes in the suspended particulates as they move across the confinement. Once disposal operations are completed, dredged material consolidation will continue to force pore water up and out of the dredged material, and it will drain toward the discharge weir. This drainage water will continue to become oxidized and lighter in color. Once the surface pore water has been removed from the confinement, the surface of the dredged material will become oxidized and lighter in color such as changing from black to light gray. The dredged material will begin to crack as it dries out. Accumulation of salts will develop on the surface of the dredged material and especially on the edge of the cracks. Rainfall events will tend to dissolve and remove these salt accumulations in surface runoff. Recent research on contaminant mobility from dredged material placed in an upland disposal site indicated that certain metal contaminants can become dissolved in surface runoff as dredged material dries out. During the drying process, organic complexes become oxidized and decomposed. Sulfide compounds also become oxidized to sulfate salts. These chemical transformations could release complexed contaminants to surface runoff, soil pore water, and leachate through the material. In addition, plants and animals that colonize the upland site could take up and bioaccumulate these released contaminants. Contaminant mobility will be significantly controlled by the physicochemical changes that occur during drying and oxidation of the dredged material.

2.2.2.2 Contaminant Mobility Determinations

Upland disposal of contaminated dredged material must be planned to contain the dredged material within the site and restrict contaminant mobility out of the site, in order to control or minimize potential environmental impacts. There are five possible mechanisms for transport of contaminants from upland disposal sites:

1. Release of contaminants in the effluent during disposal operations.
2. Surface runoff of contaminants in either dissolved or suspended particulate form following disposal.
3. Leaching into groundwater and surface waters.
4. Plant uptake directly from sediments, followed by indirect animal uptake from feeding on vegetation.
5. Animal uptake directly from sediments.

The environmental impact of upland disposal of contaminated dredged material may be more severe than aquatic discharge (Gambrell et al. 1978; Jones and Lee 1978).

Any test protocol used to predict contaminant mobility should account for the physicochemical changes occurring in the dredged material when placed in the specific disposal environment. The following discussion of test protocols will address each of the above aspects in detail.

Effluent Quality. Water-quality effects of upland disposal effluents (water discharged during active disposal operations) have been identified as one of the greatest deficiencies in the knowledge of the environmental impact of dredged material disposal (Jones and Lee 1978). Dredged material placed in an upland disposal area undergoes sedimentation, while clarified supernatant waters are discharged from the site as effluent during active dredging operations. The effluent may contain levels of both dissolved and particulate-associated contaminants. A large portion of the total contaminant level is particulate-associated.

The standard elutriate test is sometimes used to evaluate effluent water quality, but this test does not reflect the conditions existing in confined disposal sites that influence contaminant release. A modified elutriate test procedure, developed under the CE Long-Term Effects of Dredging Operations (LEDO) Research Program (Palermo 1984), can be used to predict both the dissolved and particulate-associated concentrations of contaminants in upland disposal area effluents (water discharged during active disposal operations). The laboratory test simulates contaminant release under disposal conditions and reflects sedimentation behavior of dredged material, retention time of the containment, and chemical environment in ponded water during active disposal.

The modified elutriate test procedure is illustrated in Fig. 2. Sediment and dredging-site water are mixed to a slurry concentration equal to the expected influent concentration under field conditions. The mixed slurry is aerated in a 4-liter cylinder for 1 h to ensure that oxidizing conditions will be present in the supernatant water. Following aeration, the slurry is allowed to settle under quiescent conditions for a period equal to the expected mean retention time, up to a maximum of 24 h. A sample is then collected from the supernatant water and determined for total suspended solids, then dissolved and analyzed for total concentrations of contaminants of concern. The contaminant fractions of the total suspended solids may then be calculated. Column-settling tests, similar to those used for design of disposal areas for effective settling (Palermo et al. 1978; Palermo 1984), are used to define the concentration of suspended solids in the effluent for a given operational condition, i.e., ponded area, depth, and inflow rate. Using results from both of these analyses, a prediction of the total concentration of contaminants can be made. The predictive technique is illustrated in Fig. 3. Detailed procedures are given in Palermo (1984).

The acceptability of the proposed upland disposal operation can be evaluated by comparing the predicted dissolved contaminant concentrations with applicable water-quality standards while considering an appropriate mixing zone and the quality of the receiving water body. Where the primary administrative goal is maximum containment of contaminants, appropriate controls and restrictions may be required to first meet water-quality criteria without a mixing zone or, secondarily, to ensure that an acceptable mixing zone is maintained.

Surface Runoff Quality. After dredged material has been placed in an upland disposal site and the dewatering process has been initiated, contaminant mobility in rainfall-induced runoff is considered in the overall environmental impact of the dredged material being placed in a confined disposal site. The quality of the runoff water can vary depending on the physicochemical process and the contaminants

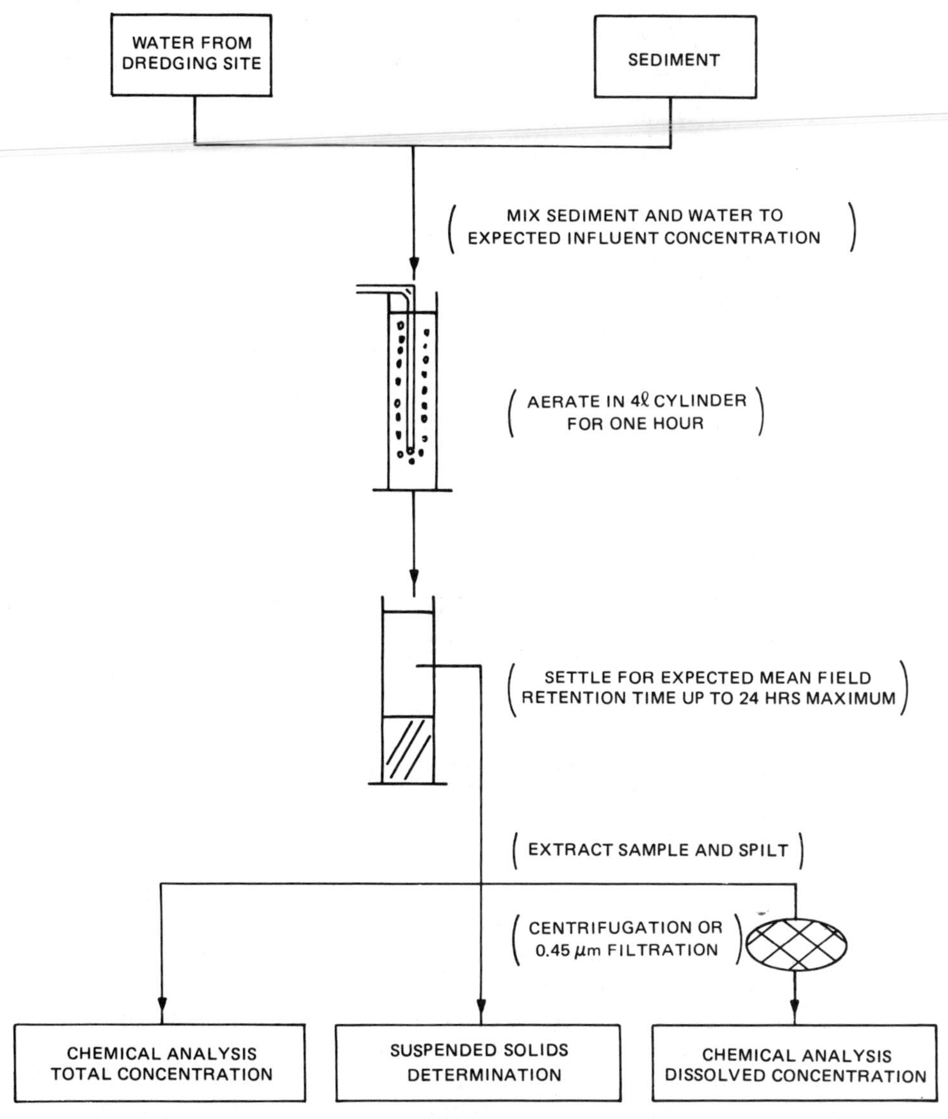

Fig. 2. Modified elutriate test procedure

present in the dredged material. Drying and oxidation will promote aerobic microbiological activity, which more completely breaks down the organic component of the dredged material and oxidizes sulfide compounds to more soluble sulfate compounds. Concurrently, reduced iron compounds will become oxidized and iron oxides will be formed that can act as metal scavengers to adsorb soluble metals and render them less soluble. The pH of the dredged material will be affected by the amount of acid-forming compounds present as well as the amount of basic compounds that can buffer acid formation. Generally, large amounts of

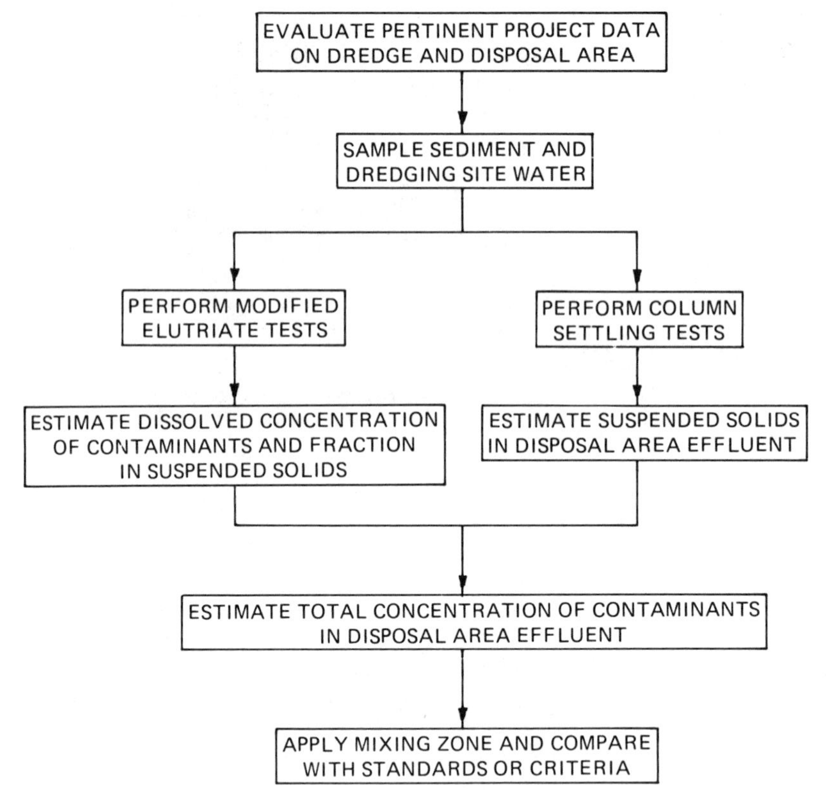

Fig. 3. Effluent quality predictive technique

sulfur, organic matter, and/or pyrite material will generate acid conditions. Basic components of dredged material such as calcium carbonate will tend to neutralize the acidity produced. The resulting pH of the dredged material will depend on the relative amounts of acid-forming and basic compounds present.

Runoff-water quality will depend on the results of the above processes as the dredged material dries out. For example, should there be more acid formation than the amount of bases present to neutralize the acid, then the dredged material will become acidic in pH. Excessive amounts of pyrite when oxidized can reduce pH values from an initial pH 7 down to pH 3. Under these conditions, surface runoff-water quality can be acid and could contain elevated concentrations of trace metals.

An appropriate test for evaluating surface runoff-water quality must consider the effects of the drying process to adequately estimate and predict runoff-water quality. At present, there is no single simplified laboratory test to predict runoff-water quality. Research was initiated in November 1984 to develop such a test. A laboratory test using a rainfall simulator has been developed and is being used to predict surface runoff-water quality from dredged material as part of the CE/EPA Field Verification Program (FVP) (Lee and Skogerboe 1983a; Westerdahl and Skogerboe 1981). This test protocol involves taking a sediment sample from a waterway and placing it in a soil-bed lysimeter in its original wet-reduced state. The sediment is allowed to dry out. At intervals during the drying process, rainfall events are applied to the lysimeter, and surface runoff-water samples are collected and analyzed for selected water-quality parameters. Rainfall simulations are repeated on the soil-bed lysimeter until the sediment has completely dried out. Results of the tests can be used to predict the surface runoff-water quality that can be expected in a confined disposal site when the dredged material dries out. From these results, control measures can be formulated to treat surface runoff water if required to minimize the environmental impact to surrounding areas.

An example of the use of this test protocol can be cited (Lee and Skogerboe 1983b). An estuarine-dredged material highly contaminated with the metals zinc, copper, cadmium, nickel, and chromium was evaluated using this test procedure. An acid rainfall simulating typical rainfall quality at the upland disposal site was used. Test results indicated significant solubilization of these metals in surface runoff water after the dredged material dried out. The pH of the dredged material became acid because of the limited base-neutralizing compounds present and the acid rainfall applied. The oxidation of sulfide compounds and organic complexes apparently released metals into more soluble and mobile forms. Based on these test results, control measures were designed to neutralize acidity and remove these metals in surface runoff water.

Leachate Quality. Subsurface drainage from disposal sites in an upland environment may reach adjacent aquifers or may enter surface waters. Fine-grained dredged material tends to form its own disposal area liner as particles settle with percolation drainage water, but the consolidation may require some time for self-sealing to develop. In addition, diffusion of contaminants through fine-grained materials will continue even after the self-sealing has stopped much of the water convection. It is surmised, but not demonstrated, that hydrophobic organic contaminants associate with naturally occurring dissolved organic carbon and thus can diffuse into groundwater beneath a site. Further work is needed to substantiate this theory. Since most contaminants potentially present in dredged material are closely adsorbed to particles, primarily the dissolved fraction will be present in leachates. A potential for leachate impacts exists when a dredged material from a saltwater environment is placed in an upland site adjacent to freshwater aquifers or to surface waters. The site-specific nature of subsurface conditions is the major factor in determining impact (Chen et al. 1978).

An appropriate leachate quality testing protocol must predict which contaminants may be released in leachate and the relative degree of release. There is presently no routinely applied testing protocol to predict leachate quality from dredged material disposal sites. An evaluation of available leaching procedures is needed before a leaching test protocol for confined dredoed material can be recommended. Although a wide variety of leaching or extraction tests have been proposed for hazardous waste (Lowenbach et al. 1977), none have been field verified for use to evaluate leaching of dredged material placed in upland disposal sites.

A review of the literature has indicated that theoretical models and data on the leaching potential of dredged material are needed in order to evaluate alternative strategies for the treatment and containment of contaminants in upland disposal sites. Theoretical developments that are needed involve pertinent transport rate equations that describe the leaching of chemicals from dewatered and consolidated dredged material. Data gaps include lack of sufficient information on: (1) bulk transport of contaminants by seepage; (2) contaminant leachability under various environmental conditions; and (3) long-term geochemical consequences that alter contaminant leachability. Leaching tests that can assist in the development of an appropriate predictive protocol for dredged material are currently being developed at the WES.

Development of leachate prediction models using mass-transport equations will require information on the relative significance of intraparticle diffusion, surface desorption, film diffusion, and other possible rate-controlling mechanisms for contaminant leaching (e.g., irreversible chemical reactions). Serial batch leach tests (Houle and Long 1980) can indicate whether leaching of a sediment is an equilibrium or kinetically controlled process. Theoretical considerations indicate that, with proper interpretation, results from serial batch leach tests can yield coefficients suitable for modeling contaminant leaching in a confined disposal site. Predicative techniques, including serial batch leach tests, are presently being evaluated at the WES (Hill et al. 1985).

Column leach tests using specially constructed parameters can provide information needed for modeling bulk transport of contaminants in an upland site (Goerlitz 1984). The disposal site environment is simulated in a test column by passing a reference liquid or site water through the dredged material. Comparison of batch leach test and column leach test results can indicate the relative significance of bulk transport and diffusive transport within a column of dredged material, and the relative importance of film effects and nonequilibrium processes on contaminant desorption mechanisms. The potential use of column and batch leaching tests for predicting leachate quality in an upland disposal site is presently under investigation at WES. Routine testing procedures cannot be recommended at this time.

Long-term geochemical changes influencing leachate quality can only be assessed directly by long-term testing procedures. Use of large pilot-scale leaching columns similar to those described by the Buffalo District (US Army Engineer District, Buffalo 1983) maintained under the environmental conditions that exist in a confined disposal facility will provide such information. This leaching procedure will determine the nature of long-term contaminant releases and the amount of release of each contaminant over time. Information on changes in leachate quality as a function of sediment geochemical alteration under the prevailing environmental conditions will also be provided. From this information, specific treatment of the dredged material and/or placement of an appropriate liner can be formulated and designed into the disposal management strategy. Alternate leaching procedures that address long-term concern are presently under investigation and will be recommended after appropriate testing and verification.

Plant Uptake. After dredged material has been placed in either an intertidal, wetland, or upland environment, plants can invade and colonize the site. In most cases, fine-grained dredged material contains large amounts of nitrogen and phosphorus, which promote vigorous plant growth. Elevations in confined disposal sites can range from wetland to upland terrestrial environments. In many cases, the dredged material was placed in upland disposal sites because contaminants were present in the dredged material. Consequently, there is potential for movement of contaminants from the dredged material into the environment through plants and then eventually into the food chain.

An appropriate test for evaluating plant uptake of contaminants from dredged material must consider the ultimate environment in which the dredged material is placed. The physicochemical processes become extremely important in determining the availability of contaminants for plant uptake.

There is a plant bioassay test protocol that was developed under the LEDO program based on the results of the DMRP. This procedure has been applied to a number of contaminated sediments and dredged material (both freshwater and estuarine). Results obtained from these plant bioassays have provided sufficient information to confirm the usefulness of the technique for predicting the potential for plant uptake of contaminants from dredged material (Folsom and Lee 1981, 1983; Folsom et al. 1981; Lee et al. 1982). The procedure is presently being field verified under the CE/EPA FVP and is being applied to a wide variety of contaminated materials such as sewage sludge amended soils in the United States and metal-mining, waste-contaminated soils in Wales, UK.

The plant bioassay procedure requires taking a sample of sediment from a waterway and placing it either in a flooded wetland environment or an upland terrestrial environment in the laboratory. Index plants, *Spartina alterniflora* and *Sporobulus virginicus* for estuarine sediments and *Cyperus esculentus* for freshwater sediments, are then grown in the sediment under conditions of both wetland and upland disposal environments. Plant growth, phytoxicity, and bioaccumulation of contaminants are monitored during the growth period. Plants are harvested and analyzed for contaminants. The test results indicate the potential for plants to become contaminated when grown on the dredged material in either a wetland or upland terrestrial environment. From the test results, appropriate management strategies can be formulated as to where to place a dredged material to minimize plant uptake or how to control and manage plant species on the site so that desirable plant species that do not take up and accumulate contaminants are allowed to colonize the site, while undesirable plant species are removed or eliminated.

There is another laboratory test being developed under the LEDO program that utilizes an organic extractant of dredged material to chemically predict plant uptake of certain trace metals such as zinc, cadmium, nickel, chromium, lead, and copper. This test procedure attempts to simulate the capacity of a plant root to extract metals from a dredged material. Field verification of test protocol is being conducted under the CE/EPA FVP. This test procedure takes a sample of dredged material in the flooded reduced wetland condition and another sample that has been air dried for an upland condition. The samples are extracted for 24 h in a modified diethylenetriamine-pentaaceticacid (DTPA) extraction solution according to Lee et al. (1983). This solution is then filtered through a millipore filter and the filtrate is analyzed for soluble contaminants. This procedure has been successful in predicting plant leaf tissue contents of certain metals. There is no existing extraction procedure that predicts plant availability of organic contaminants.

Animal Uptake. Many animal species invade and colonize upland dredged material disposal sites. In some cases, prolific wildlife habitats have become established on these sites. These habitats are usually rich in waterfowl and often become the focus of public interest through local ornithologists, sportsmen, and the environmentally aware public. Concern has developed recently over the potential for invertebrate animals inhabiting upland terrestrial disposal sites to become contaminated and contribute to the contamination of food webs associated with the site.

An appropriate test for evaluating animal uptake of contaminants from dredged material must consider the ultimate environment in which the dredged material is placed, the anticipated ecosystem developed, and the physicochemical processes governing the biological availability of contaminants for animal uptake.

There is a recommended test protocol being tested under the CE/EPA FVP that utilizes an earthworm as an index species to indicate toxicity and bioaccumulation of contaminants from dredged material. In the procedure, earthworms are placed in sediment maintained in moist and semimoist air-dried environments. The toxicity and bioaccumulation of contaminants are monitored over a 28-day period (Marquenie and Simmers 1984; Simmers, Rhett, and Lee 1983). This procedure is a modification of a procedure developed by C.A. Edwards in England for determining the hazardous nature of manufactured chemicals to be sold in the European Economic Community. Test results to date indicate that the terrestrial earthworm test procedure can indicate potential environmental effects of dredged material disposal in upland environments. The evaluative portion of the test is mainly tissue analysis rather than strictly mortality. While the test is being established, those treatments necessary to ensure survival for the test period (such as washing or dilution) can indicate potential field-site management strategies. The earthworm contaminant levels can also be related to the food web that could exist on the site after disposal. This type of test can be conducted simulatenously under optimum conditions in the laboratory and in the field at or near the proposed disposal site, to further assess the extent of contaminant mobility. This test can identify bioavailable metals and organic contaminants in the material to be dredged.

2.3 Decision-Making Framework

A decision-making framework has been described by Peddicord et al. (1986) that utilizes the results from the suite of test protocols described previously into eleven flowcharts (Fig. 4–14). Terms that will be used in the framework include:

1. Reference site: Location from which biological and sediment or water chemistry data are used for comparison to test results from contaminated dredged material. This may vary from an existing disposal site to an existing background site and will be determined by a local authority decision.
2. Local Authority Decision (LAD): A decision made by local regulatory authorities having jurisdiction over the project in question.

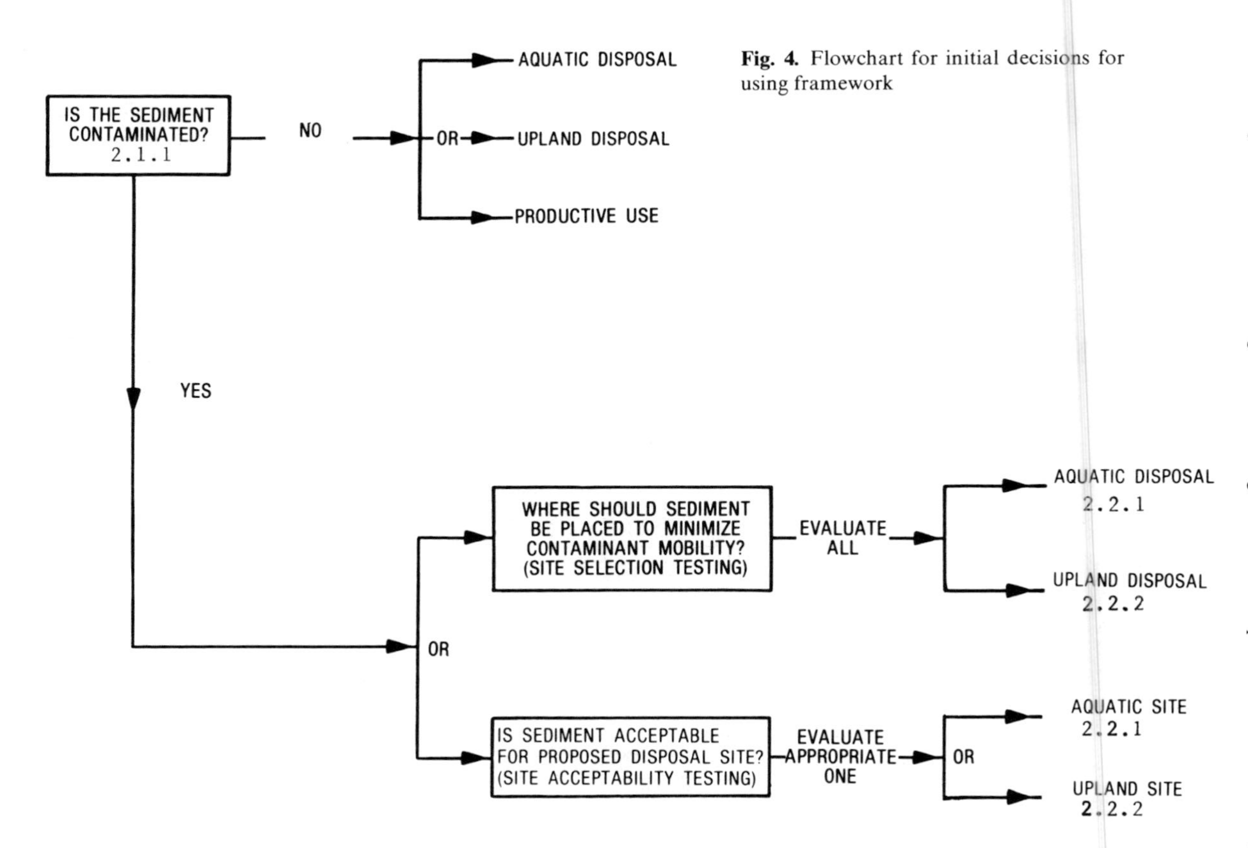

Fig. 4. Flowchart for initial decisions for using framework

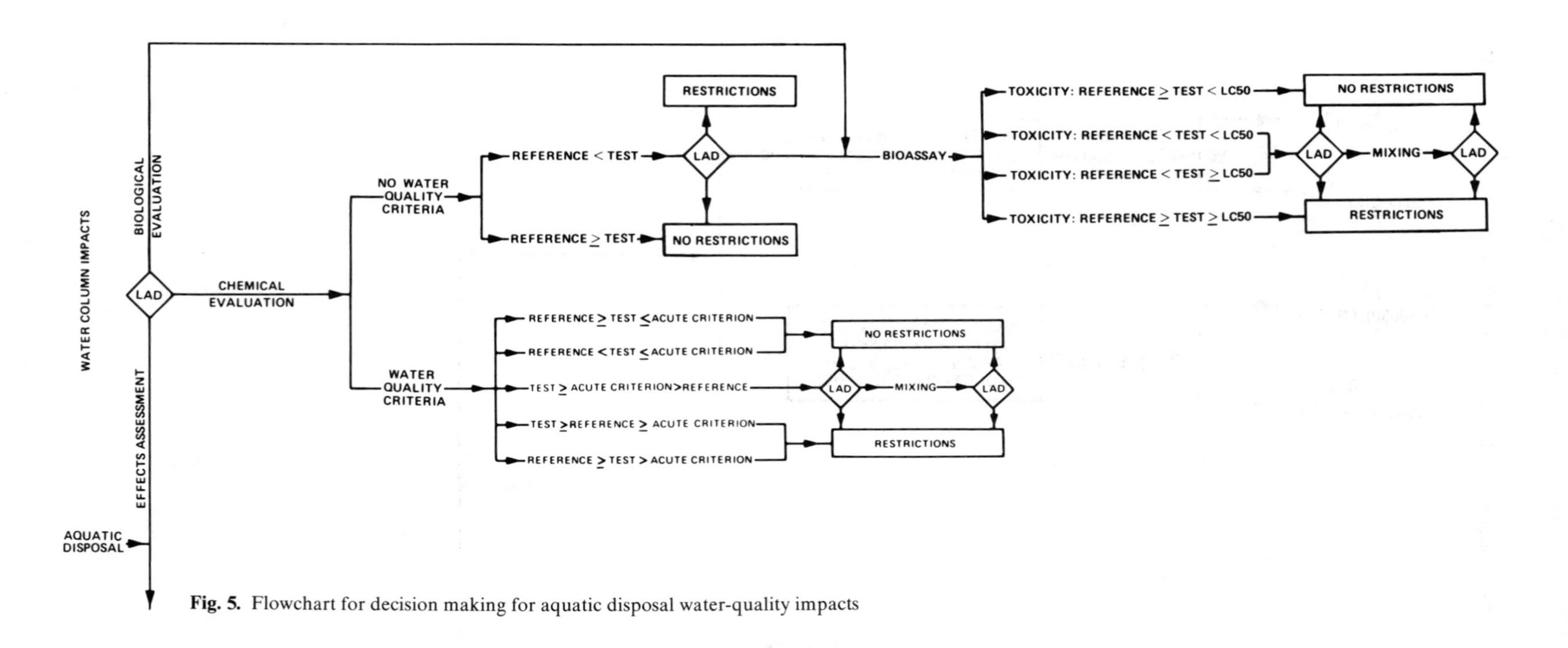

Fig. 5. Flowchart for decision making for aquatic disposal water-quality impacts

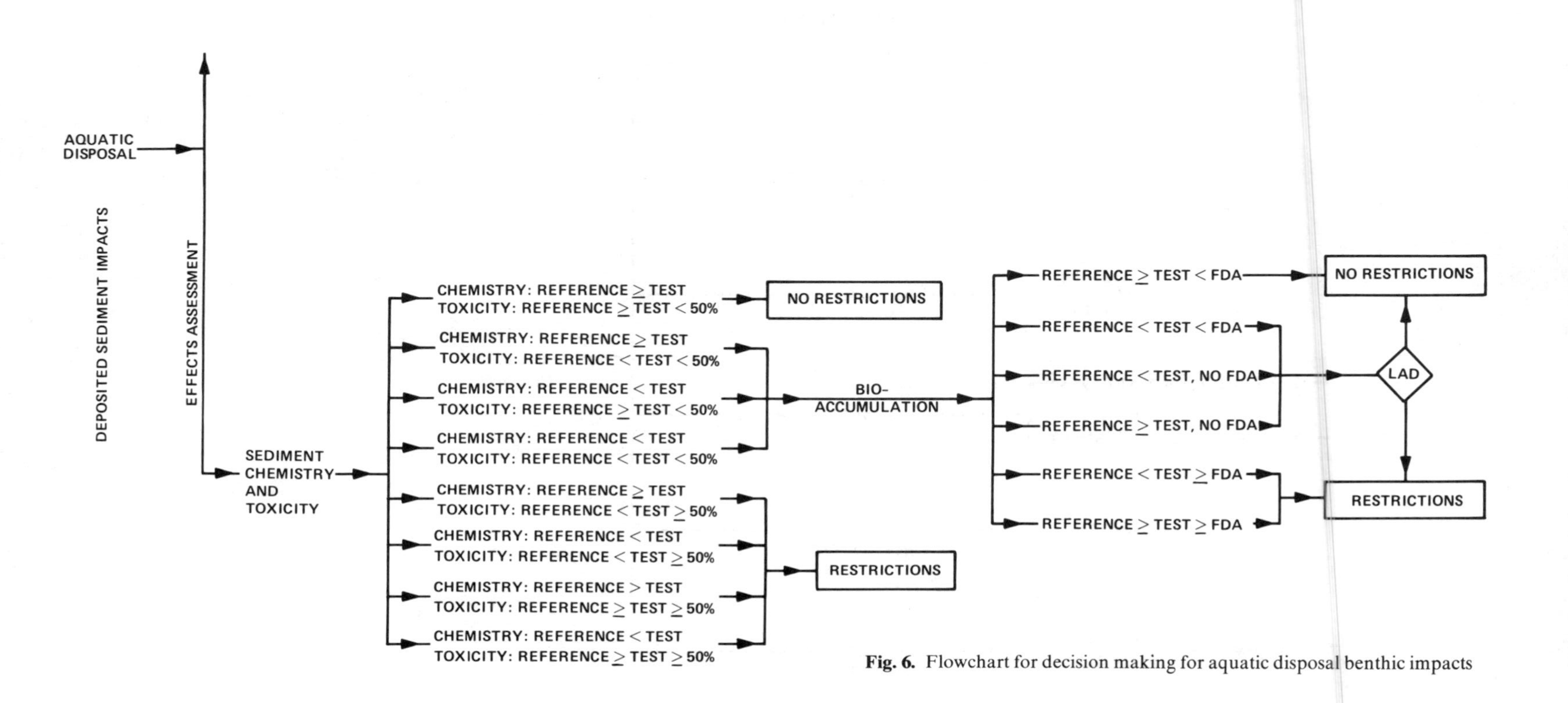

Fig. 6. Flowchart for decision making for aquatic disposal benthic impacts

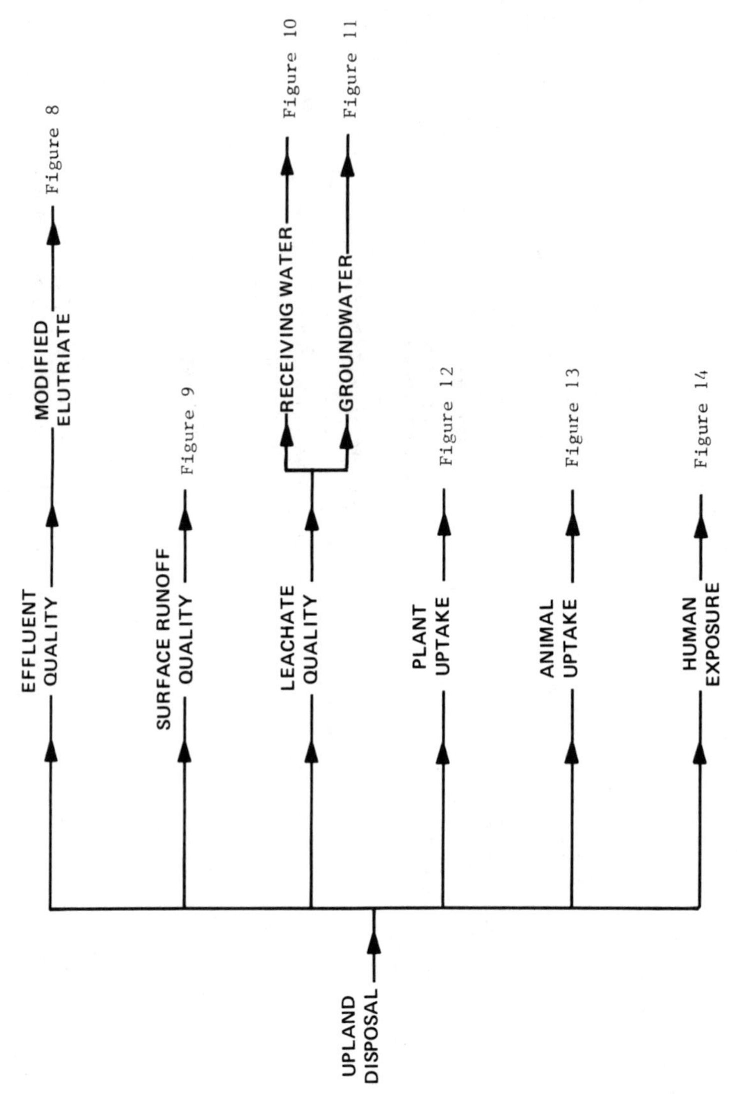

Fig. 7. Summary flowchart for decision making for upland disposal

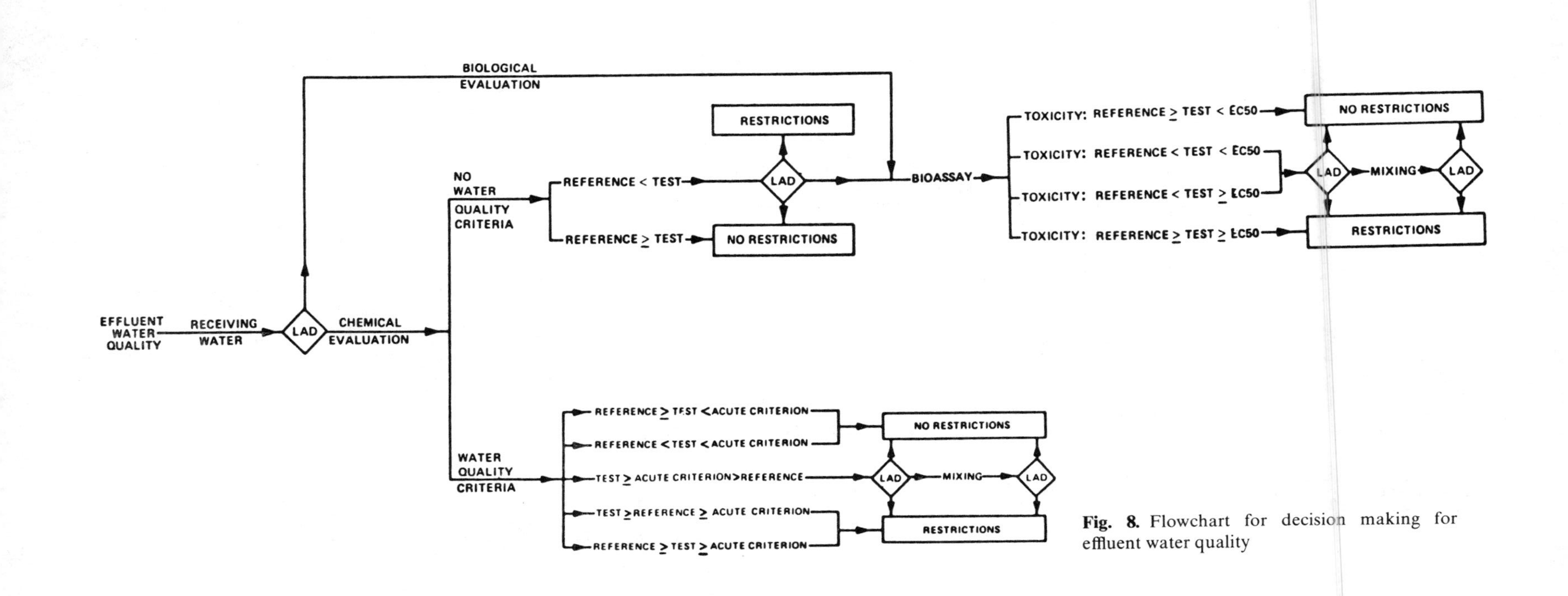

Fig. 8. Flowchart for decision making for effluent water quality

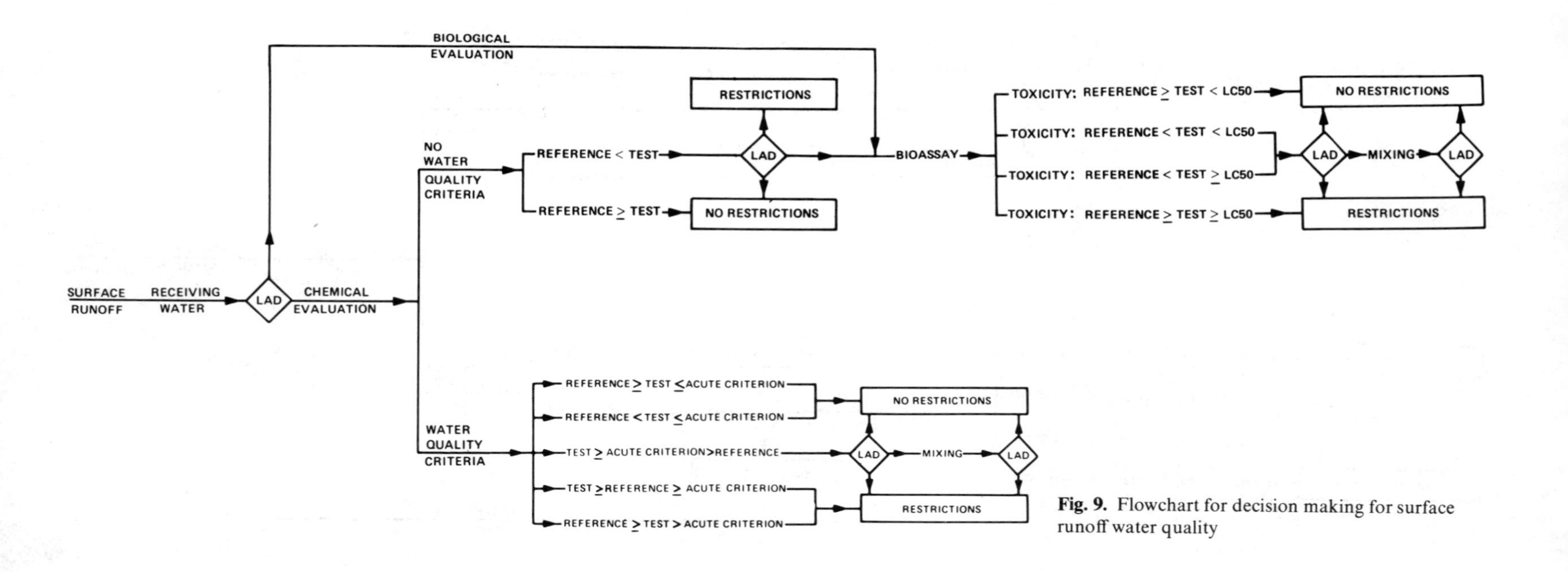

Fig. 9. Flowchart for decision making for surface runoff water quality

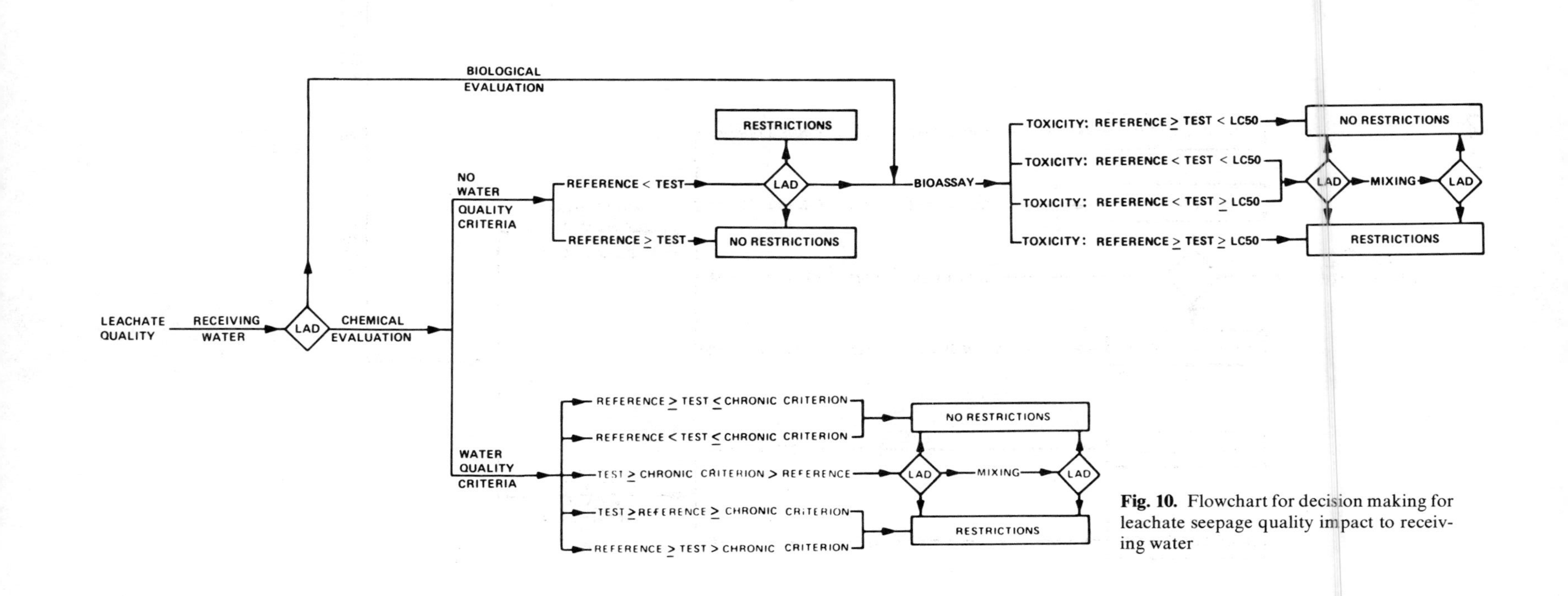

Fig. 10. Flowchart for decision making for leachate seepage quality impact to receiving water

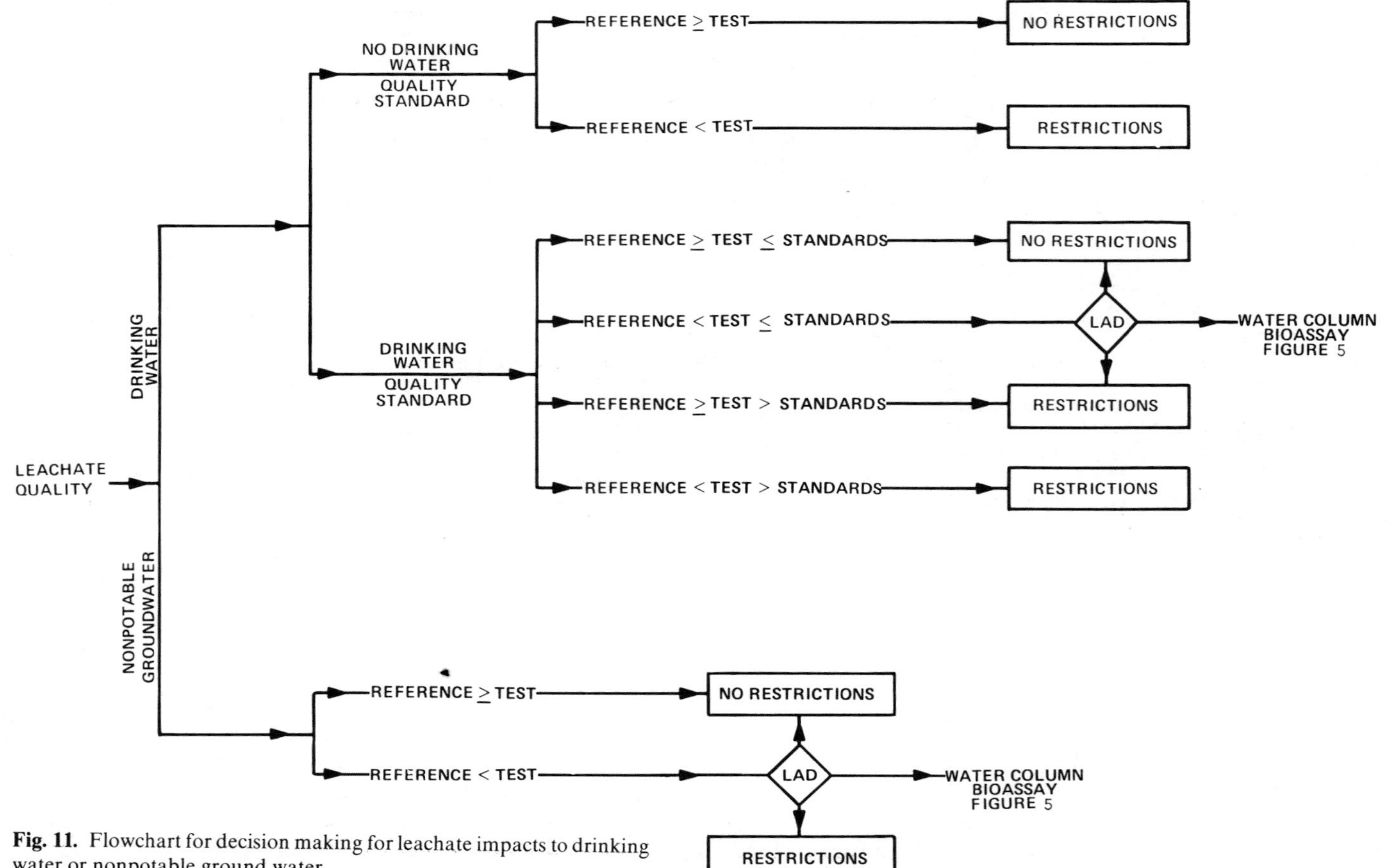

Fig. 11. Flowchart for decision making for leachate impacts to drinking water or nonpotable ground water

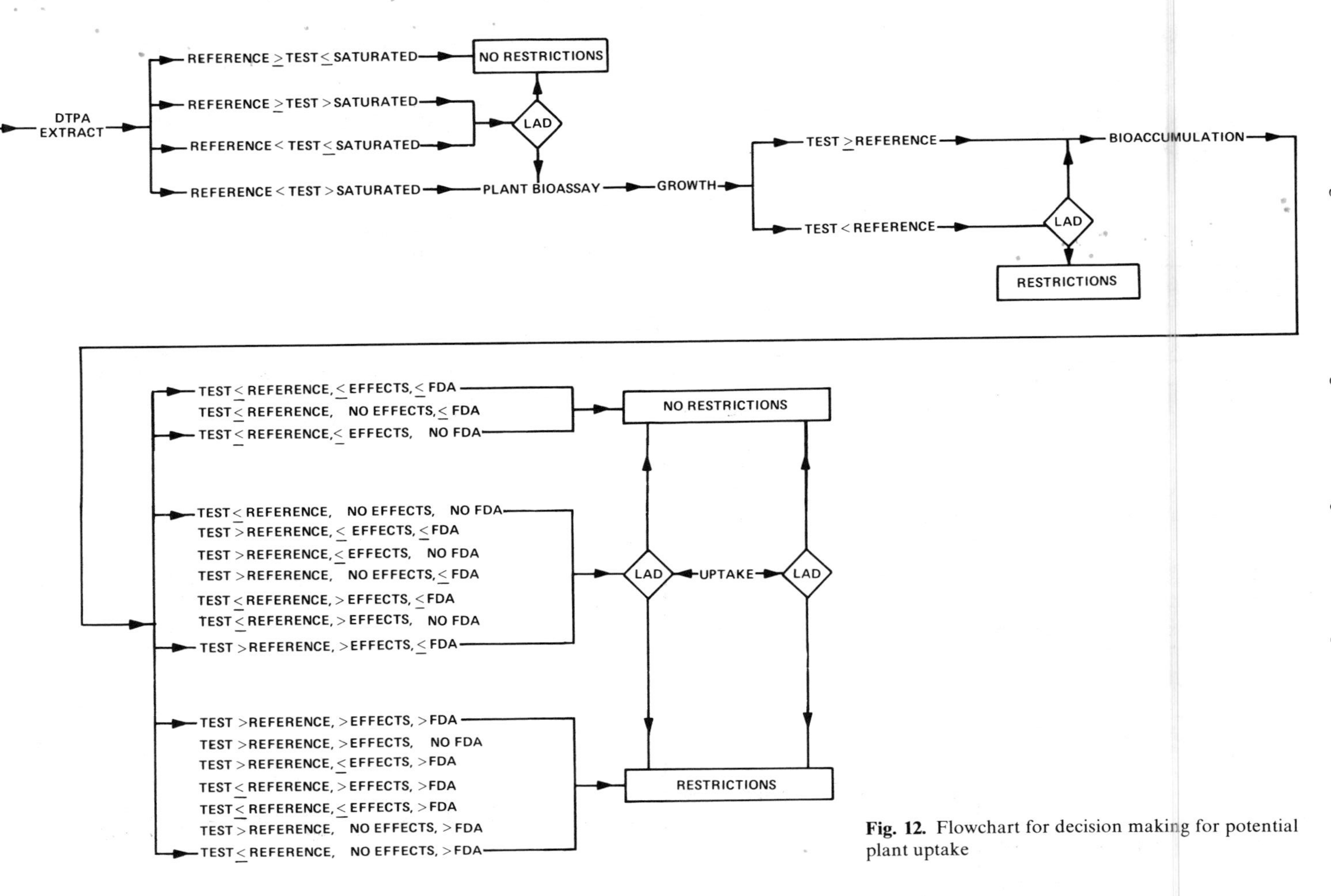

Fig. 12. Flowchart for decision making for potential plant uptake

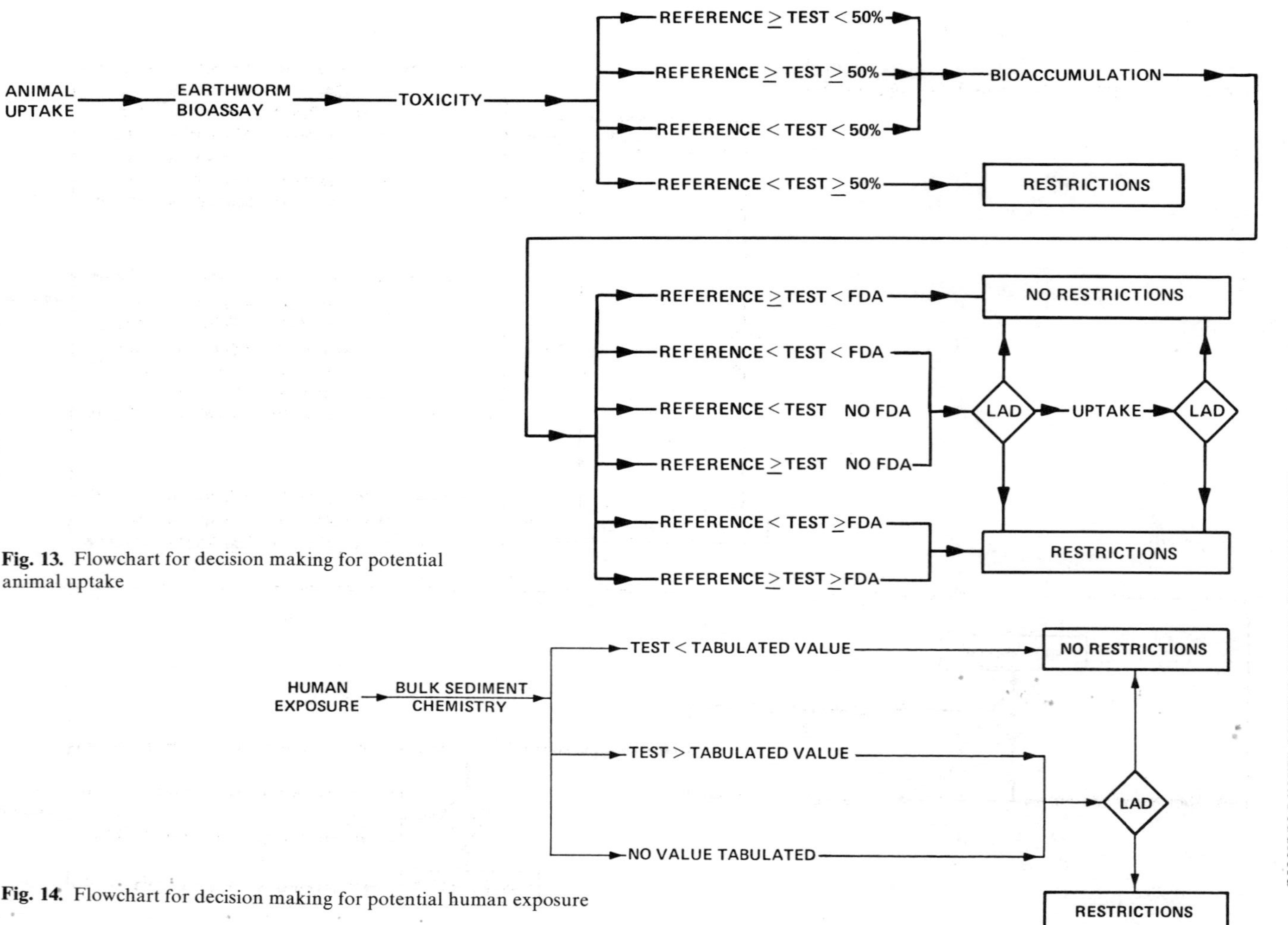

Fig. 13. Flowchart for decision making for potential animal uptake

Fig. 14. Flowchart for decision making for potential human exposure

2.3.1 Responsibility for Local Authority Decisions

There are certain decisions that must be made initially and then periodically within the decision-making framework that are the sole responsibility of the local authorities. These LADs are required to initially set specific goals to be achieved. For example, a LAD must establish the environmental quality ultimately desired at the site and the rate at which this goal is to be achieved. A LAD must determine the appropriate reference site(s) for test result comparisons in the decision-making framework in order to achieve the ultimate and intermediate goals. As described previously, the selection of reference sites can vary from the actual disposal site to a pristine background site. This selection is dependent on the goal sestablished for the area such as a goal of nondegradation (reference site is disposal site) or cleaner-than-present condition (reference site is pristine background site) or some other goal. *The clear identification of the ultimate and intermediate goals and the selection of appropriate references to achieve them is a crucial responsibility of the local authorities and will influence the outcome of all test result interpretations.* In addition, LADs must be made whenever technical knowledge and understanding are inadequate to support a scientific judgement and administrative considerations. For example, a LAD must determine whether or not to consider mixing zones when test results exceed reference-site values of water-quality criteria. Should the LAD be to consider mixing zones and an acceptable mixing zone is available, a decision for no restrictions on that particular aspect of the disposal might be made. In contrast, should the LAD *not* consider mixing zones, then a decision for restrictions might be made which will generally be more conservative, but may prove to be more costly upon implementation of the restrictions. Many of these LADs are shown in the flowcharts as ◇. Scientific guidance for making each LAD is provided at the appropriate points in the text. This general guidance is appropriate for nationwide use, but the actual implementation of the general guidance must vary in different areas to meet different goals, objectives, and concerns.

2.3.2 Initial Evaluation of Contaminants

The initial evaluation determines if the sediment to be dredged is likely to be contaminated (Fig. 1). This decision is based on consideration of available information. The information considered in the initial evaluation also allows identification of the specific contaminants of concern in each sediment being considered.

It is recommended that all potential dredging projects collect at least one composited sediment sample from the project. This sample should be representative of the entire depth of dredging as well as the reach of waterway to be dredged. An example of a composited sample might be the collection of a sediment core for each 4,000 yd^{-3} of sediment along the waterway. This would be the equivalent of one typical barge load of dredged material (one management unit). These cores are then divided in half lengthwise. One-half of all the cores are kept separate, while the other half of all cores are mixed to get a homogeneous composited sample. This sample is then analyzed for the entire list of EPA priority pollutants. If the composite sample indicates elevation of one or more contaminants, then each separate remaining half-core can be analyzed to determine which sample or samples along the waterway contains contaminants. Likewise, a composited sediment sample should be obtained from an appropriate LAD reference site and analyzed for the entire list of EPA priority pollutants. Further details on sediment sampling and processing procedures are reported by Plumb (1981).

2.3.2.1 Decision of No Contamination[4]

If sufficient information is available and provides no substantial reason to believe contaminants are present above reference-site levels and based on the chemical analysis of a composite sediment sample, a DECISION FOR NO FURTHER TESTING is made. The sediment can be dredged and disposed in an aquatic site, in an upland site, or used productively such as for marsh creation or enhancement of agricultural land with *no restrictions and no contaminant impacts on the environment.* In such cases, the selection of a disposal site is based on considerations other than potential contaminant impacts on the environment.

2.3.2.2 Decision of Sediment Contamination

If the available information is inadequate or provides a substantial reason to believe contaminants are present above reference site levels, then a DECISION FOR FURTHER TESTING is made. The testing of the sediment depends on which of the two questions in Fig. 4 is being addressed. The question *"In what type of disposal environment should the sediment in question be placed to minimize contaminant mobility?"* is SITE SELECTION TESTING and represents the situation where aquatic and upland (and nearshore) disposal sites are available. The emphasis is on selecting the disposal environment to minimize the potential for adverse contaminant impacts from the dredged material. The second question, *"Is this sediment suitable from a contaminant perspective for placement in a particular disposal environment?"*, could be considered as SITE ACCEPTABILITY TESTING and addresses the situation that there are limitations on available disposal sites. Therefore, the sediment is tested to determine the acceptability of a given disposal environment for the disposal of the sediment. For example, if the only disposal sites available are upland sites, then testing should focus on upland disposal and not on aquatic disposal. Ultimately, the testing should be tailored to the available disposal site. Once the appropriate question is identified, a decision to consider AQUATIC DISPOSAL (Figs. 5 and 6) or UPLAND DISPOSAL (Figs. 7 and 14) can be made. Test results are compared to established numerical values tables where these are available and appropriate for test interpretation (Tables 2–8). When such values do not exist, interpretation of test results is made in

---

[4]All decisions reached on the basis of test results and interpretations are indicated in capital letters.

Table 2. Environmental protection agency water-quality criteria for protection of aquatic life[a]

Chemical	Criterion for protection of aquatic life ($\mu g\ l^{-1}$)			
	Salt water		Fresh water	
	24-h Average (chronic)	Maximum at any time (acute)	24-h Average (chronic)	Maximum at any time (acute)
Aldrin	—	1.3	—	3.0
Arsenic (total trivalent)	—	—	—	440
Cadmium[b]	4.5	59		
Hardness h, mg l^{-1} $CaCO_3$ = 50			0.012	1.5
100			0.025	3.0
200			0.051	6.3
Chlordane	0.0040	0.09	0.0043	2.4
Chromium (total trivalent)[b]	—	—		
Hardness h, mg l^{-1} $CaCO_3$ = 50			—	2,200
100			—	4,700
200			—	9,900
Chromium (total hexavalent)	18	1,260	0.29	21
Copper[b]	4.0		5.6	
Hardness h, mg l^{-1} $CaCO_3$ = 50		23		12
100				22
200				43
Cyanide (free)	—	—	3.5	52
Dieldrin	0.0019	0.71	0.0019	2.5
DDT	0.0010	0.13	0.0010	1.1
TDE	—	—	—	—
DDE	—	—	—	—
Endosulfan	0.0087	0.034	0.056	0.22
Endrin	0.0023	0.037	0.0023	0.18
Heptachlor	0.0036	0.053	0.0038	0.52
Lindane	—	0.16	0.080	2.0
Lead[b]	—	—		
Hardness h, mg l^{-1} $CaCO_3$ = 50			0.75	74
100			3.8	170
200			20	400
Mercury	0.025	3.7	0.00057	0.0017
Nickel[b]	7.1	140		
Hardness h, mg l^{-1} $CaCO_3$ = 50			56	1,100
100			96	1,800
200		160		3,100
PCB (total)	0.030	0.030	0.014	0.014
Selenium (inorganic selenite)	54	410	35	260
Silver[b]	—	2.3		
Hardness h, mg l^{-1} $CaCO_3$ = 50			—	1.2
100			—	4.1
200			—	13
Toxaphene	—	0.070	0.103	1.6
Zinc[b]	58	170	47	
Hardness h, mg l^{-1} $CaCO_3$ = 50				180
100				320
200				570

[a]Federal Register, Vol. 45, No. 231, Friday, November 28, 1980, pp. 79318–79357.

[b]Criteria for some metals in fresh water are hardness-dependent and are derived from the following equations, where h is hardness in mg l^{-1} as $CaCO_3$ and e is the natural logarithm base:

Metal	24-hr Average	Maximum at any time
Cadmium	$e^{1.05 (\ln h) - 8.53}$	$e^{1.05 (\ln h) - 3.73}$
Chromium (total trivalent)	Not established	$e^{1.08 (\ln h) + 3.48}$
Copper	(Main table)	$e^{0.94 (\ln h) - 1.23}$
Lead	$e^{2.35 (\ln h) - 9.48}$	$e^{1.22 (\ln h) - 0.47}$
Nickel	$e^{0.76 (\ln h) + 1.06}$	$e^{0.76 (\ln h) + 4.02}$
Silver	Not established	$e^{1.72 (\ln h) - 6.52}$
Zinc	(Main table)	$e^{0.83 (\ln h) + 1.95}$

Table 3. Standards for contaminant concentrations in drinking water

Parameter, mg l^{-1} (unless otherwise noted)	Drinking water standards Federal	State of Washington
Arsenic	0.05	0.05
Barium	1.0	1.0
Cadmium	0.010	0.010
Chromium	0.05	0.05
Lead	0.05	0.05
Mercury	0.002	0.002
Selenium	0.01	0.01
Silver	0.05	0.05
Fluoride	1.4–2.4	1.4–2.4
Nitrate (as N)	10.0	10.0
Endrin	0.0002	0.0002
Lindane	0.004	0.004
Methoxychlor	0.1	0.1
Toxaphene	0.005	0.005
2,4-D	0.1	0.1
2,4,5-TP Silvex	0.01	0.01
Trihalomethanes	0.1	0.1
Turbidity (JU)	1.0	1.0
Coliform bacteria membrane filter test (No./100 ml)	1.0	1.0
Gross alpha (pCi l^{-1})	15.0	15.0
Combined radium 226 and radium 228	5.0	5.0
Beta and photon particle activity (Mrem yr^{-1})	4.0	4.0
Sodium	Monitor	250.0
Chloride	250.0	250.0
Color (units)	15.0	15.0
Copper	1.0	1.0
Corrosivity	Noncorrosive	Noncorrosive
Foaming agents	0.5	0.5
Iron	0.3	0.3
Manganese	0.05	0.05
Odor (threshold No.)	3.0	3.0
pH (units)	6.5–8.5	6.5–8.5
Sulfate	250.0	250.0
Total dissolved solids	500.0	500.0
Zinc	5.0	5.0

Table 4. Demonstrated effects of contaminants on plants

	Plant growth effect-contaminated content, mg kg^{-1}leaves				
Contaminant	Normal[a]	Critical content[b]	10% Yield reduction[b]	25% Yield reduction[b]	Phytotoxic[a]
Arsenic	0.1–1	–	–	–	3–10
Boron	775	–	–	–	75
Cadmium	0.1–1	8	15	Varies	5–700
Cobalt	0.01–0.3	–	–	–	25100
Chromium (III), oxides	0.1–1	–	–	–	20
Copper	3–20	20	20	2040	25–40
Fluorine	1–5	–	–	–	–
Iron	30–300	–	–	–	–
Manganese	15–150	–	–	500	400–2,000
Molybdenum	0.1–3.0	–	–	–	100
Nickel	0.1–5	11	26	50–100	500–1,000
Lead	2–5	–	–	–	–
Selenium	0.1–2	–	–	–	100
Vanadium	0.1–1	–	–	–	10
Zinc	15–150	200	290	500	500–1,500

[a]From Chaney (1983).
[b]From Davis, Beckett, and Wollan (1978); Davis and Beckett (1978); Beckett and Davis (1977).
[c]From Chaney et al. (1978).

Table 5. Maximum recommended application of municipal sludge-applied metals to medium-textured cropland soils to prevent phytotoxicity[a]

	Maximum application		
Contaminant	Kg ha^{-1}	lb acre^{-1}	mg kg^{-1}
Lead	1,000	891	500[b]
Zinc	560	446	250
Copper	280	223	125
Nickel	112	111	62
Cadmium	11.2	4.5	2.5

Note: Soil bulk density 1.33; potentially acidic soil. Recommended limits to prevent yield reduction in sensitive vegetable crops at PH $\geq$6.2, or most crops and cover crops at pH$\geq$5.5.
[a]EPA, USDA, USFDA (1981).
[b]Maximum allowable lead content in soil human child exposure as related to direct soil ingestion in the United Kingdom and in the United States.

Table 6. Additional action levels for contaminants in foodstuffs used by various countries

Source	Contaminant	Commodity	Content (mg kg^{-1})	References
Britain	Lead	All foods	1.0 (fresh wt.)	M.A.F.F. (1972)
World Health Organization (WHO)	Lead	Root vegetables	0.1 (fresh wt.)	WHO (1972)
		Cereal	0.1 (fresh wt.)	
		Leafy vegetables	1.2 (fresh wt.)	
	Cadmium	Root vegetables	0.05 (fresh wt.)	WHO (1972)
		Leafy vegetables	0.1 (fresh wt.)	
		Potatoes, cereal	0.1 (fresh wt.)	
Dutch	Copper	Animal feed	20.0 (dry wt.)	DMAFCMN (1973)
Dutch (unofficial)	Cadmium	Single animal feed	0.5 (dry wt.)	European Community (1974)
		Mixed animal feed	1.0 (dry wt.)	
		Roughage	1–2 (fresh wt.)	
European Economic Community	Lead	Single animal feed	10.0 (dry wt.)	Van Driel et al. (1985)
		Mixed animal feed	5.0 (dry wt.)	
		Roughage	40.0 (fresh wt.)	
FDA (as of Sept. 82)	Mercury	Wheat seed	1.0 (dry wt.)	FDA (1982)
	PBB	Animal feed	0.5 (dry wt.)	
	Various pesticides	Vegetables, grains, and feeds	0.03–0.1	

Table 7. Background levels and allowable applications of several heavy metals for US cropland soils[a]

	Background concentration in Surface soils (mg kg^{-1})			No effect allowed[b] addition[a]	Median + allowed [b] application
Parameter	5 Percentile	Median	95 Percentile	(kg ha^{-1})	(mg kg^{-1})
Lead	4.0	11	27	1,000	511
Zinc	7.3	54	129	500	304
Copper	3.7	19	96	250	144
Nickel	3.8	19	59	125	82
Cadmium	0.035	0.20	0.78	5	2.7
pH	4.6	6.1	8.1	–	–

[a]Holnigren et al. (1987) and Table 5.
[b]Allowed application is mixed into the 0–15 cm (0–6 in.) surface layer of soil.

comparison to results of the same test performed on a reference sediment. For each test the framework provides guidance for determining whether or not restrictions on the discharge are required to protect against contaminant impacts, or whether further evaluation is required to determine the need for restrictions. In some cases, there is inadequate scientific knowledge to reach a decision solely on the basis of test results, and LADs that incorporate both scientific and administrative judgements are required to reach a decision. In such cases, guidance is given on evaluating the scientific considerations involved. In this manner, a systematic

Table 8. Recommended or regulated limitations on potentially toxic constituents in surface (0–15 cm) soils

Basis for limitation	Contaminant	Soil concentration	Reference
Soil ingestion	Lead	500 mg kg^{-1}	EPA (1977)
	Mercury	5 mg kg^{-1}	
	PCBs, etc.	2.0 mg kg^{-1}	Fries (1982)
Plant uptake	Cadmium	2.5 mg kg^{-1} (pH5.5)	EPA (1979)
Phytotoxicity	Zinc	250 mg kg^{-1}	Logan and Chaney (1983)
	Copper	125 mg kg^{-1}	
	Nickel	62 mg kg^{-1}	
	Cobalt	62	
Leaching	Cr (VI)	0.05 mg l^{-1}	EPA drinking water standard (Table 3)

interpretation of the results of each test required to evaluate potential impacts of aquatic disposal and upland disposal is made. Applying the systematic interpretation of test results will lead to a decision that restrictions are, or are not required for aquatic disposal and/or upland disposal.

2.3.3 Aquatic Disposal with Restrictions

In cases where testing protocols indicate that water-column or benthic effects will be unacceptable when conventional aquatic disposal techniques are used, aquatic disposal with restrictions may be considered. This alternative involves the use of dredging or disposal techniques that will reduce water-column and benthic effects. Such techniques are discussed in detail in US Army Engineer District, Seattle (1984) and include the use of submerged discharge points and diffusers, subaqueous confinement of material, or capping of contaminated material with clean material, and treatment techniques. The same basic considerations for conventional aquatic disposal-site designation, site capacity, and dispersion and mixing also apply to aquatic disposal with restrictions.

2.3.3.1 Submerged Discharge

The use of a submerged point of discharge reduces the area of exposure in the water column and the amount of material suspended in the water column and susceptible to dispersion. The use of submerged diffusers also reduces the exit velocities for hydraulic placement, allowing more precise placement and reducing both suspension and spread of the discharged material. Considerations in evaluating feasibility of a submerged discharge and/or use of a diffuser include water depth, bottom topography, currents, type of dredge, and site capacity. The DMRP

(Barnard 1978) developed a conceptual design for a submerged diffuser that has been successfully demonstrated by European dredging interests and is now being considered for more detailed study in the United States under the CE Dredging Operations Technical Support (DOTS) Program.

2.3.3.2 Subaqueous Confinement

The use of subaqueous depressions or burrow pits or the construction of subaqueous dikes can provide confinement of material reaching the bottom during disposal operation, thereby reducing both physical benthic effects and the potential for release of contaminants. Considerations in evaluating feasibility of subaqueous confinement include type of dredge, water depth, bottom topography, bottom sediment type, and site capacity. Subaqueous confinement has been utilized in Europe and to a limited extent by US Army Engineer District, New York. Precise placement of material and use of submerged points of discharge increase the effectiveness of subaqueous confinement.

2.3.3.3 Capping

Capping is the placement of a cleaner material over material considered contaminated. Considerations in evaluation of the feasibility of capping include water depth, bottom topography, currents, dredged material and capping material characteristics, and site capacity. Both the Europeans and the Japanese have successfully used capping techniques to isolate contaminated material in the aquatic disposal environment. Capping is also currently used by the US Army Engineer District, New York and US Army Engineer Division, New England as a means of offsetting the potential harm of aquatic disposal of contaminated or otherwise unacceptable dredged material. The London Dumping Convention has accepted capping, subject to careful monitoring and research, as a physical means of rapidly rendering harmless contaminated material disposed in the ocean. The physical means are essentially to seal or sequester the unacceptable material from the aquatic environment by covering with acceptable material.

The efficiency of capping in preventing the movement of contaminants through this seal and the degradation of the biological community by leakage, erosion of the cover (cap), or bioturbation are being addressed by research under the LEDO Program. The engineering aspects of cap design and placement are also being addressed under this program. It is possible that techniques and equipment can be developed that will provide a capped dredged material disposal area as secure from potential environmental harm as upland confined disposal areas. The capping technique for disposal of dredged material has potential for relieving some pressure on acquiring sites for confined disposal areas in locations where land is rapidly becoming unavailable.

2.3.3.4 Chemical/Physical/Biological Treatment

Treatment of discharges into open water may be considered to reduce certain impacts. For example, the Japanese have used an effective in-line dredged material treatment scheme for highly contaminated harbor sediments (Barnard and Hand 1978). However, this strategy has not been widely applied and its effectiveness has not been demonstrated for solution of the problem of contaminant release during aquatic disposal.

2.3.4 Upland Disposal with Restrictions

Conventional confined upland disposal methods can be modified to accommodate disposal of contaminated sediments in new, existing, and reusable disposal areas. The design of modification to these areas must consider the problems associated with contaminants and their effects on conventional design. Many of the following design considerations apply to all of the implementation options.

2.3.4.1 Site Selection and Design

Site location is an important consideration since it can mitigate many contaminant mobilization problems. Proper site selection may reduce surface run-on and therefore contaminated runoff and contaminant release by flooding. Groundwater contamination problems can be minimized through selection of a site with a natural clay foundation instead of a sandy area and through avoidance of aquifer recharge areas (Gambrell et al. 1978).

Careful attention to basic site design as discussed previously will aid in implementing many of the controls outlined. Retention time can be increased to improve suspended solids removal and therefore contaminant removal. Additional ponding depth can also improve sedimentation. Decreasing the weir-loading rate can also reduce the suspended solids and contaminant concentration of the effluent.

Dewatering should be examined carefully before selecting a method since dewatering promotes oxidation of the material and thereby increases the mobility of certain contaminants (Gambrell et al. 1978). Care must also be taken to reduce loss of contaminated sediment by erosion during drainage and storm events.

2.3.4.2 Available Options

Depending on the particular dredging operation, one or more types of restrictions may be required. The particular restriction or combination of restrictions may eliminate certain disposal options. For the purposes of developing a management strategy, four options are considered available for upland disposal with restrictions. These options include:

1. *Containment*: Dredged material and associated contaminants are contained within the disposal site.
2. *Treatment*: Dredged material is modified physically, chemically, or biologically to reduce toxicity, mobility, etc.
3. *Storage and rehandling*: Dredged material is held for a temporary period at the site and later removed to another site for ultimate disposal.
4. *Reuse*: Dredged material is classified and beneficial uses are made of reclaimed materials.

Obviously, combinations of the above options are available for a particular dredging operation.

Containment of contaminated dredged material can be either in an existing or a new facility. These facilities can be designated or modified to handle a wide variety of contaminants. Most contaminated sediments can be disposed of in an existing site where special controls have been incorporated in consideration of the restrictions discussed in Section 2.3.4.4. In the case of highly contaminated sediments, a more secure disposal facility would be required, and, in all probability, such a disposal facility would dictate the design of a new facility.

The treatment option can be associated with either existing or new facilities. Some form of physical, and chemical, or biological treatment would probably be associated with the disposal of highly contaminated dredged material. Treatment may also be combined with other options for disposal of slightly to moderately contaminated dredged material in confined disposal sites.

Of the four available options, storage and rehandling can serve two beneficial functions: continued use of upland sites located close to dredging areas and use as a rehandling facility for contaminated dredged material prior to later disposal off-site.

Finally, the concept of a reuse option would incorporate beneficial uses of materials and could include sand and gravel or slightly contaminated construction fill to be used for raising dikes or acceptable off-site uses.

2.3.4.3 Design Considerations

Contaminated dredged material management includes methods for dewatering, transporting, storing, treating, and disposing of contaminated material. The most technically and economically effective strategy to handle contaminated dredged material will depend on many site-specific variables, which include the following:

1. Method of dredging used – hydraulic versus mechanical.
2. Method of dredged material transport – pipeline versus truck, hopper, or barge.
3. Physical nature of removed material – consistency (solids/water content) grain-size distribution.
4. Volume of removed material.
5. Nature and degree of contamination; physical and chemical characteristic of contaminants.
6. Proximity of acceptable treatment, storage, containment, or reuse facilities.
7. Available land area for construction of new or expansion of existing facilities.

2.3.4.4 Restrictions

Conventional confined upland disposal methods may be modified to accommodate disposal of slightly to highly contaminated dredged material. Many of the restrictions on upland disposal that may be required are common to the available options. Among these restrictions are:

1. Effluent quality controls during dredging operations.
2. Runoff water quality controls after dredging operations.
3. Leachate controls during and after dredging operations.
4. Control of contaminant uptake by plants and animals during and after dredging operations.
5. Control of atmospheric contaminants after dredging operations.

Many of the contaminant controls described in the following paragraphs are directly applicable to the control of highly contaminated sediments. These controls will be extremely site-specific. Special considerations that are based on the physical nature and chemical composition of the dredged material will be required to effectively design a confined disposal facility. For example, some contaminated dredged material may require in-pipeline treatment prior to discharging the material into the containment facility. Similarly, if the facility requires a bottom-liner system, the liner material (synthetic membrane or clay) must be chemically compatible (resistant) with the dredged material to be placed on them. Special compatibility testing will be needed for selection of appropriate liner material. Other requirements such as leachate detection and monitoring are likely due to the potentially adverse environmental effects of the liner leaking.

Effluent Controls. Effluent controls at conventional upland disposal areas are generally limited to chemical clarification. The clarification system is designed to provide additional removal of suspended solids and associated adsorbed contaminants as described in Schroeder (1983). Additional controls can be used to remove fine particulates that will not settle or to remove soluble contaminants from the effluent. Examples of these technologies are filtration, adsorption, selection ion exchange, chemical oxidation, and biological treatment processes. Beyond chemical clarification, only limited data exist for treatment of dredged material (Gambrell et al. 1978).

Runoff Controls. Runoff controls at conventional sites consist of measures to prevent the erosion of contaminated dredged material and the dissolution and discharge of oxidized contaminants from the surface. Control options include maintaining ponded conditions, planting vegetation to stabilize the surface, liming the surface to prevent acidification and to reduce dissolution, covering the surface with synthetic geomembranes, and/or placing a lift of cleaner material to cover the contaminated dredged material (Gambrell et al. 1978).

Leachate Controls. Leachate controls consist of measures to minimize groundwater pollution by preventing mobilization of soluble contaminants. Control measures include proper site selection, dewatering to minimize leachate production, chemical admixing to prevent or retard leaching, lining the bottom to prevent leakage and seepage, capping the surface to minimize infiltration and thereby leachate production, using vegetation to stabilize contaminants and to increase drying, and leachate collection, treatment, or recycling (Gambrell et al. 1978).

Control of Contaminant Uptake. Plant and animal contaminant uptake controls are measures to prevent mobilization of contaminants into the food chain. Control measures include selective vegetation to minimize contaminant uptake, liming or chemical treatment to minimize or prevent release of contaminants from the material to the plants, and capping with cleaner dredged material or excavated material (Gambrell et al. 1978).

Control of Atmospheric Contaminants. The control of gaseous emissions or dust that might present human health hazards can consist of physical measures such as covers or vertical barriers. Control of contaminated surface material is another type of management or operating control to minimize transport of contaminants off-site. Techniques for limiting wind erosion are generally similar to those employed in dust control and include physical, chemical, or vegetative stabilization of surface soils (Corps of Engineers, CE 1983, Lee et al. 1985, Peddicord et al. 1986).

3 Conclusions and Recommendations

A technically feasible and environmentally sound management approach to the disposal of dredged material has been developed and presented. This strategy is based on results of many years of research and dredging experience by the Corps of Engineers and others. The evaluation procedures allow specific, potential problem areas to be defined and addressed. A number of variations are presented for each of the major alternatives of open water (aquatic) and upland confined disposal, each having a significant influence on the fate of contaminants at disposal sites. The management strategy provides a framework for assessing and choosing an appropriate alternative for disposal based on specific problem areas. It is applicable to materials ranging from clean sand to highly contaminated sediments. It is recommended that the strategy be implemented for managing all dredged material disposal. Application of the strategy should be thoroughly documented to allow refinement based on experience.

Although there has been much research and some field experience gained in handling and control of contaminated materials generated by industrial and chemical manufacturing operations (USEPA 1982), few applications to dredging can be cited. Considerable effort is needed to apply these control technologies to dredging operations. Research sponsored by the CE, EPA, and others will continue to provide input into the described management strategy for dredged material disposal that will reduce potential environmental impacts.

Acknowledgments. This chapter was prepared from information obtained from research reports prepared under the US Army Corps of Engineers Dredging Operations Technical Support Program (DOTS) and under mission support for the US Army Engineer District, Seattle.

References

American Public Health Association (1980) Standard Methods for the Examination of Water and Wastewater. 15th Ed. APHA; New York

Barnard WD (1978) Prediction and control of dredged material dispersion around dredging and open-water pipeline disposal operations. US Army Eng Waterways Exp Stn, Vicksburg, Miss, Tech Rep DS-78-6

Barnard WD, Hand TD (1978) Treatment of contaminated dredged material. US Army Eng Waterways Exp Stn, Vicksburg, Miss Tech Rep DS-78-14

Beckett PHT, Davis RD (1977) Upper critical levels of toxic elements in plants. New Phytol 79:95-106

Biddinger GR and Gloss SP (1984) The Importance of Trophic Transfer in the Bioaccumulation of Chemical Contaminants in Aquatic Ecosystems, Residue Rev, Vol 91, pp 104-130

Brannon JM (1978) "Evaluation of Dredged Material Pollution Potential," Technical Report DS-78-6, US Army Engineer Waterways Experiment Station, Vicksburg, Miss

Branson, DR, Blau GE, Alexander HC, and Neely WB (1975) "Bioconcentration of 2,2′, 4,4′ – tetrachlorobiphenyl in Rainbow Trout as Measured by an Accelerated Test," Transactions of the American Fishery Society, Vol 4, pp 784-792

Burks SA, Engler RM (1978) Water quality impacts of aquatic dredged material disposal (laboratory investigations). US Army Eng Waterways Exp Stn, Vicksburg, Miss, Tech Rep DS-78-4

Chaney RL, (1983) Potential effects of waste constituents on the food chain in land treatment of hazardous wastes. Noyes Data Corp., NJ, pp 152-240

Chaney RL, Hundemann PT, Palmer WT, Small RJ, White MC, Decker AM (1978) Plant accumulation of heavy metals and phytoxicity resulting from utilization of sewage sludge and sludge composts on cropland. Proc Nat Conf Composting municipal residues and sludge, information transfer, Inc., Rockville, Md, pp 86-97

Chen KY, Eichenberger D, Mang JL, Hoeppel RE (1978) Confined disposal area effluent and leachate control (laboratory and field investigations). US Army Eng Waterways Exp Stn, Vicksburg, Miss, Tech Rep DS-78-7

Davis RD, Beckett PHT (1978a) Upper critical levels of toxic elements in plants. II. Critical levels of copper in young barley, wheat, rape, lettuce, and ryegrass, and of nickel and zinc in young barley and ryegrass. New Phytol 80:23-32

Davis RD, Beckett PHT (1978b) Critical levels of twenty potentially toxic elements in young spring barley. Plant Soil 49:395-408

Dillon TM (1984) Biological Consequences of Bioaccumulation in Aquatic Animals: As Assessment of the Current Literature, Technical Rep D-84-2, US Army Eng Waterways Exp Stn, Vicksburg, Mississippi

Driel W van, Smilde KW, Luit B van (1985) Comparison of the heavy-metal uptake of *Cyperus Escluentus* and of agronomic plants grown on contaminated dutch sediments. US Army Eng Waterways Exp Stn, Vicksburg, Miss, Miscellaneous Pap D-83-1

Dutch Ministry of Agriculture and Fisheries, Committee on Mineral Nutrition, The Hague (1973) Tracing and treating mineral disorders in dairy cattle. Centre Agric Publ Doc, Wageningen

Environmental Effects Laboratory (1976) Ecological evaluation of proposed discharge of dredged or fill material into navigable waters – Interim guidance for implementation of section 404(b)(1) of public law 92-500. US Army Eng Waterways Exp Stn, Vicksburg, Miss, Miscellaneous Pap D-76-17

Environmental Protection Agency/Corps of Engineers (EPA/CE) (1977) Ecological Evaluation of Proposed Discharge of Dredged Material into Ocean Waters, Implementation Manual for Section 103 of Public Law 92-532 (Marine Protection, Research, and Sanctuaries Act of 1972), US Army Eng Waterways Exp Stn, Vicksburg, Miss

Environmental Protection Agency (EPA) (1980) Guidelines and Testing Requirements for Specification of Disposal Sites for Dredged or Filled Material, *Federal Register*, Vol 45, No. 249, pp 85, 336-85, 367

Environmental Protection Agency (EPA), FDA, USDA (1981) Land application of municipal sewage sludge for the production of fruits and vegetables. A statement of federal policy and guidance. US Environ Protect Ag Joint Policy Statement SW-905

European Community (1974) Richtlijn Raad EEG (17/12/73) tot vaststelling van de macimale gehalten aan angewenste stoffen en produkten ub duervieders. Publ Blad EEG 17, L38:31-36

Folsom BL, Jr, Lee CR (1981) Zinc and cadmium uptake by the freshwater plant *Cyperus esculentus* grown in contaminated sediments under reduced (flooded) and oxidized (upland) disposal conditions. J Plant Nutrit 3:233–244

Folsom BL, Jr, Lee CR (1983) Contaminant uptake by *Spartina alterniflora* from an upland dredged material disposal site – application of a saltwater plant bioassay. Proc Int Conf Heavy metals in the environment, Heidelberg, Germany

Folsom BL, Jr, Lee CR, Preston KM (1981) Plant bioassay of materials from the blue river dredging project. US Army Eng Waterways Exp Stn, Vicksburg, Miss, Miscellaneous Pap EL–81–6

Francingues NR, Palermo MR, Lee CR, and Peddicord RK (1985) "Management Strategy for Disposal of Dredged Material: Test Protocols and Contaminant Control Measures," Miscellaneous Paper D–85–1, US Army Eng Waterways Exp Stn, Vicksburg, Miss

Gambrell RP, Khalid RA, Patrick WH (1978) Disposal alternatives for contaminated dredged material as a management tool to minimize adverse environmentaleffects. US Army Eng Waterways Exp Stn, Vicksburg, Miss, Tech Rep DS–78–8

Goerlitz DF (1984) "A Column Technique for Determining Sorption of Organic Solutes on the Lithological Structure of Aquifers, " *Bull of Environ Contamination and Toxicology,* Vol 32, pp 37–44

Haliburton TA (1978) Guidelines for dewatering/densifying confined dredged material. US Army Eng Waterways Exp Stn, Vicksburg, Miss, Tech Rep DS–78–11

Herner and Company (1978) Dredged material research program publication index and retrieval systems. US Army Eng Waterways Exp Stn, Vicksburg, Miss Tech Rep DS-78–23

Hill DO, Myers TE, and Brannon JM (1985) Development and Application of Techniques for Predicting Leachate Quality in Confined Disposal Facilities, in preparation, US Army Eng Waterways Exp Stn, Vicksburg, Miss

Holliday BW, Johnson BH, Thomas WA (1978) Predicting and Monitoring Dredged Material Movement. US Army Eng Waterways Exp Stn, Vicksburg. Miss, Tech Rep DS–78–23

Holnigren GGS, Meyer MW, Daniels RB, Chaney RL, Kubota J (1987) Cadmium, lead, zinc, and nickel in agricultural soils of the United States. J Environ Qual (in press)

Houle MJ and Long DE (1980) "Interpreting Results from Serial Batch Extraction Tests of Wastes and Soils," *Disposal of Hazardous Waste, Proceedings of the Sixth Annual Research Symposium, Chicago, March 17–20, 1980,* Grant No. R807121, EPA-600/9–80–100, Municipal Environ Research Laboratory, Office of Research and Development, US Environ Protection Agency, Cincinnati, pp 60–81

Jones RA, Lee GF (1978) Evaluation of the elutriate test as a method of predicting contaminant release during open water disposal of dredged sediments and environmental impact of open water dredged material disposal. US Army Eng Waterways Exp Stn, Vicksburg, Miss, Tech Rep D–78–45

Kay SH (1984) "Potential for Biomagnification of Contaminants Within Marine and Freshwater Food Webs," Technical Report D–84–7, US Army Eng Waterways Exp Stn, Vicksburg, Miss

Lee CR, Skogerboe JG (1983a) Quantification of soil erosion control by vegetation on problem soils. Proc Int Conf Soil erosion and conservation society of America, soil conservation of America

Lee CR, Skogerboe JG (1983b) Prediction of surface runoff water quality from an upland dredged material disposal site. Proc Int Conf Heavy metals in the environment, Heidelberg, Germany

Lee CR, Folsom BL, Jr, Engler RM (1982) Availability and plant uptake of heavy metals from contaminated dredged material placed in flooded and upland disposal environments. Environ Int 7:65–71

Lee CR, Folsom BL, Jr, Bates DJ (1983) Prediction of plant uptake of toxic metals using a modified DTPA soil extraction. Sci Tot Environ 28:191–202

Lee CR, Skogerboe JG, Eskew K, Price RA, Page NR, Clar M, Kort R, Hopkins H (1985) Restoration of problem soil materials at corps of engineers construction sites. US Army Eng Waterways Exp Stn, Vicksburg, Miss, Instruct Rep EL–85–2

Lowenbach WF, King E, Cheromisinoff P (1977) Leachate testing techniques surveyed. Water Sew Worksh, pp 36–46

Marquenie JM and Simmers JW (1984) Bioavailability of Heavy Metals PCB and PCA Components to the Earthworm (Eisenia foetidus), *Proceedings, International Conference on Environmental Contamination*, London, UK, In Press

McFarland VA, Gibson AB, and Meade LE (1984) "Application of Physicochemical Estimation Methods to Bioaccumulation from Contaminated Sediments, II. Steady State from Single Time-Point Observations," *Applications in Water Quality Control-Proceedings*, RG Willey ed., US Army Hydrol Eng Center, Davis, Calif, pp 150–167

Montgomery RL, Ford AW, Poindexter ME, Bartos MJ (1978) Guidelines for dredged material disposal area reuse management. US Army Eng Waterways Exp Stn, Vicksburg, Miss, Tech Rep DS-78-12

Neff JW, Foster RS, and Slowey JF (1978) Availability of Sediment-Adsorbed Heavy Metals to Benthos with Particular Emphasis on Deposit-Feeding Infauna, Technical Report D-78-42, Texas A&M Res Found, College Stn, Tex. (NTIS No. AD-A061 152)

Palermo MR (1984) Interim guidance for predicting the quality of effluent discharged from confined dredged material disposal areas. US Army Eng Waterways Exp Stn, Vicksburg, Miss, Tech EEDP-04-1,2,3 and 4

Palermo MR, Montgomery RL, Poindexter ME (1978) Guidelines for dredging, operating, and managing dredged material containment areas. US Army Eng Waterways Exp Stn, Vicksburg, Miss, Tech Rep DS-78-10

Peddicord RK, Lee CR, Palermo MR, Francingues Jr (1986) General decision-making framework for management of dredged material – Example application to commencement Bay, Washington. US Army Eng Waterways Exp Stn, Vicksburg, Miss, Miscellaneous Pap D-86

Plumb RH, Jr. (1981) Procedure for Handling and Chemical Analysis of Sediment and Water Samples, Technical Report EPA/CE-81-1, US Army Eng Waterways Exp Stn, Vicksburg, Miss

Schroeder PR (1983) Chemical clarification methods for confined dredged material disposal. US Army Eng Waterways Exp Stn, Vicksburg, Miss, Tech Rep D-83-2

Simmers JW, Rhett RG, Lee CR (1983) Application of a terrestrial animal bioassay for determining toxic metal uptake from dredged material. Proc Int Conf Heavy metals in the environment, Heidelberg, Germany

US Army Corps of Engineers (1983) Preliminary guidelines for selection and design of remedial systems for uncontrolled hazardous waste sites. Draft Eng manual 1110-2-600, Washington, DC

US Army Engineer District, Buffalo (1983) Analysis of Sediment from Ashtabula River, Ashtabula, OH, Technical Report No. G0072-02, prepared by Aquatech Environmental Consultants, Inc., Buffalo, NY

US Army Engineer District, Seattle (1984) Evaluation of Alternative Dredging Methods and Equipment, Disposal Methods and Sites and Site Control and Treatment Practices for Contaminated Sediments, Draft report

US Environmental Protection Agency (USEPA) (1977) Ocean dumping, final revision of regulations and criteria. Fed Register 42, January 11, 1977

US Environmental Protection Agency (USEPA) (1982) Handbook – remedial action at waste disposal sites, EPA-625/6-82-600, Cincinnati, Ohio

Vierveijzer HC, Lepelaar A, en Dijkstra J (1979) Analysemethoden voor grand, rioolslibi, gewas en Vloeistof. Inst Bodemvruchtbaarheid, WHO 1972, p 259

Westerdahl HE, Skogerboe JG (1981) Realistic rainfall and watershed response simulations for assessing water quality impacts of land use management. Proc Int Symp Rainfall runoff modeling. Miss State Univ Water Resourc Publ, pp 87–104

Wright TD (1977) "Aquatic Disposal Field Investigations, Galveston, Texas, Offshore Disposal Site: Evaluative Summary," Tech Rep D-77-20, US Army Eng Waterways Exp Stn, Vicksburg, Miss

Environmental Management of New Mining Operations

ROGER J. HIGGINS[1]

1 Introduction

The environmental management of a mining operation is an integral part of that operation. Environmental management can be planned and organized so that it contributes to the achievement of company goals including productivity, product quality and profitability. Environmental performance is also a prime and often highly visible indicator of corporate citizenship. In short, the standard of environmental management can play an important role in determining the overall success of an operation as seen by its shareholders, employees, local community and the public at large.

Conversely, poor environmental management or its absence altogether can be responsible for poor company performance, or even financial failure when extensive clean-up or add-on installations are required to meet environmental obligations.

Because of their potential impact on the success or otherwise of a new mining venture, environmental issues should be incorporated into project planning from the outset. This incorporation requires plans for environmental monitoring of project impacts and the ways and means of dealing with unacceptable impacts. Frequently, it also requires an assessment of the adequacy and relevance of the prevailing legislative framework and its potential development. Hopefully with the full cooperation of the regulatory authority, a clear definition of environmental criteria to be met, and what constitutes "acceptable" impact will be determined.

Incorporation of environmental management issues into the planning of a new mining venture should recognize two precepts. Firstly, the construction and operation of a mine, related processing facilities and infrastructure will not be accomplished free of any impacts on the environment; secondly, the level of impact which will be acceptable will be strongly governed by the specifics of the project and its setting, i.e. the geographic, ecological, economic and political environment.

The natural environment has many legitimate uses to man, one of which is the assimilation of mining and industrial wastes. The task of environmental management is to make sensible use of the environment's capacity to do this, in full consideration of other demands on the environmental resources at the same time and place.

The following sections set out a framework for incorporating environmental considerations into the planning of a mining project at its various stages and also

[1]Ok Tedi Mining Ltd., PO Box 1, Tabubil, Papua New Guinea

for determining an appropriate basis on which to formulate the project's environmental monitoring and management program. Emphasis is placed on the disposal of mine overburden and the process plant tailing, as these are the dominating waste products of most mining operations.

2 The Phases of a Project

Before achieving full operational status, a new mining venture undergoes a number of distinct phases of development. Each phase is categorized by the emphasis placed on the activities being conducted, so that appropriate levels of information are provided for sound decisions regarding the future of the project. The phases may overlap and need not be continuous, but do follow in an established sequence. Figure 1 illustrates the relative timing and duration of project development phases. The time scale shown is arbitrary. However, for a new large-scale mining and processing venture, the scale could roughly represent elapsed months.

The decisions required from the first four phases shown on Fig. 1 are "whether to" decisions, i.e. on the basis of progressively improved and broader information, it is decided whether the project should or should not proceed. Subsequent phases enable "how to" decisions, i.e. how to develop the resource, how to build the facilities and how to operate them.

The distinction between the whether to and how to phases is also important with respect to environmental issues. If environmental management is to be integrated with overall project planning, neither slipping behind nor pushing ahead of the whole, then it must develop comparable levels of information for comparable decisions.

Exploration. A mining project's exploration phase may last for many years. During most of this phase, there is no project as such, and no project-related environmental activity. It is necessary that the exploration activities are conducted in a manner which is environmentally sound, and in sensitive environments an impact assessment of the exploration phase may be desirable before any field parties or drill crews enter the area.

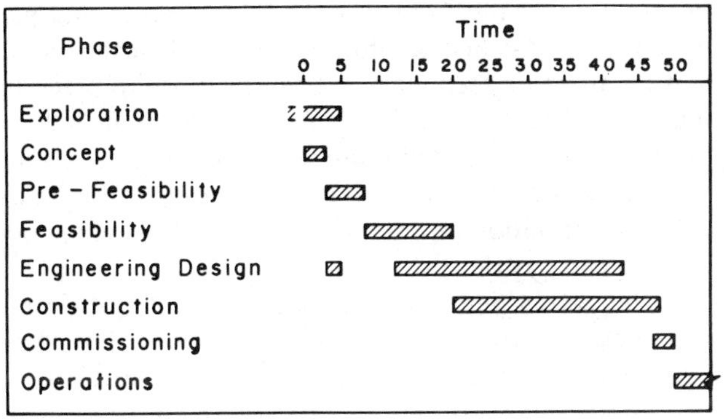

Phase
Time
0 5 10 15 20 25 30 35 40 45 50
Exploration
Concept
Pre - Feasibility
Feasibility
Engineering Design
Construction
Commissioning
Operations

Fig. 1. Phases of project development

Concept. Towards the end of the exploration phase, a conceptual study will be carried out to establish the project's order of magnitude. This study is where the project is conceived, and on Fig. 1 corresponds with time zero. This study is conducted at desk-top, with little field information other than that gleaned from the field notes and log books of the exploration team. Ore reserve and capital cost estimates are probably no better than ±50%, but the first outlines of the project are drawn. The report of the concept study should contain a listing of likely environmental issues, in order to provide direction for subsequent phases. These will be broadly stated: for example, high annual rainfall is likely to govern the design parameters for tailing retention dams; the project site is located in the midst of an established agricultural community; the project will require access through the traditional lands of a subsistent population. It is important at this early stage that issues are not too readily classified as constraints. In fact, the early identification of issues permits the project-planning process to minimize the emergence of environmental constraints. These can often be avoided by the same innovative engineering and management that overcome the geotechnical, logistic, financing or other potential problems of any project.

Prefeasibility. A project's prefeasibility phase is marked by two significant events. The mineralization discovered during exploration is assessed for reclassification as an orebody, where an orebody is defined as a mineralized deposit which is economic to exploit. This is the first attempt to determine the net value of the resource. The second event is the commencement of engineering endeavour, generally equally divided between mining, mineral processing and infrastructure studies. This first proposal for a specific rather than conceptual development plan can reduce the uncertainty surrounding capital cost estimates to around ±30% overall, although particular project areas may remain poorly defined.

Two significant environmental activities should also be undertaken during the prefeasibility phase. Firstly, a thorough search for and comprehensive compilation of existing environmental information on the project area is advisable. For a new project in a well-developed region this may be quite straightforward and yield extensive information, while previous scientific documentation of remote areas may be scattered and very limited. In either case, government agencies, research institutions, local clubs and societies and interested and informed individuals are likely to provide in total more pertinent information than was first expected. This information, in the form of reports, books, journal papers, maps and interview notes can be classified according to both its bearing on specific aspects of the project and its direct or secondary relevance to the project area. This activity requires diligence, but is unlikely to amount to more than 1–2% of the effort devoted to the prefeasibility phase. It maximizes the amount of existing data that is available for use, and avoids the embarrassment of suggesting an expensive field program to collect information that is later discovered in the log book of the local canoe club or village magistrate!

The second environmental activity of the prefeasibility phase is a field reconnaisance by environmental specialists. This "testing of the water", both figuratively and literally, can place in priority order the issues identified during the conceptual phase, bring to light any further issues and highlight the practical aspects of environmental monitoring and management which are specific to the project. The need for an immediate start to the collection of key field data is also likely to become apparent and the collection of such data should be initiated. Often, a record of data for environmental assessment is equally important to project construction and operations, for example in respect of river discharges and flood heights, and climatic variables. A coordinated environmental monitoring program makes the most efficient use of resources.

Feasibility. A project's feasibility phase is the last which is devoted predominantly to a whether to decision. Three conclusions may emerge from a feasibility study: a clear decision to proceed, a clear decision not to proceed or the project may be held in abeyance, pending some expected or hoped for change in local or world economic conditions or other controlling factors. In the event of a clear decision to proceed, engineering and design are likely to commence almost immediately in order to maximize the economic advantage from the favourable conditions which led to that decision. The project may of course be stopped at some later stage due to a change in conditions, but work will concentrate on how to issues in a positive climate. In the event of a clear decision not to proceed, work will cease and the owners are likely to divest themselves of their interest in the project.

Should the project be placed in suspension as a result of the feasibility study, limited activities will continue related to monitoring business aspects of the project: metal prices, competitive suppliers and financing opportunities for example, for changes which might lead to a clear decision either way. No significant refinement of the feasibility study or engineering studies will take place.

Environmental activities during the feasibility phase should recognize the possible outcomes described above. The principal requirement of environmental investigations in this phase is to determine whether or not emerging environmental issues could negate the success a project would otherwise enjoy. This requires that an impact assessment be made of the issues which have been identified, so that an informed and rational decision can be made on whether these impacts can be managed within the overall objectives of the project. It must also be determined feasible to abandon the project facilities in an environmentally sound manner, at the end of the mine life. As with the feasibility phase generally, this is the last real opportunity for a whether to decision on environmental matters. Future environmental work will concentrate on how to manage impacts, and should a project be stopped at a later stage on environmental grounds, then feasibility-phase environmental studies were inadequately conducted and evaluated.

Environmental activities during the feasibility phase will require both intensive and extensive collection of baseline data on the project's setting, to the extent necessary to complement existing information. This data collection should conform to a design for long-range environmental monitoring of the setting and later project impacts, so that a continuous data base is provided. Environmental assessments with discontinuous or spot data are tenuous at best.

Engineering Design. Details of how the orebody will be mined, the ore processed, the product shipped and support services provided are determined during the engineering design phase. Construction techniques and program are also established. This is a period of intensive activity in design offices, which are sometimes widely separated, and also in the field at the project site as geotechnical, survey and similar design data are gathered. Schedules are drawn and target dates are set, missed, revised and met as solutions to many how to problems are found. Construction contracts are bid and negotiated, procurement sources are established and long lead-time equipment is ordered. In related but separate activities, project financing is finalized and sales contracts for the project's products are arranged.

During this period of intensive and diversified activity, environmental issues can easily be passed over. In fact, they deserve as much attention at this time as in any phase of a project's development because at this time future environmental problems can be designed out with careful care and attention, or designed in by omission. The early incorporation of environmental controls need not be expensive and will almost always avoid costly capital and operating costs later. Where major expenditure is required for environmental control, this will be minimized by coordinated project design and development incorporating environmental criteria. Designs must also recognize the need to leave the project site stable and free from sources of on-going pollution after the orebody is mined out.

Design of environmental control facilities requires upgrading of impact predictions as further field data is generated. The design phase is also the last opportunity to gather environmental data which is demonstrably free from significant impacts of the project. These "natural", "background" or "baseline" conditions will be referred to continually throughout the project's life when actual impacts are observed or claimed.

Also during this phase of project development, the environmental monitoring and management program should be designed, so that appropriate hardware can be installed and so that monitoring operations are compatible with construction activities and with mining and processing operations.

Environmental activities during the engineering design phase are crucial to efficient and successful environmental management later. They should be given their due attention.

Construction. Construction of project facilities involves a similar level of intensive and diversified activity on the project site as engineering design involves off-site. Activities during the construction phase include prestripping of the orebody, and building of haul roads, ore-treatment facilities, workshops, warehouses and offices. In remote areas the establishment of towns, roads, ports and/or airports may also be necessary. In making this omelette, some eggs will be broken, and construction activities will inevitably have environmental impacts. These impacts are not a cause for concern per se, provided they were anticipated during earlier phases and remain within the limits which were agreed to be acceptable when the "whether to" decision for the project was taken.

As in the exploration phase, construction impacts can be severe in sensitive ecosystems and construction contracts should contain particular and explicit reference to practices to be adopted in minimizing environmental disturbance.

Construction impacts should be monitored, wherever possible within the framework of the program designed for observing the project's long-term impacts. In some cases, however, this will not be possible as construction impacts will be different both in nature and in geographical extent from operational impacts. Impacts may be observed which are different from or more extensive than those anticipated prior to the start of construction. Because of the pace at which construction usually proceeds on a major project, rapid assessment of any such impacts must be undertaken to determine whether the impacts are unacceptably large, whether they are irreversible or can be remedied on completion of construction, and what, if any, changes in construction methods or programs should be implemented. The project manager's brief should include the need to maintain control over environmental effects along with control over budgets and schedules.

Environmental monitoring of construction impacts also provides the first check on the reliability of earlier impact prediction techniques. The process of refining and narrowing impact predictions can commence, based often for the first time on the observed response of the local environment to significant activities of the project. During construction monitoring, field techniques, staffing arrangements and procedures are further developed.

Commissioning. A project's commissioning phase is when installed equipment is test run, first by individual components and sections and then for the total system. During this phase, project facilities are progressively handed over from the construction group, who complete their job, to the operations group, who accept that the facilities have been built according to their design. Final modifications and improvements are made and operators receive hands-on training. All environmental control facilities undergo test runs and operational adjustment during this phase. While initial discharges from the process plant occur during commissioning, they are inevitably highly variable in rate and episodic as the plant stops and starts. This greatly confuses any interpretation of environmental monitoring results, but such monitoring should nonetheless take place.

Operations. Full operational status is a major event in the history of a new mining venture. With it come a dramatic shift in management interest and project emphasis, and also the stability necessary for the trouble-free performance of a continuous operation. The stability is only relative, however, to the continual changes of earlier phases of the project. In all aspects of the project from mining and milling to environmental control, operational performance is monitored and results fed back in a continuous effort to improve that performance. In addition, cut-off grades and throughput rates are subject to revisions as the project matures, very often with cut-off grades being lowered and throughputs raised. Projects may go through major expansions and associated construction phases while in operation.

Important feed-back mechanisms should be driven by environmental activities during the operations phase. As a result of routine monitoring of environmental indicators, according to the predetermined program, and assessment of monitoring results by appropriate specialists, impacts should be assessed against predictions and level of acceptability. A clear mechanism should exist in the organization structure and procedures of the operating company, whereby necessary changes can be introduced to the operations in order to meet environmental standards. These changes need not be detrimental to the operations of the company and indeed if actual impacts are less than those predicted and agreed to be acceptable, the company may be in a position to relax its environmental control procedures accordingly.

Also, the monitoring and management program itself should be under regular periodic review, to ascertain that appropriate environmental indicators are being observed, that impacts are being adequately recorded and that effort and cost are not being wasted collecting data which does not materially assist in assessing impacts. Environmental monitoring programs now being designed are likely to be too broad rather than too narrow, in order to ensure that significant effects are not missed. As experience is gained with data from the program, some segments of it can usually be scaled down in areal extent, monitoring frequency and intensity, while other segments may deserve increased attention.

Abandonment. Operations monitoring also allows definitive plans to be made for ensuring that at the end of the project's life, facilities can be left in an environmentally sound and stable condition. Feasibility phase plans for abandonment should be reviewed and updated during operations, and at an appropriate time towards the end of the operations phase, designs finalized and implementation procedures agreed upon.

By the time mining and milling operations cease, site areas of the project should be as close as possible to their stable abandonment condition. This avoids expensive clean-up activities at a time when income from the project has ceased. Periodic check monitoring may be required after abandonment until the physical and chemical stability of the site is confirmed.

Table 1 summarizes the full life of a project into the nine phases described in this section. In addition to describing the dominant project activity in each phase, related environmental activities are specified.

3 Environmental Monitoring and Management

3.1 Development of an Environmental Monitoring Program

An environmental monitoring program for any project can be built up from a knowledge of the project's components and the characteristics of the environmental setting. Differences between these parameters for different projects mean that foci and levels of effort will not be directly comparable from one monitoring program to another. In these circumstances, consistency can be achieved by adopting a standard approach to the development of monitoring programs, wherein project and environmental factors are assessed and balanced in like manner. Such an approach is outlined below and illustrated in Fig. 2.

Effluent Sources. The nature of a mining project and the component facilities necessary for the mining and processing of its ore will establish the waste materials which must be handled. From completion of construction, effluents can be broadly sourced to the mine, process plant and infrastructure areas of the project. More

Table 1. Emphasis of activities

Phase	Project Activity	Environmental Activity
Exploration	Identify significant mineralization	Nil
Concept	Order-of-magnitude assessment of benefits, costs, development strategy and problems	Desk identification of probable issues
Prefeasibility	Classification of mineralization as economic orebody; preliminary examination of economics, logistics and marketing	Field reconnaissance; compilation of existing data; initiation of limited, long lead-time data collection
Feasibility	Establish extent and grade of oreody and framework for development; decide that there are no insurmountable obstacles	Identify major issues and assess with available data; predict impacts as manageable or otherwise; initiate baseline data collection
Engineering design	Field investigations for design (such as geotechnical and survey); detailed design of all facilities; initiate contracts and procurement	Establish details of environmental baseline; predict impacts in detail; design monitoring and management program; incorporate environmental controls in design
Construction	Build facilities; establish operations staffing and procedures	Monitor construction impacts; establish operations team, structure, program and equipment
Commissioning	Testing and handover of facilities to operator; modifications and improvements; operations training	Performance testing of environmental control facilities; monitoring of initial discharges
Operations	Operate plant; produce, ship and market product	Operations monitoring; feedback and control of impacts, audit and review of monitoring and management program; prepare for abandonment
Abandonement	Close-down of operations and salvage or sale of facilities	Clean-up of site; ensure long term stability and drainage; arrangements for periodic check monitoring

specific sourcing is possible for any particular case, as well as a weighting among sources of likely importance regarding environmental contaminants.

Mining activity results in movement of rock and earth with three dominant environmental consequences. Material is available for erosion, increasing the sediment load of mine-area streams. This effect is greatest for high tonnage, open-pit mines, where large areas of ground surface are disturbed and rock below cutoff grade is moved to overburden dumps. Where erosion is uncontrolled or difficult to control, the sediments themselves can present an environmental hazard irrespective of their mineralogical composition. Because material exposed to erosion is likely to be mineralized, a further hazard is presented by the potential for heavy metals to be released into solution from the sediments, either in the vicinity of the source or at some downstream depositional environment. Metals bound in

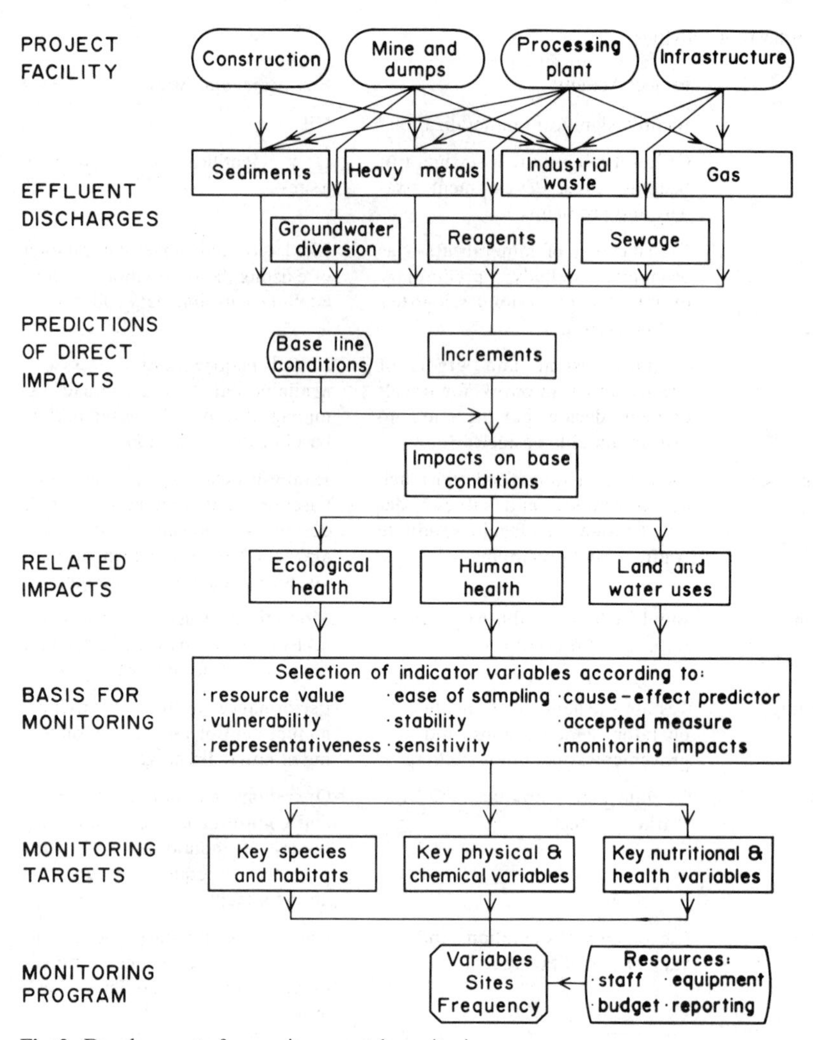

Fig. 2. Development of an environmental monitoring program

or adsorbed onto sediments are of little or no environmental consequence unless the geochemistry of the receiving waters permits their release.

A further significant environmental consequence of mining activity, whether surface or underground, is interference with natural groundwater tables and in-situ permeability. This frequently releases groundwater to surface streams, although diversion of surface water to groundwater can also occur. In either case, drainage from mineralized rock can result in contamination of receiving waters with high concentrations of metals in solution.

The highest volume discharge from a mineral processing plant is the tailing, incorporating finely ground ore from which the economic minerals have been removed. The proportion of metal recovered can be extremely small and in a gold recovery operation, for example, as little as several grams per tonne of ore may be extracted. The production of a concentrate by froth flotation may extract 1–10% of the weight of ore, depending on head grades. Sand- and silt-sized materials in the tailing may present environmental hazards both as sediment and as a further source of heavy metals. The plant tailing will also contain residual reagents from the process. These may be highly toxic and at significant concentrations, as in the case of cyanide from gold leaching.

A mineral-processing plant may also generate gaseous emissions, from ore roasting for example, and other industrial effluents such as waste oil and containers which have transported toxic reagents.

Infrastructure supporting mining and milling operations also produces industrial wastes, including sewage and domestic garbage, waste lubricants from workshops and equipment servicing and gases from thermal power generation. Where such infrastructure is constructed specifically for a new mining project, the possible environmental hazard presented by these discharges should be considered and incorporated into the project's environmental monitoring program.

Environmental Baseline. The second parameter set which determines the detail of an environmental monitoring program is that which describes the receiving environment. The relevant environmental baseline is the status of the project's setting in the absence of the project but including the influences of earlier projects in the vicinity and of other man-induced disturbances. Whether the first projects in a region have first rights to the capacity of the natural environment to assimilate wastes depends on the legislative framework and the approach of regulatory authorities.

The environmental baseline should be established during a project's feasibility phase, as outlined in Section 2 above. Data collected to complement that from external sources can be focussed towards expected impacts from potential contaminant sources, as these sources will have been broadly defined in earlier phases.

Predictions of Impact. A scientific description of the receiving environment and a knowledge of the quantity and quality of waste discharges allow predictions to be made of direct impacts on air and water quality. Such predictions are required to establish that environmental considerations will not jeopardize the project's feasibility and, during engineering design, to determine the extent of environmental protection facilities which must be incorporated in order to meet regulatory criteria. These same predictions can be utilized in developing a monitoring program which is effective and efficient in picking up both early-warning signs of environmental effects and actual impacts on the ecosystem and other users of the environmental setting.

Impact predictions will be undertaken by a variety of means and with varying reliability of results. The value of the predictions to the monitoring program lies in defining the changes in environmental parameters in both the spatial and temporal dimensions.

Direct impacts of effluent discharges, changes in air and water quality for example, can be projected to ecological health, human health and limitations on alternative land and water users. These related changes are true environmental impacts as they take cognizance of the environment's capacity to absorb some physical and chemical change without adverse consequences. Projections from direct impacts to environmental consequences are based on assumed system responses, and an environmental monitoring program must be designed in such a way that these assumptions are confirmed or can be refined.

Basis for Monitoring. The natural environment comprises a complex system of dependent and independent variables. It is neither desirable nor feasible to monitor every element of the system and a rational basis is required for selecting those elements to be recorded by the program. Selection criteria fall into three categories:

1. "Significance criteria". Some elements of the environmental system should be monitored because of some special value attached to them. Significance criteria will result in selection for monitoring of a resource having commercial, subsistent or cultural value. Other elements might be selected due to particular vulnerability, being either endangered species for example, or known to be close to limits of environmental stress under background conditions. A further significance criterion is representativeness, whereby a particular environmental parameter adequately represents a part of the natural system, such as a group of chemical characteristics, a habitat or a system response.
2. "Feasibility criteria". It may not be technically or logistically feasible to monitor all significant elements of the environmental system, and further selection is necessary. Elements (such as aquatic species or chemical variables) which are easy to sample and easy to interpret are obviously preferred, other things being equal. A parameter which is stable or varies consistently under background conditions will more readily demonstrate induced change than one subject to high "noise" levels. Likewise, a parameter with a known sensitivity to expected environmental stress, or an expected systematic and graduated response, will be a preferred indicator.
3. "Technical criteria". Environmental parameters which are of significance and which are feasible to monitor should be screened against one further set of criteria. Impact predictions for any project rely to a large extent on the cumulative experience of scientists and engineers, gained during previous work at other locations and reported in technical literature. It is necessary that data to be collected be comparable with that available from elsewhere. This means that established cause-effect predictors and accepted environmental parameters must be considered for monitoring, even if not judged to have particular significance or if such monitoring is not easily managed in a particular case.

One further criterion is that the monitoring program should not itself result in environmental impact, such as might occur for example by too frequent sampling of a limited population of an aquatic species.

Monitoring Targets. The selection criteria described above and the baseline description of the project environment allow the selection of specific target elements for monitoring. Key physical and chemical variables may include rainfall and air temperature, streamflow, suspended sediment concentrations, wind speed and direction, dust levels, river water pH and concentration of dissolved heavy materials and biological oxygen demand (BOD). A specific list of these or other parameters will depend on the particular project and the receiving environment.

Biological parameters of project-area flora and fauna will form another set of monitoring targets. Monitoring locations should be closely related to the sites where physical and chemical parameters are monitored, to assist the development of cause-effect relationships. However, frequencies of sampling often can be independent. The biological parameters being observed generally exhibit a response which is an integration of the time series of changes in ambient environmental quality. On the one hand, this means that sampling can be carried out relatively less often. Biological monitoring might be conducted quarterly or biannually, for example, where chemical monitoring takes place daily or weekly. On the other hand, a much longer elapsed time is required to detect trends in the collected data, or to establish that a definite change has occurred against a background of natural variability.

Environmental impacts on the mental and physical health of the people of the project area are a further integration of impacts on the ambient environment. Factors far removed from the project, to and including the development of the society as a whole, also influence health status, so that the project's impact can be difficult to discern. However, there are obvious exceptions to this. Particular environmental contaminants can be responsible for direct impacts on human populations, (e.g. asbestos dust, radioactive waste) and in subsistent societies impacts on biological resources lead to changes in the available food supply and nutritional status of the people.

Effects on the health of project area peoples are rarely a major target of specific environmental monitoring programs and are usually addressed by the social services provided to the community. They nonetheless merit at least general consideration in an environmental monitoring program and may require a more specific component of the program where a clear cause-effect link is established.

Monitoring Program. With selection of the variables to be monitored, the locations, frequency and routine of observations necessary to quantify environmental impacts, and the resources necessary to carry out the program can be determined. Monitoring may be carried out directly by staff of the mining company and/or by contract to specialist firms. Field equipment, laboratory facilities and data handling arrangements will be required, leading to a budget for capital equipment and an annual operating budget. Arrangements will be made for regular reporting of the data and the interpretations to be drawn from it. This reporting is the interface between the monitoring program and responsive environmental management, both internally to the mining project and with the regulatory authority.

3.2 Responsive Environmental Management

The data collected by an environmental monitoring program describes the response of the environmental system to activities of the project. The integration of environmental management into the project's operations requires an established mechanism, whereby the results of the monitoring program contribute to the achievement of company goals including that of responsible environmental performance. An approach to environmental management is shown in Fig. 3.

Data Analysis and Review. An environmental monitoring program can quickly produce a large and unwieldy data bank. These data are rarely direct measures of environmental impact and require manipulation and analysis to reveal trends, shifts and patterns which may be masked by the natural variability of the system parameters. It is important that the data base be easily accessed and updated regularly with newly collected information. Good quality field data is usually collected at considerable expense of time and resources but is of very limited worth if it is held only in field books and desk drawers.

A critical review of the data should take place periodically to ensure that the data is providing the information expected and needed for assessment of environmental impacts. Are the most useful parameters being observed? Are observations being carried out at the best locations, at too many locations or at too few locations? Is the frequency of observations appropriate? No particular detail of a monitoring program should be regarded as beyond review and the effort and resources expended on field data collection should be justified regularly by the production of meaningful information pertinent to the objectives of the program.

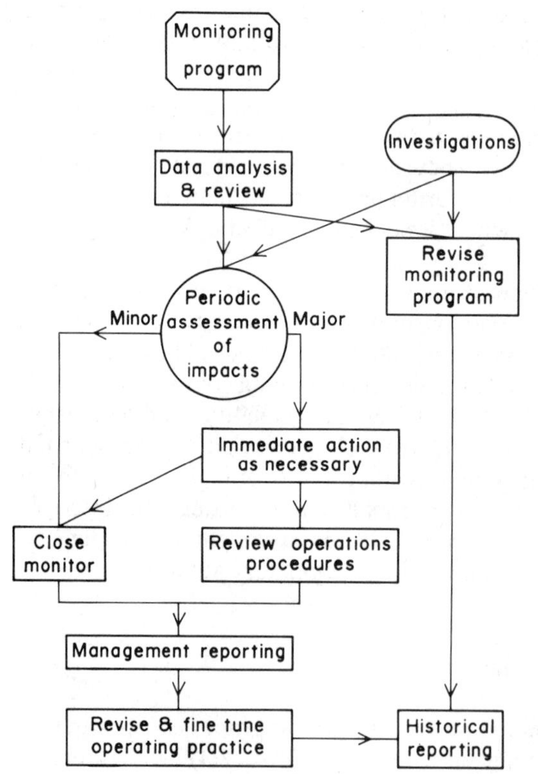

Fig. 3. Environmental management flows

Investigations. The term "environmental monitoring" carries the implication of scheduled and routine observations. There is no doubt that this monitoring is of great value in developing an understanding of the environmental systems and in detecting the onset of changes. Environmental management decisions can also be assisted by once-off studies, herein grouped together as "investigations".

Investigations within an environmental program are designed to clarify a particular area of uncertainty and thereby enable more effective monitoring or establish more sensitive indicators of environmental response.

Investigations pertinent to the environmental program of a mining operation could include the following:

1. Erosion potential of waste rock dumps and the transport of eroded material by mine-area streams;
2. Decay behaviour of naturally degrading chemicals contained in effluent discharges;
3. Mixing experiments to determine changes in chemical equilibria when tailing effluents are mixed with natural stream waters;
4. Bioavailability of complexed heavy metals;
5. Biotoxicity tests on locally resident species of aquatic fauna subjected to a range of effluent concentrations;
6. Testing of "what if" scenarios using hydrologic, geochemical or biologic models of the receiving environment;
7. Opportunistic monitoring of natural and operational events, such as severe storm and discharge peaks, droughts and temporary plant shutdowns.

Such investigations should of course be planned, executed and documented in a proper scientific manner, so that their conclusions can be defended and can contributed to the objective of sound environmental management.

Revision of the Monitoring Program. Regular review of the collected data to assess its quality and pertinence, and the results of investigations into components of the environmental system, will provide cause for periodic revisions to the monitoring program. Revisions to the suite of parameters being observed should be approached with some caution, particularly with respect to dropping parameters from the suite. While there is no point in collecting worthless or excessive data, continuity of records over extended periods is of great importance in obtaining an understanding of the behaviour of a natural system and its response to change. Modifications to an environmental monitoring program therefore need to be fully justified by an analysis of the existing data base and should on all occasions be directed towards improving the relevance and continuity of the total record of observations, bearing in mind also the inevitable restraints imposed on resources available for data collection.

Where any set of parameters is being replaced by another, judged to be more relevant, easier to monitor, more sensitive indicators of change or for any other reason preferable, an attempt should be made to incorporate a period of overlap. During this period both parameter sets are observed and relationships between them used to avoid a sharp discontinuity in the time series of data.

Assessment of Impacts. Rarely will the assessment of environmental impacts be clear-cut and objective on the evidence of collected data. Whatever tools are used in making these assessments, from statistical analysis of field data through laboratory experiments to simulation modelling of components of the environmental system, ultimate assessments will involve subjective albeit professional judgements. Judgements and debate will focus on impact thresholds, or what constitutes an impact, and impact magnitudes, or what extent of impact is acceptable in a particular case.

Simple decision rules are sometimes proposed. For example, water quality standards can be specified such that daily or longer-term average concentrations of various constituents must not exceed fixed numerical values. If such a water quality standard is exceeded, then an impact is assumed to have occurred.

Ideally, such simple numeric standards could be established for the specific environmental characteristics of a project area and the specific nature of effluent discharges. There might then be a clearly-defined relationship between the standard and actual environmental impacts. In practice, environmental systems are complex and simple decision rules are at best an operating guide to environmental impact. Standards are generally established by extrapolation or broad generalizations on a national or regional scale, with an in-built tendency towards ensuring environmental protection for the more sensitive system components of the more sensitive regions.

The actual impact of a particular project may not be well defined by such water or air quality standards which are broadly applied. In addition, such standards are generally designed to delineate the onset of environmental effects, and may be of limited assistance in quantifying an acceptable degree of impact.

Response to Observe Impacts. Like all management functions, the quality of environmental management can be assessed in large measure by the ability to respond quickly and effectively to actual situations as they are encountered. The appropriate response to any observed environmental impact depends on the nature and severity of the impact observed.

If the impact is minor, generally within the range of predictions and acceptable in the context of the project, then the response is largely routine. A close monitor is maintained, probably with a heightening of emphasis towards any areas approaching the limits of range or with the potential to do so. Such observed minor impacts are noted in routine management and compliance reporting.

If severe impacts are observed or if environmental monitoring suggests that they are likely, then two steps follow. Immediate action is required to remove or suppress the cause of the impact; this may involve such actions as diversion of effluent discharges to a holding area for treatment, reduction of processing throughput rates or in extreme cases the temporary suspension of operations. These immediate steps are taken until such time as conditions stabilize or until revised operating procedures can be designed and implemented so that the observed severe impacts are mitigated or avoided.

Any temporary disruption of operations and/or revision of operating procedures to avoid environmental impacts can have severe implications for the smooth running and profitability of a mining and ore-processing project. However, such disruptions should occur very rarely if project planning has incorporated sound environmental assessment, if project engineering has incorporated appropriate environmental controls and a stable framework of environmental legislation prevails.

In cases of severe environmental impacts being observed and counteracting measures being taken, management reporting must take place outside the routine reporting structure, in such a way that the importance of the situation is highlighted.

Fine Tuning and Historical Reporting. Two further activities should be incorporated into the environmental management process. The fine tuning of operating practice, which activity is a continuous part of any operations management function, can sometimes be driven by environmental considerations. On other occasions fine tuning activities may have their own environmental implications for which allowance must be made. Fine tuning takes account of factors such as process efficiency, ease of maintenance, work force organization and resource optimization. Environmental impact has a legitimate place among such considerations.

In addition to the reporting of environmental monitoring and impact assessment for management purposes, a conscious effort should be devoted to maintaining a historical record of environmental performance. This includes not only the raw data collected by the monitoring program, but also operational conditions which influenced effluent quality or other sources of environmental stress, and the reactions and responses observed or initiated. This record can be of great importance in predicting the effects of planned changes or unexpected events and is also necessary to minimize the disruption to environmental programs which can occur with inevitable changes in program staffing and management approach.

3.3 Feedback

There are three predominant closed loops which can create effective interaction between environmental management and monitoring programs. These are illustrated in Fig. 4.

The inner loop ensures the validity of the nuts and bolts aspects of the monitoring program. Revision of the program on the basis of an analysis of collected data and the findings of new research feeds back to the mechanics of data collection: staff, equipment, sites, data analysis, budgets and reporting. This is the tightest loop, with a close association between cause and effect and shortest response time.

The middle feedback loop provides a response to actual project impacts, vis-a-vis predicted impacts. As described in Section 3.1 and illustrated in Fig. 2, impact predictions early in the life of a project are utilized in developing the environmental program, including the selection of monitoring targets and detailed program design. Where monitored impacts differ from those predicted, it is appropriate to re-examine the basis of the monitoring program, as well as its emphasis and detail. The feedback mechanism represented by this loop is slow-acting. Before feedback information is available, time is required to collect data on actual impacts and to assess this data, distinguishing project-related changes in

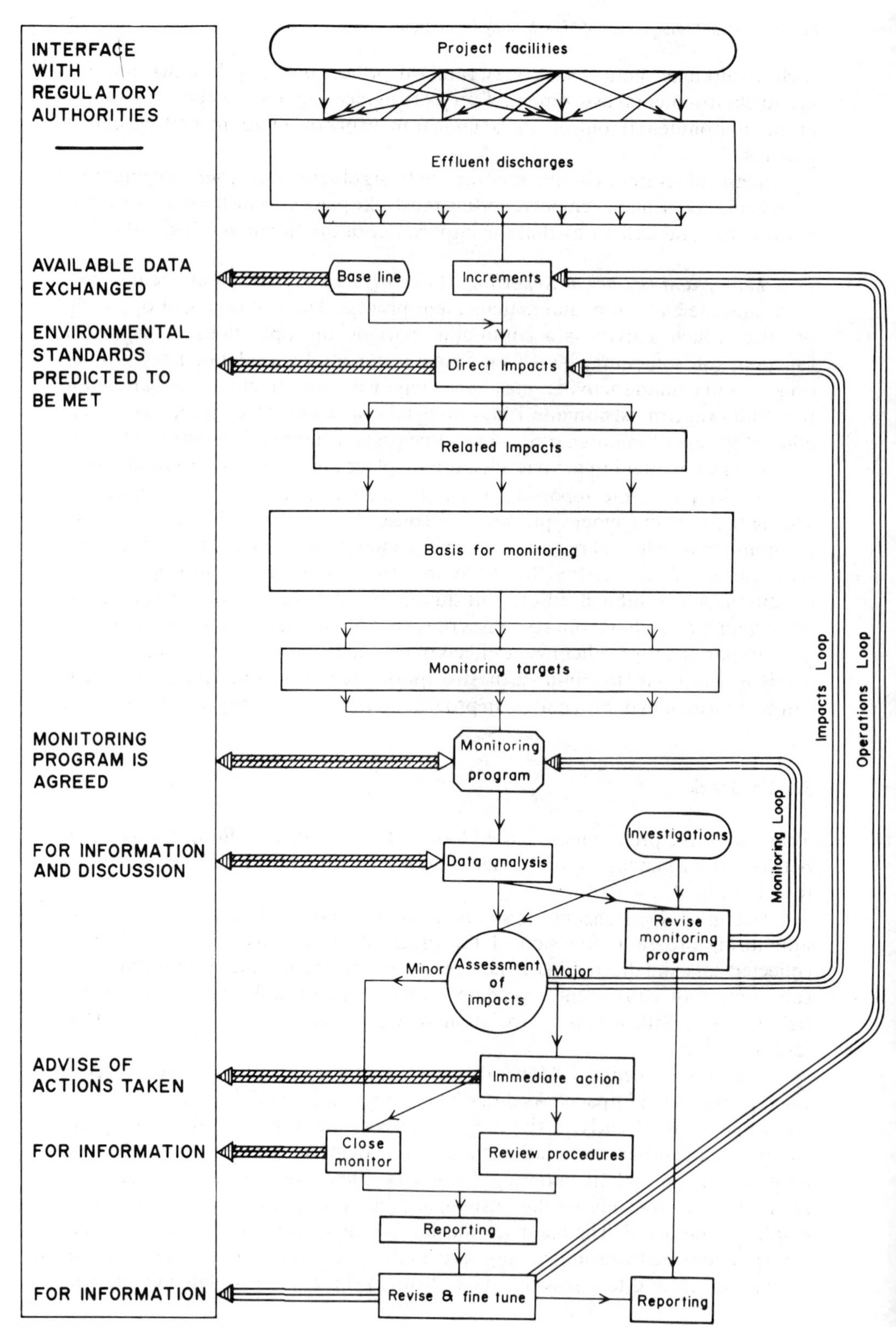

Fig. 4 Environmental monitoring, management and government liaison

environmental parameters from natural variability and any other changes not related to the project. Further time is then required to assess whether differences between predicted and observed impacts warrant revision of the monitoring program. The time frame of this information and response system can vary from weeks to years, depending principally on the time necessary to confirm actual impacts on the environment.

The outer loop provides the mechanism for responsive management of the overall project to its environmental performance. Where environmental impacts lead to a need to revise or fine tune operating practices, then component effluents at their source will have been modified in either quantity or quality.

Likely impacts from such revised discharges to the environment can be re-estimated, perhaps also justifying a re-examination of the monitoring program design. In addition to the time required to monitor and assess actual impacts, this response loop contains the time required to design and implement changes to the operating project. Where such changes involve equipment modifications or additions, extensive total elapsed times may be invoked.

3.4 Interface with Regulatory Authorities

Environmental control facilities are incorporated into project design in order that government-imposed environmental criteria can be met; monitoring is carried out to check performance against those criteria; and environmental management procedures and mechanisms are put in place to ensure appropriate environmental performance under operating conditions. There is clearly a need for an established interface between the mining project and the regulatory authority having environmental responsibilities.

Government authorities will require information from the project at various specified points in the development sequence, and at specified intervals during operations. It is particularly important that the regulatory body is familiar with and understands baseline conditions so that the project is not encumbered with some assumed ideal background. In addition, government authorities are likely to hold or have access to preproject data on the environment and a full exchange of information between the project and government bodies is in the best interest of all parties.

Environmental standards are set by the regulatory authorities and it is important to the project that these be clearly defined and are not subject to frequent or arbitrary change. It is legitimate for proponents of the project to express their opinion on the validity of environmental standards and to propose that they be altered in a way that is favourable to the project. Such proposals, properly backed by scientific data and argument, can be of great assistance to regulatory bodies in their assessment of appropriate standards, both generally and for a particular project. While complete agreement may not always (or often) be reached on appropriate standards, the greatest disparity of views is likely to occur if a strongly adverse relationship exists between the project and its regulatory body.

When impact predictions are made during the course of an environmental assessment, regulatory authorities will wish to compare them against their imposed standards, and may of course make thier own predictions. Particularly at early stages of a project's life, comparisons between standards and predictions should take into account other relevant aspects of both the project and the environment, including the importance of the project to regional or national development, presence or lack of competing users of the local environment and future options for parallel or sequential developments or uses.

The basis of the project's monitoring program should also be agreed between the project and the regulatory body. Such agreement should include key variables for monitoring, sampling frequency and locations. Details of how monitoring will be carried out, including staffing and budgeting, are internal decisions of management and should not be offered "for approval", although they may be discussed if a productive working relationship between the parties is established.

It is generally in the project's best interests for various other information to be provided to the regulators. This includes periodic reviews and analysis of collected data, advice on actions taken to investigate any observed impacts, changes in operating procedures which have an influence on effluent discharges and the historical record of the project's environmental monitoring and performance. Legislation may require that some of this information be provided. However, a broader view should be taken, and any information which demonstrates the management's commitment to a high standard of environmental performance and a responsive management style should be made available. Feedback from the regulators should be encouraged at every opportunity.

4 Conclusions

Sound environmental management, initiated and directed from within and pursued as an explicit corporate goal, is sound business practice for a mining venture. Integration of environmental planning and management into the development and operation of a project enables project decisions to reflect the complete scope of the project, in the same way as does the integration of such components as mine planning, mineral processing, maintenance planning and cash forecasting. Realistic environmental planning recognizes a need to balance technical, economic and political constraints on the development of the project. A mine cannot be developed as a "zero-impact" project, but development at an acceptable level of impact can be achieved at an acceptable cost if environmental planning is incorporated at the initial phases of the project.

At any point in the development and operation of a project, the appropriate level of environmental activity can be assessed by comparison with the levels of activity on other aspects of the development. Environmental management programs should be conceptualized as the exploitation of the ore resource is undergoing conceptual design; during a feasibility study, it should be established that environmental factors will not prevent a project from being carried through; monitoring programs and control facilities should be designed during the engineering phase; and these programs and facilities should be up and running by the time the project is commissioned. If environmental activities lag project activities generally, then there is the risk of unforeseen impacts and potentially high remedial costs. If environmental activities lead project activities generally, then considerable effort and expense will be wasted by the project proponents, or effort will be spread so thingly as to be of little value to whatever future course is selected. In neither of these cases has sound project planning been pursued.

By keeping in step with project planning, environmental monitoring and management programs for a project can be tailored to the unique combination of the project and its environmental setting. A monitoring program can be built-up from a logical assessment of effluent sources, baseline conditions, impact predictions and environmental criteria. When the program is in place and operating, a responsive management approach with defined feedback loops can ensure that monitoring is producing appropriate management information, that the project's operations can be controlled and fine-tuned in a sound manner which recognizes environmental constraints, and that unexpected impacts are quickly detected and any necessary corrective action taken. Environmental management is well served also by the exchange of information with regulatory bodies, both for purposes of compliance with legislation and for the enhanced understanding of the nature and magnitude of environmental constraints.

Subject Index